T0361799

SUSTAINABILITY, INNOVATION AND PARTICIPATORY GOVERNANCE

Sustainability, Innovation and Participatory Governance

A Cross-National Study of the EU Eco-Management and Audit Scheme

Edited by
HUBERT HEINELT
Darmstadt University of Technology, Germany
RANDALL SMITH
University of Bristol, UK

Routledge
Taylor & Francis Group

LONDON AND NEW YORK

First published 2003 by Ashgate Publishing

Reissued 2019 by Routledge
2 Park Square, Milton Park, Abingdon, Oxon, OX14 4RN
52 Vanderbilt Avenue, New York, NY 10017

Routledge is an imprint of the Taylor & Francis Group, an informa business

Publisher's Note
The publisher has gone to great lengths to ensure the quality of this reprint but points out that some imperfections in the original copies may be apparent.

Disclaimer
The publisher has made every effort to trace copyright holders and welcomes correspondence from those they have been unable to contact.

A Library of Congress record exists under LC control number:

ISBN 13: 978-1-138-70816-7 (hbk)
ISBN 13: 978-1-315-19870-5 (ebk)

Contents

PART I: POLICY OVERVIEW AND ANALYTICAL FRAMEWORK

PART II: COUNTRY STUDIES

PART III: REFLECTIONS

List of Tables and Figures

Tables

Figures

List of Contributors

Zafeiroula Dimadama is a regional and environmental economist with a PhD in Environmental Policy and Management. She is a researcher at the Research Institute of Urban Environment and Human Resources of Panteion University (Athens) where she is currently working on a project on "EU Enlargement and Multi-level Governance in European Regional and Environmental Policies/ADAPT" funded by the EU 5th Framework Programme in Research and Development.

Femke Geerts was a Research Assistant at the School for Policy Studies, University of Bristol. She is now working in a planning consultancy in the Netherlands.

Brigitte Geißel is currently the co-ordinator of a research project on "Local Political and Administrative Elites in Eastern and Western Germany", Martin-Luther-University Halle-Wittenberg. Her main research interests are: citizen participation, environmental policy, gender studies, political culture, German unification and sustainability.

Panagiotis Getimis is a Professor in the Department of Economics and Regional Development and Director of the Research Institute of Urban Environment and Human Resources of Panteion University (Athens). He is co-editor of the Greek journal "TOPOS: Review of Urban and Regional Studies". His main publications include "The Welfare State and Social Policy in Greece" (with D. Gravaris, Themelio: Athens, 1996), "Urban and Regional Policies" (with G. Kafkalas and G. Maravegias, Themelio: Athens, 1998). His current research interests are focused on urban and metropolitan governance, regional innovation and institution building, structural funds policy, environmental and spatial planning policy.

Georgia Giannakourou was a researcher at the Research Institute of Urban Environment and Human Resources of Panteion University (Athens).

Hubert Heinelt is Professor of Political Science at Darmstadt University of Technology. Apart from co-ordinating the project of which this volume is a product, he is currently working on several EU policies (such as environment and cohesion policy). Among his recent publications are "European Union Environment Policy and New Forms of Governance" (co-editor with T. Malek, R. Smith and A. Töller) Aldershot: Ashgate, 2001, and "Policy Networks and European Structural Funds. A Comparison between Member States" (co-editor with R. Smith) Aldershot: Avebury, 1996.

Tanja Malek is Assistant Professor for Political Science at the University of Bielefeld. She is currently co-ordinating a project on organisational learning processes within the European Commission and is undertaking research into the societal dimensions of the European integration process. In 2000 she completed her PhD on the development of the European Structural Funds. Among her recent publications are "European Union Environment Policy and New Forms of Governance" (co-editor with H. Heinelt, R. Smith and A. Töller) Aldershot: Ashgate, 2001.

Britta Meinke is a researcher at the Institute of Political Science at Darmstadt University of Technology. She is currently working on a project on civil society in the EU. In February 2000 she completed her PhD on multi-level regulation through international environmental regimes. Recent publications include "Civic society and interest mediation in the context of the EU. Reflections about the Flora Fauna Habitats Directive" in: Wolfram Lamping et al. (eds.): "Democracies in Europe", Opladen: Leske and Budrich 2002 (with Hubert Heinelt).

Randall Smith is Senior Research Fellow at the School for Policy Studies, University of Bristol. Recent publications include "European Union Environment Policy and New Forms of Governance" (co-editor with H. Heinelt, T. Malek and A. Töller) Aldershot: Ashgate, 2001, "Policy Networks and European Structural Funds: A Comparison between Member States" (co-editor with H. Heinelt) Aldershot: Avebury, 1996 and "Cities in Transition" (co-editor with B. Blanke) Basingstoke: Macmillan, 1999.

Georgios Terizakis is a Research Fellow at the Institute of Political Science at Darmstadt University of Technology. He is currently working on his doctoral thesis on "Democratic Governance and Civil Society in Greece".

Annette Elisabeth Töller is Assistant Professor of Political Science at the University of the Federal Armed Forces (Hamburg), Institute of Public Administration. She is working on a research project on "Voluntary agreements as an instrument for environmental policy in Germany and the Netherlands". In her PhD thesis she analysed the functioning of comitology in EU environmental policy (Leske and Budrich: Opladen, 2002). Among her recent publications are "European Union Environment Policy and New Forms of Governance" (co-editor with H. Heinelt, T. Malek and R. Smith) Aldershot: Ashgate, 2001.

Acknowledgements

The editors of this book are indebted to a large number of people who made the research on which it is based possible and who helped to bring it to fruition. They include, of course, the authors of the individual chapters, but less visibly our hardworking research assistant in Darmstadt, Georgios Terizakis, and our patient secretary in Bristol, Claudia Bittencourt. Members of advisory groups in Germany, Greece and Britain and participants at a workshop in Brussels also helped to shape the final outcome. Our publishing editor, at Ashgate has provided invaluable, and again patient, support. Subventions to underpin the production of this book have come from the Vice Chancellor's budget at the University of Bristol and from the budget of the Centre for Urban Studies in the School for Policy Studies at the University of Bristol. We are most grateful for this support. Finally, in acknowledging the key role of the Research Directorate of the European Commission in funding this project under the Fifth Framework Programme, we pay particular thanks to Angela Liberatore for her infectious enthusiasm and wise guidance. The text, of course, is the responsibility of the editors, not the Commission.

Hubert Heinelt
Randall Smith

List of Acronyms and Abbreviations

ACCA	Association of Certified and Chartered Accountants (UK)
ALARM UK	Network of Local Anti-Roads Groups in the UK
ANDYP	Local Development Agency of Piraeus Region (Greece)
ASAOS	Greek Eco-Labelling Council
BAM	Federal Agency for the Testing of Materials (Germany)
BAT	Best Available Technique
BATNEEC	Best Available Technique Not Entailing Excessive Costs
BDI	German Federal Industry Association
BEN	Business Eco Network (UK)
BGBl	Federal Law Gazette (Germany)
BMU	German Ministry for the Environment, Nature Conservation and Nuclear Safety
BMWi	German Ministry for Economic Affairs
BPEO	Best Practicable Environmental Option
BREEAM	Building Research Establishment's Environmental Assessment Method (UK)
BSI	British Standards Institution
BV	Best Value
BVQI	Bureau Veritas Quality International
CBI	Confederation of British Industry
CEC	Commission of the European Communities
CEI	Centre for Environmental Initiatives (Sutton, UK)
CEN	European Standardisation Committee
CHP	Combined Heat and Power
CNC	Computerised Numerical Control
COD	Chemical Oxygen Demand
DAU	German Association for the Accreditation and Supervision of Environmental Verifiers
DEFRA	Department of the Environment, Food and Rural Affairs (UK)
DETR	Department of the Environment, Transport and the Regions (UK)
DGB	German Federation of Trade Unions
DIHT	German Organisation for Trade and Commerce
DIN	German Standardisation Institute
DNVQA	Det Norske Veritas Quality Assurance
DOE	Department of the Environment (UK)
EA	Environment Agency (UK)
EAC	European Accreditation Organisation
EBEAFI	European Better Environment Awards for Industry

ECG	Environmental Co-ordination Group (Huntsman Polyurethanes, UK)
ECOSOC	Economic and Social Committee
EEB	European Environmental Bureau
EFA	European Foundation for Accreditation
EFQM	European Foundation for Quality Management
EKBY	Greek Biotope Wetland Centre
ELBA	East London Business Alliance
ELOT	Greek Organisation for Standardisation
ELP	East London Partnership
EMAS	Eco-Management and Audit Scheme
EMS	Environmental Management System
EP	European Parliament
EPA	Environmental Protection Act (UK)
EPB	Financial Programme for the Implementation of EMSs (Greece)
EPET II	Operational Programme for Research and Development (Greece)
EPP	European People's Party
ESYD	Greek Accreditation Council
ETI	Ethical Trading Initiative
ETUC	European Trade Union Confederation
FHG	Airport Hamburg GmbH
GGET	General Secretary of Research and Technology (Greece)
GM	Genetically Modified
H&S	Health and Safety
HASP	Health and Safety Plans
HERRA	Greek Recovery and Recycling Association
HMIP	Her Majesty's Inspectorate of Pollution (UK)
ICC	International Chamber of Commerce
ICLEI	International Council for Local Environment Initiatives
IDeA	Improvement and Development Agency (UK)
IEA	Institute of Environmental Assessment (UK)
IEM	Institute of Environmental Management (UK)
IEMA	Institute of Environmental Management and Assessment (UK)
IIP	Investor in People
IÖW	Institute for Research in Environmental Economics (Germany)
IPC	Integrated Pollution Control
IPPC	Integrated Prevention and Pollution Control
ISO	International Standardisation Organisation
IUKE	Department of Trade and Industry's Inside UK Enterprise Programme
KEDKE	Central Union of Municipalities of Greece
LGMB	Local Government Management Board (UK)
LPG	Liquefied Petroleum Gas
MEP	Member of the European Parliament
NACCB	National Accreditation Council for Certification Bodies (UK)
NEEC	Not Entailing Excessive Costs
NEF	New Economics Foundation (UK)

NGO	Non-Governmental Organisation
NIFES	Consultancy (UK)
NRA	National Rivers Authority (UK)
NTUA	National Technical University of Athens
NVQs	National Vocational Qualifications (UK)
OPRA	Operator Pollution Risk Appraisal
PCB	Polychlorinated Biphenyl
PIRA	Printing Industry Research Association (UK)
PR	Public Relations
PRISM	Award for Environmental Improvement Company of the Year (UK)
PuSH	Public Sector Helpdesk (UK)
R&D	Research and Development
RDC	Regional Distribution Centre (Sainbury's, UK)
RSPB	Royal Society for the Protection of Birds (UK)
SA	Societé Anonyme
SAGE	Strategic Advisory Group on the Environment (ISO)
SCEEMAS	Small Company Energy and Environmental Management Assistance Scheme (UK)
SDU	Sustainable Development Unit (UK)
SEA	Single European Act
SERM	The Safety and Environmental Risk Management rating agency (UK)
SEV	Greek Employers Association
SLUNC	Sustainable Land Use and Nature Conservation (Sutton, UK)
SME	Small and Medium Sized Enterprise
SPDR	Staff Performance and Development Review
SRU	Standing Committee on the Environment (Germany)
T&G	Transport and General Workers Union (UK)
TCO	Swedish Labour Union
TEC	Training and Enterprise Council (UK)
TEE	Greek Technical Chamber
TGA	German Accreditation Organisation
TPM	Total Production Maintenance
TPU	Thermoplastic Polyurethanes
TQM	Total Quality Management
TUSDAC	Trade Union and Sustainable Development Advisory Committee (UK)
UBA	Federal Environment Agency (Germany)
UDP	Unitary Development Plan (UK)
UGA	Committee for Environmental Verifiers (Germany)
UKAS	United Kingdom Accreditation Service
UNICE	Union of Industrial and Employers Confederation of Europe
VALPAK	A collective scheme to capture, recycle and handle packaging waste
VDI	Association of German Engineers
VOCs	Volatile Organic Compounds

WTO/TBT	World Trade Organisation/Technical Barriers to Trade
WWF	World Wildlife Fund
YHOP	Ministry for Planning, Housing and the Environment (Greece)
YPEHODE	Ministry for the Environment, Spatial Planning and Public Works (Greece)
ZERI	Zero Emissions Research Initiative

PART I
POLICY OVERVIEW AND
ANALYTICAL FRAMEWORK

Chapter 1
Introduction

Hubert Heinelt, Britta Meinke, Randall Smith and
Georgios Terizakis

1.1   The Purpose of the Book

It was the purpose of the research project "Achieving sustainable and innovative
policies through participatory governance in a multi-level context" (funded by the
European Commission under the 5th Framework Programme on Research and
Development) to find out *under what circumstances participatory governance leads
to sustainable and innovative outcomes.* It is assumed that there is a link between
participation on the one hand and sustainability and innovation on the other, such
that participation leads to a higher degree of sustainable and innovative outcomes.

The theoretical basis for such an assumption and its further conceptual
elaboration were discussed at a conference on "Democratic and Participatory
Governance: From Citizens to 'Holders'" held in September 2000 at the European
University Institute in Florence.[1] The papers focused on the current debate about
governance, and one of the core questions was: What does the shift from
"government to governance" imply in respect to participation? "If one is trying to
design an arrangement for participatory governance, one has to provide convincing
answers to two questions: (1) Who should participate? And, (2) how should they
participate?" (Schmitter 2002, 58). This is not "that simple – especially since the
rules that are most likely to facilitate the mechanics of governance may not be the
ones that are the most likely to conform to democratic principles" (Schmitter 2002,
68). Is the traditional idea of citizenship appropriate for legitimising participation in
governance arrangements – or do we have to go beyond such traditional notions?
Schmitter offers an answer to this question with his "holder" concept, where
"holders" are individual or collective actors who possess specific qualities or

---

[1]   The papers from this conference have been published separately (see Grote/Gbikpi 2002).
Furthermore, the empirical results from the research project and some further theoretical
and conceptual issues were discussed at another conference held in October 2001 at the
Panteion University in Athens. The papers from this second conference are also published
separately (see Heinelt et al. 2003).

resources which entitle them to participate or which require that they should participate. But holder involvement does not lead per se to participatory governance, let alone sustainable and innovative outcomes. Therefore, specific (governance) arrangements have to be considered.[2]

1.2   Why EMAS?

Although these ideas guided the overall work of the project, the empirical research has a place of its own. We tried to identify opportunities for participatory governance which support a shift towards sustainable and innovative policy developments.[3] We therefore identified different governance mixtures in Germany, Greece and the UK, which reflected organisationally determined as well as socially and culturally embedded particular arrangements, from where policy change has to start. Furthermore, we took account of different territorial levels of government within which these arrangements were moulded by the political process.

The empirical work of the project covered two different policy areas: (a) water supply[4] and (b) enterprise oriented environmental management systems, specifically the Eco-Management and Audit Scheme (EMAS). The differences between these two policy areas can be found in the forms of participation which are linked to particular characteristics of the structure of policy networks. Water supply, on the one hand, is an issue which potentially concerns everyone. This implies an open network structure with unclear boundaries. Apart from a relatively clearly defined set of actors officially responsible for water supply, the spectrum of holders can be widened depending on the perception, articulation and organisation of interests. But the way interested parties are actually involved in "governing" a water supply system depends on the political options which enable them to participate. EMAS, on the other hand, suggests a closed network structure, because, by its nature as a management tool in organisations, it involves only a defined set of actors and a clear boundary at the level of the organisation. So participation is inherently restricted.[5] The two types of case studies are also indicative of different types of governance, different modes of interest intermediation and different types of actors. The water supply case studies indicate (at least in Great Britain) a shift towards the privatisation of governance and the exclusion of some political actors, resulting in less transparency and diminution of accountability. By way of contrast, EMAS is intended to foster environmental self-regulation by companies (and also public

---

[2]   Interestingly, similar expressions – like "stakeholder involvement" and "involvement of interested parties" – have played an important role in the decision making process on the EMAS II regulation at the EU level, as described in Chapter 3.

[3]   This was in part done in order to identify opportunities for EU intervention to promote these developments by setting up specific governance arrangements institutionally through EU legislation.

[4]   For the results of the empirical work on water supply, see www.geog.ox.ac.uk/cgi-bin/ dir.pl?~bpage/files.

[5]   Conceptual reflections on the different features of policy networks are addressed in Jordan/Schubert (1992) and van Waarden (1992).

authorities) which are controlled indirectly by the state and answerable to the general public and certain external and internal actors (from verifiers and customers to the neighbourhood and employees), thereby leading to greater involvement by actors and to transparency of environmental effects.

It is this characteristic of fostering environmental self-regulation which makes EMAS an interesting example of what can be called "modern governance". Without going into detail about the main elements of this scheme (which are described in section 3.2.3) it is important to stress that EMAS encourages enterprises to voluntarily adopt policies dedicated to legal compliance and continuous improvement in environmental performance[6] beyond what is required by law. According to the decision-making process at the EU level (see Chapter 3), EMAS is seen as an instrument to address the complexity of environmental impacts at the site level. As such, it was designed to deal with failures of environmental protection resulting from command and control instruments. In addition, it is an aim of the instrument that environmental benefits should be achieved, which would otherwise not be addressed.

Participating organisations establish a management system for the site that is designed to evaluate its environmental impact, to set goals for future improvement, and to carry out regular audits of environmental protection measures and their effects. To ensure that each site's management system conforms to the EMAS standard, they are subject to an external verification procedure that is carried out by an independent, accredited verifier, who has to validate a published environmental statement. Through the requirement to publish the validated statement, it is possible for the general public to act as an additional control agent on the enforcement of environmental law as well as on the environment policy developed for the site itself.

Environmental systems in general, and more specifically EMAS, can be seen as a part of a broader trend that is fundamentally changing the way businesses and policies are regulated. More precisely, EMAS is creating *new forms of governance* in three different ways.

First, EMAS can be interpreted as creating in some countries a new form of *Government-Market Governance*. The objective of EMAS is to reduce regulation by the state and to further self-regulation on the part of companies/organisations. Traditionally, command and control policies, which make up the core of most countries' environmental regulations, have sought to bring about environmental improvements by setting strict emission limits as well as prescribing the industrial technologies and processes needed to meet these limits. These policies are typically applied uniformly across industry regardless of local economic or environmental conditions, and they impose stiff penalties on violators. Business in general is more open to the use of such instruments (and especially market instruments such as eco-taxes or emissions trading). Governments intend to create stronger capacities for self-regulation by playing a more limited role in the EMAS "beyond compliance policy".

Second, EMAS can be seen as creating a new form of *Market-Civil Society Governance*. In principle, by putting the general public, or more precisely the

---

6   The standard of "the best available technology not entailing excessive costs" (Article 3a of the regulation) indicates the purpose of the desired improvement.

consumer or the neighbourhood, in an "observatory position" in relation to the individual site through the publication of the environmental statement, EMAS tries to increase the transparency of the registered sites in general. One could argue that it is the aim of EMAS to enforce better environmental performance in the light of public scrutiny.

Third, the revision of EMAS, EMAS II (see sections 3.5 and 3.6), focused more on the participatory aspects of internal decision-making and the implementation of environmental management systems within a firm or organisation. EMAS can here be seen as creating a new form of *Governance within a site*. The intention of EMAS is to change environmental performance on a site by the increased participation of interested parties/actors (employees, experts).

Table 1.1 lists the number of registered EMAS sites in April 1999 and October 2001. However, this rough overview can be misleading. First, Germany is the front runner in terms of the absolute number of sites, but if one relates the figures for example, to GDP (i.e. to the economic development of a country) or to population,[7] the picture is different. For both indicators Austria gets a higher ranking than Germany and Denmark is nearly the same as Germany. France gets a lower ranking than Greece when GDP is taken into account. Second, one can get the impression that EMAS is unimportant. Even in the case of Germany less than 6 per cent of all sites in the manufacturing sector were registered in 2001. But one has to take into account that EMAS is taken up unevenly by companies, not least according to their size (as will be shown in the case studies in this book). To put it precisely: bigger firms are taking up EMAS more than smaller ones. This means that thinking about the importance of EMAS has to take into account (i) differing distribution by the size of sites and (ii) the importance bigger sites have economically in respect to the workforce and total turnover. An earlier survey (see section 8.3.1) undertaken at the end of 1997, revealed that in Germany "only" 699 sites (i.e. 1.6 per cent of all sites in the manufacturing sector) were registered, but 22.9 per cent of the registered EMAS sites had more than 1,000 employees and only 0.4 per cent between 1 and 49 employees. We have not calculated how many people were employed in total by the EMAS sites with more than 1,000 employees or those with less than 50 employees. However, it is clear from the official German statistics (German Statistical Office 1997) that in 1997 sites with more than 1,000 staff employed about 30 per cent of all employees and were responsible for about 36 percent of the total turnover of the manufacturing sector in Germany. Yet these large sites comprised only 1.6 per cent of all manufacturing sites. On the other hand, sites with less than 50 employees comprised more than 50 per cent of all manufacturing sites but employed only 10 per cent of the workforce and were responsible for only 7 per cent of the turnover.

---

[7] See for such a ranking of ISO 14001 certifications www.inem.org/htdocs/iso/speedometer/speedometer-4_2001.html (accessed January 2002). Data for GDP and population have been taken from this source.

Furthermore, the development of EMAS in different counties has to be considered in relation to the growing importance of management schemes in general and especially of ISO 14001 as another environmental management system.[8] The differences between EMAS and ISO 14001 as well as attempts at the EU level to "bridge" both schemes are addressed in sections 3.3 to 3.6.

Table 1.1   The Development of EMAS and ISO 14001 by Country

Country	Registered EMAS Sites		Certified ISO 14001 Companies	
	October 2001	April 1999	January 2001	August 1999
Germany	2523	2020	2400	1400
UK	78	70	1400	1009
Greece	4	1	57	0
Austria	359	169	223	200
Sweden	211	139	1370	645
Denmark	174	99	580	350
Spain	151	30	592	234
Italy	68	13	724	150
Finland	36	18	526	191
France	35	29	802	285
Netherlands	27	21	800	443
Belgium	11	6	130	130
Ireland	8	6	150	82
Japan	-	-	5338	2338
USA	-	-	1340	480

Sources: Official EMAS register and
www.inem.org/htdocs/iso/speedometer/speedometer-4_2001.html (accessed January 2002).

The emphasis here – as shown in Table 1.1 – is that most of the frontrunners in relation to EMAS are also the frontrunners for ISO 14001, although some countries (Sweden and Denmark) are more in favour of ISO 14001.[9] This is due to a strong awareness of the usefulness of environmental management systems and the fact that a lot of companies in these countries have implemented both schemes. However, there are other cases – like the UK – where a strong awareness of the usefulness of environmental management systems can be assumed, but sites/organisation are showing a marked preference for the ISO standard. Finally, there are countries – like Greece – where neither scheme has been well developed.

---

[8]   The growing importance of management schemes in general and in relation to EMAS is addressed in the case studies.
[9]   This was measured by the numbers of registered or certified sites/organisations in relation to GDP and population; see www.inem.org/htdocs/iso/speedometer/speedometer-4_2001.html (accessed January 2002)

Table 1.2   EMAS sites in the UK and Germany by Sector, January 2002

Economic sectors (NACE Code)		UK	Germany
11	Oil and gas	21	-
14	Mining	-	20
15	Food	-	252
17-19	Textiles/Clothing/Leather	2	62
20	Wood	-	57
21	Paper	2	63
22	Printing	4	95
23/24	Chemicals/Petroleum products	16	252
25	Rubber/Plastic	2	149
26	Glass	4	61
27	Basic metal material	3	63
28	Metal products	4	259
29	Machinery	2	159
30	Office machinery	1	11
31	Electronic machinery	-	76
32	Telecommunication	-	39
33	Med./Optical Instruments	-	39
34/35	Cars, vehicles, trailers	4	137
36	Furniture etc.	-	73
37	Recycling	-	27
40/41	Energy and water supply	7	75
51/52	Wholesale/Retail trade	-	53
55/63	Tourism and travel agencies	-	35
60-62	Transport	-	14
65/66	Banks/Insurance	-	19
75	Local authorities*	2	66
80	Schools/Universities	-	21
85	Hospitals	-	55
90	Waste/Waste water	-	29
XX	Waste Recycling	4	188
YY	Others	-	53
	Total	78	2523

*   It is worthy of note that just two local authorities in the UK are listed. The UK EMAS
    Register lists another 52 sites (44 of them in the London Borough of Sutton) under Article
    14 of Regulation 1836/93. See Table 4.2 below.

Source: www.emas.org.uk (accessed January 2002).

Beyond the number and size of sites, there are, as shown in Table 1.2 – referring to Germany and the UK[10] – also some considerable differences between the economic sectors. Leaving aside local government, in the UK EMAS sites are concentrated in the oil and gas and chemical industries (with 26.9 and 19.3 per cent respectively) followed by energy supply (with 9.0 per cent). In Germany the chemical industry is again important (9.5 per cent and together with the rubber and plastic sector 15.4 per cent of all sites). In contrast to the UK, the food sector as well as the sectors of metal products, electronic machinery and waste recycling are important in Germany (with 10.0, 10.3, 6.3 and 7.5 per cent respectively).

Given these figures the question arises: why has EMAS been implemented so differently – and with what consequences for participation, sustainability and innovation?

1.3 Outline of the Book

This book starts with a clarification of key expressions and related theories and concepts (Chapter 2). A number of questions are addressed. What is the core notion of sustainability? What is the specific notion of sustainability as something that should and can be reached by EMAS? The idea of innovation is not developed in detail, because it is relatively self-evident that EMAS, as a management scheme developed for and adopted by enterprises or sites mainly in the manufacturing and to a lesser extent in the service sector, should improve products, production processes and organisational structures in general as well as address their environmental impacts. The second chapter ends with reflections on participation, and more precisely, on the question of who can participate in EMAS and with what possible outcomes.

The third chapter focuses on the development of EMAS at the EU level. The crucial questions are: Who influenced the concrete policy output, i.e. the EMAS regulations of 1993 and 2000? What were the key "design principles" (Schmitter 2002) and obstacles for participatory governance arrangements (especially at the site level) that were discussed and promoted by those involved in the decision-making process and agreement on the joint text of the EMAS regulations? This chapter also deals with the development of EMAS in terms of the implementation of the regulation through the committee established by Article 19 of the EMAS regulation of 1993. Special attention is paid to the decision-making process on EMAS II and its implementation through the Article 19 committee.

Chapters 4, 6 and 8 present the findings based on scrutiny of the administrative systems which were set up in the three member states – the UK, Germany and Greece – to make EMAS work. Because of differences in national institutional arrangements and in environmental policy, Articles 6 and 18 of the 1993 EMAS I regulation allowed member states to set up country-specific systems to implement the regulation. The member states made use of this opportunity, and therefore EMAS has had to work within quite different systems. Do institutions matter? This

---

[10] The four Greek EMAS cases are distributed over the sectors of clothing, printing, chemicals and furniture (see Chapter 7).

is the basic question addressed in these three chapters, including a short description of the historical development and main features of environmental policy in the three member states. The question "Do institutions matter?" is linked to the issue of the number of participating sites. The chapter also focuses on the important issue of the options available to particular actors (holders) to participate in the implementation of EMAS in the different member states and whether or not (and to what extent) these options have an impact on the acceptability of EMAS to enterprises, the general public and specific interested parties (holders).

Chapters 5, 7 and 9 deal with case studies at the site/company level in the three member states and scrutinise at the "operational level" (Kiser/Ostrom, 1983) how environment management systems work and what conditions foster or hinder participation as well as lead to sustainable and innovative outcomes.[11] The empirical work was based on analysis of documents (especially the environmental statement but also other publications and written material about the sites) and on interviews. The interviews at each site were conducted with the same kinds of "holders" using a common topic guide. The interviewees were the owner or the general manager of each company, the environmental manager, an employee and the verifier. Some consultants and knowledge holders (from universities, research institutes etc.) were interviewed where they had played an important role. Additionally, representatives of local government and/or the neighbourhood were contacted in those cases where environmental conflicts or problems had emerged.

In summary, we combined case studies at the site level with analysis of EU level decisions on legislation and its implementation at the national level, because we wanted to see how "higher" level decisions as well as differing national institutional settings affect particularly governance arrangements at the "operational level" (Kiser/Ostrom 1983), and consequently also affect available options for participation which can lead to (specific degrees of) sustainability and innovation.[12]

---

[11] The sample of sites has been chosen on the basis of surveys of all EMAS sites in the three countries, using the criteria of different sectors and sizes (number of employees) to reflect the overall features of EMAS sites. Furthermore, we have chosen "champions of change" because the limited number of cases allowed us only to look at "good practice" where we could expect to find the conditions fostering participation as well as sustainable and innovative outcomes. For further reflections on case study selection, see the last two paragraphs of section 2.1.

[12] Furthermore, by this methodological multi-level approach we have also used Jan Kooiman's (2002) distinction of three governing orders (meta governing, second order governing and first order governing) with their different forms of participation or participatory governance (as summarised by Grote/Gbikpi 2002). *Meta governing* and linked forms of participatory governance (i.e. participation in deliberation/dialogues about the problem definition and about the "appropriate" political solution) were addressed by analysis of the debates at the EU level on the EMAS directives. Additionally, the debates at the national level in the three countries were analysed on how to implement and organise eco-management and audit schemes. *Second order governing* was addressed through analysis of the decisions in the institutional settings and on the legal basis of EMAS at both the national and EU levels. *First order governing* and the related forms of participatory governance involving those possessing some quality or resource to solve a concrete problem or to resolve a specific conflict (see Schmitter 2002, 63) were analysed through the case studies at site level.

Chapter 2

Sustainability, Innovation, Participation and EMAS

Hubert Heinelt and Annette Elisabeth Töller

2.1   The Theoretical Debate

The aim of the research project was to examine the conditions *under which participatory decision-making leads to sustainable and innovative outcomes.* It is assumed that there is a link between participation on the one hand and sustainability and innovation on the other.

Theoretical debate on participation in a broader context of democratic theory and on sustainability and innovation suggest that there is some kind of interrelationship (see Schmitter 2002). Participation is traditionally seen as a means to legitimise the decision-making procedure. In addition, according to some traditions in democratic theory, it is assumed that participation produces "better results" (for an overview see Schmalz-Bruns 2002). This is argued on several grounds. First, one basic assumption of traditional democratic theory, starting from the idea of the natural rights of man, is that those who are affected by a decision also have to be given a right to participate in the decision. Thus, even if the final decision is not based on their ideas, they have had a say and a chance to make their argument heard. Second, participation by a broad range of interests, if undertaken in an open and free way, makes everybody give good reasons for their position. This can help to eliminate both egotistical and logically wrong positions. Third, Charles Lindblom (1965) developed an argument called the "intelligence of democracy": those who are given the right to participate might have the relevant knowledge to help produce better results (see Heinelt 2002b).

This last argument is important with respect to innovation. The involvement of those possessing relevant knowledge is crucial for detecting the potential for innovation and to make innovation work. Furthermore, and related to the second argument, the involvement of or participation by relevant actors is important for the social acceptability of innovation, and therefore its applicability. This needs to be considered in the broader context that innovation is related to social change, and social change may imply a negative sum game, i.e. an unequal distribution of gains and losses resulting from innovation. Provided that through participation all those who are relevant to the solution of problems or the resolution of conflicts are involved, the chances of a positive sum game increase, whereby everybody gains or the losers are compensated through benefiting from overall gains.

In relation to sustainability, there is a tendency to see it not only as a desirable outcome, but as a broad process in which a wide range of resources (including the

11

resources of legitimacy) are brought into play. Because the sustainability ideal is seen to be inclusive, empowering and transparent, it is closely linked to participation (see O'Riordan/Voisey 1998, 3, 16).

However, whilst a close link or even a causal interrelationship between participation and sustainability may seem obvious, in the empirical research – especially in the EMAS case studies – one has to analyse *whether* they are linked or not. This means that although one can find best cases at the site level, one might also be confronted with cases where participation takes place without sustainable outcomes or there are sustainable outcomes without participation. Furthermore, even if the two coincide, it is necessary to examine closely whether one is the causal consequence of the other. This is likely to be extremely difficult to prove. Thus, in the empirical research one must be open to contingent results. In summary, the empirical research has to deal with three aspects: (i) participatory decision-making; (ii) sustainable and innovative outcomes; and (iii) a number of unanticipated situations.

The aim of the rest of this chapter is to operationalise the ideas of sustainability, innovation and participation in order to apply the theoretical framework to the empirical field of EMAS. The "conditions" under which they come together can, however, be derived only after completing the case studies at the level of the organisation and through careful comparative analysis.

2.2 Sustainability

2.2.1 Some Reflections on Sustainability

Sustainability as articulated at the Rio Conference in 1992 is generally seen as a concept which tries to accommodate the needs of present and future generations. Sustainability as an ideal aims at a situation in which current habits, activities and developments can be indefinitely continued in the future without endangering the fundamental conditions of life on planet earth. In doing so, the notion of sustainability covers the dimensions of *ecology, economy* and of *social needs*. This means that the perspective from which current habits, activities and developments are judged, has to be broadened. It needs to cover space (geographical extension), time and third parties whose interests have to be taken on board. Moreover, sustainability introduces a strong moral notion into areas where this has not hitherto featured.

Sustainability is so wideranging that it can be realised neither by simple top-down policies nor by following the conventional paths of decision-making. Rather, all kinds of players, local and global, those in politics and those in every sector of society have to be brought on board. As O'Riordan and Voisey note: "At the heart of sustainability is self-regeneration – of the soul as well as of economy, polity and society" (O'Riordan/Voisey 1998, 6). Thus, the responsibility of everybody is emphasised.

It would not be an exaggeration to state that in the field of environmental policy the debate on sustainability has led to a *paradigm shift*. Most elements of environmental policy aim at the regulation of private economic activities, mainly industrial activities. Under the traditional *liberal approach*, these activities are in principle free to develop, unless it is shown that they damage third parties (such as the neighbourhood or the water supply). This has two consequences. First of all, only certain activities can be the target of regulation. Second, actual or potential damage to someone or something results in a high demand for regulation. Even though the precautionary principle has brought about considerable change, sustainability brings in a new dimension. First of all, sustainability – if acknowledged by business interests – introduces a moral notion into their activities, which means that practices generally have to be *justified* and measured against the yardstick of sustainability. This is the case for all activities and practices, not just those which are potentially or actually harmful. This comes close to a reversal of the burden of proof and is the opposite of the liberal approach. Second, activities have to be measured by their environmental consequences and whether they can continue for ever which is a much stricter criterion than whether they clearly damage someone or something.

Thus, under the heading of the overall survival of the planet, the paradigm of sustainability overcomes the traditional antagonism between those who regulate and those who are the object of regulation. Furthermore, by stressing common responsibility, a *common moral aim* is introduced which cannot be ignored by any party (although it would be naive to think that this aim is seriously supported by all those who pay lipservice to the idea of "sustainability"). This could therefore present an opportunity for a threefold extension of the decisional horizon of business interests beyond traditional environmental protection approaches: an extension with regard to space (environmental effects beyond the boundaries of the site or organisation become an issue); with regard to time (future generations are taken into account); and with regard to third party interests only indirectly affected by a company's activities (such as environmental or social effects on workers, endangered species or on the availability of raw materials). Thus, the notion of sustainability underpins the arguments of those who had already taken these perspectives on board (such as environmental NGOs or environmental policy makers) but who had yet to come to grips with the dominant liberal approach which does not legitimise these perspectives.

2.2.2 The Notion of Sustainability and EMAS

Although sustainability as a general concept covers the dimensions of ecology, economy and social needs, it makes sense in terms of EMAS to focus particularly on the ecological dimension of sustainability. Of course, ecological and social sustainability cannot in the long run be achieved in an economically unsustainable environment. Thus economic sustainability is a necessary but insufficient condition for ecological sustainability. Similarly, economic sustainability can flourish in the context of, or be followed by measures to improve, social sustainability. However, this varies widely from site to site and from country to country, as the case studies show. Thus economic sustainability can be seen as a prerequisite for ecological sustainability and social sustainability, and it can be seen as a possible consequence of ecological and social sustainability. We are open to both aspects, but we do not see either economic or social sustainability as a necessary prerequisite for environmental sustainability.[1]

Many of the arguments outlined above on sustainability and environmental policy are particularly true in respect of EMAS.[2] As an integrated instrument addressing all aspects of the environment, EMAS reflects a holistic approach – the kind of approach sustainability needs. While the instrument as a whole is in line with sustainability issues, there are a number of features which need to be specifically emphasised.

First of all, its voluntary and integrated nature stresses the general responsibility of industry and business to embrace the aim of environmental protection and a sustainable future. In doing so, EMAS aims to stimulate an internal capacity for change and innovation and opens up the opportunity for companies (or organisations in general) to combine environmental protection, innovation and cost efficiency.

Second, legal compliance, also associated with the idea of substitution development,[3] reflects the notion of responsibility being taken for each site. It particularly stresses the need to stimulate internal capacity in order to get closer to the very demanding expectation of systematic compliance with statutory duties, instead of waiting for the arrival of compliance inspectors.

Third, the aim of continuous improvement emphasises the paradigm shift away from the liberal approach (regulation only of identified harmful activities) and stresses that there is constant potential for improvement with regard to environmental effects.

Fourth, the stringency with which specific environmental issues have to be addressed goes beyond the traditional environmental protection approach. For example, energy management and type of energy sources have to be considered as well as the use of raw materials, the minimisation, recycling and transport of waste, the avoidance of accidents, the training of employees, and the new design of production processes and products, as well as the overall structure of the organisation. The fact that a reduction in the use of resources or the production of waste can save a lot of money – thus proving that economic and environmental improvements can go together – is an additional benefit for the companies. However, the ideal of sustainability covers more than cuts in costs through saving energy or water. The threefold extension of the perspective (space, time and third parties) affecting managers as well as shopfloor workers, goes beyond traditional environmental protection and aims at sustainability.

---

[1] Reflecting the empirical findings, it does not seem sensible to suggest that a strong social dimension of sustainability is a necessary condition of ecological sustainability.

[2] See Chapter 3 of this book on the development of EMAS I where the sustainability debate was one of the three key arguments for advancing the idea of a regulation.

[3] Substitution in the context of EMAS means that controls and inspections based on environmental statutory obligations are reduced for those sites participating in EMAS, because legal compliance as one element of EMAS has been verified externally and thus can be assumed to have been achieved.

Fifth, the idea of including the supply chain, i.e. putting pressure on suppliers and trying to influence the behaviour of customers, is a crucial element of sustainability policy since it aims at penetrating the whole market with environmental protection measures.

Sixth, the idea of institutionalising an environmental management system makes environmental protection go beyond static concerns by guaranteeing that environmental protection is continually addressed and gets a voice in decision-making by both managers and shopfloor workers.

Seventh, the duty to publish an environmental statement is not only important for the credibility of the scheme, but is also an instrument that can in principle open the company up to an interested public which can then comment on what is going on within the company/site, another sign of a move away from the liberal approach.

2.2.3 Critical Features of EMAS and Sustainability

Based on experience gained from the implementation of EMAS in industrial sites (and on the results of our empirical research), there are a number of important features which are often forgotten during EMAS implementation when the main target is sustainable outcomes. For instance, a number of environmental issues can be omitted from analysis and particularly from improvement. This is especially true for transport which has very sensitive environmental dimensions and the production of energy. In addition, the means of securing raw materials and the environmental qualities of products, especially their disposal, are often neglected. In respect of legal compliance, the translation of new legal requirements into practical work is a problem that many companies face. Continuous improvement can turn out to be difficult for those sites with relatively low environmental impacts which might, after some time, be seen to have achieved the limits of possible improvement. With regard to the management system, the co-ordination and integration of environmental protection between different departments can prove to be problematic.

2.3 Innovation

It can be argued that technological as well as organisational innovations can help to address the different dimensions of sustainability in so far as they contribute to economic growth, to better environmental performance and to improvement in working conditions (see Weizsäcker et al. 1995).[4]

Innovations can be focused on the product, the production process and the structure of an organisation. In the context of EMAS, *innovation of products* implies first of all improvements in respect to their environmental impact. This can refer to

---

[4] From this viewpoint the meaning of sustainability as well as of innovation can be linked to a particular understanding of social change (as suggested in section 2.1). An emphasis on sustainable and innovative development (in the political and theoretical debates) can imply that the process of social change is considered

   (i) to be manageable as a positive sum game through innovations, that is by increasing public welfare so that nobody loses or those who do lose can be compensated, and

the use of raw materials, options for recycling and/or limitations of emissions directly related to the product. Such improvements can have economic effects (i) by decreasing the costs of the product but also (ii) by improving competitiveness in cases where customers are looking for environmentally friendly commodities. The social impacts of such innovations can be ambiguous. They can affect working conditions at the EMAS site positively (by banning dangerous substances), but the choice of a particular raw material could favour some producers rather than others, with consequent positive or negative social effects, not least in respect to employment options.

Innovations in the production process can be linked to challenges to produce a new or more environmentally friendly product. In the context of EMAS, this kind of innovation mainly implies attempts to avoid or minimise emissions, to reduce the use of natural resources or to substitute them by renewable resources, or to enhance recycling. These objectives may be reached by the use of new technology but also by a systematic reorganisation of the production process. Such environmental improvements may lead to cost savings, but they may also require costly new investments. As with product innovations, the social impacts of innovations on the production process can be ambiguous. They can affect working conditions positively through improvements in health and safety as well as "job enrichment", i.e. by increasing the responsibilities (autonomy) and the qualifications of employees. But through new technologies and new organisation of the production process, the responsibilities of employees can be diminished and their qualifications can be devalued. It is reasonable to argue that EMAS aims at the achievement of positive impacts because otherwise active involvement of employees could not be assumed. This can be construed as a crucial precondition to realising the environmental objectives of this particular management scheme.

Innovations in the structure of an organisation can be linked to systematic and on-going scrutiny, monitoring and control of what is happening over the whole site (not only in the production process) to avoid unintended environmental effects. EMAS is in this respect one management scheme alongside others which are addressing quality management, health and safety and so on. The growing importance of such schemes can be seen in the context of new technologically determined options *and* requirements to reorganise the division of labour and to structure the organisation less hierarchically. Insofar as such options and requirements are realised in line with a new "industrial divide" (Piore/Sabel 1984) or "flexible specialisation" (Lipietz 1991) in both the industrial sector and through the introduction of New Public Management ideas (Hood 1991; Schröter/Wollman 2001), particular difficulties arise for management which can be labelled as the

---

(ii) to be controllable with respect to the impact of innovation by a process of reflection (referring to space, time and third parties), which results in avoiding negative consequences.

Such an understanding of innovative and sustainable processes of social change not only emphasises participation with respect to the social acceptance of innovations, but new forms of participation are also needed to implement innovation in circumstances where it depends on the mobilisation and use of knowledge hold by particular actors/ holders.

tension between autonomy and control. On the one hand, autonomy has to be "granted" by management to achieve full efficiency and effectiveness, and on the other hand management has to remain in control in order to know what is going on in "their" organisation. Management systems – like EMAS – can offer a solution. They promote more decentralised responsibilities, more horizontal interactions and more involvement by employees, but they also secure for management, through the systematic gathering and documentation of what is being done in different parts of the organisation, evaluation and control of what is going on. Furthermore, through systems of continuous reporting and documentation, management can access decentralised dispersed knowledge, and it can ensure by these systems that such knowledge is not lost when "holders" disappear or try to withhold information.[5]

2.4 Participation

Although it is a challenging task to clarify theoretically the meaning of new forms of participation or participatory governance,[6] it is much easier to address the question of "Who participates how?" with regard to EMAS. The aim of this section is, using Schmitter's "holder" concept (see Schmitter 2002, 62ff), to describe the range of participants in the implementation of EMAS at the site level and how they can get involved.

The "holder" concept attempts to overcome traditional approaches of democratic theory which, by focusing on the citizen, give him/her a right to participate. Instead, there are so-called "holders" who are involved in an endeavour in various ways (by holding shares, stakes, through spatial proximity and so on) and who *by their involvement* have a *right to participate*, although the specific form and extent of participation may vary considerably (for a critical commentary, see Heinelt 2002a, 15-16). The "holder" concept seems to be able to capture theoretically the idea that the sphere of governance extends far beyond a state-centred view of policymaking (and a state-centred view of citizen involvement), and that participatory practices have not followed the extension of policymaking into the broader reaches of society nor the enhanced involvement of a wider range actors in the policymaking process.

In the context of EMAS, we can distinguish participants who are internal or external to the company or who participate directly or indirectly. They can participate directly in implementing EMAS or they can participate indirectly, for example by making certain demands on the company or by reading the environmental statement and responding to it. Furthermore, participation can take place at different stages of the EMAS cycle, starting with the decision at the site to participate in EMAS, followed by policy formulation, the initial review, the definition of objectives and measures, the establishment of the management system, the drafting of the

---

[5] This issue is addressed in the growing literature on knowledge management (see for example Myers 1996; Sanchez/Heene 1997; Edvinsson/Malone 1997; Brooking 1999; Probst 1999; Mertins 2001).

[6] See the books edited by Grote/Gbikpi (2002) and Heinelt et al. (2003).

environmental statement up to the external validation and verification, the registration, the publication and distribution of the environmental statement.

Last but not least, participation in the context of EMAS can be seen not only in relation to the decision to implement the scheme as such but also in relation to the routines of running the system, because a management system and audit schemes can open up new options for participation on an everyday basis. In other words, EMAS can alter the patterns of daily interactions on the site and it can create new horizontal forms of interaction, albeit operating alongside traditional hierarchical structures.

Most categories of the "holder" concept can be applied to EMAS at site or organisation level, though empirical work is needed to demonstrate whether these holders actually participate or not.

Within the company, the most obvious group is the *share holders*, who are usually the owner(s). Whether and to what extent they are involved in EMAS varies considerably depending on the size and legal status of the company. In those cases where the owner is the boss, he/she will be heavily involved in EMAS, although involvement may vary between the different stages of implementing the scheme.

Second, there are *status holders* within companies, for example, managing director, heads of department, works council representatives and safety representatives. These people are likely to participate formally in EMAS implementation because of their status. They could well be part of an "environmental team". The degree to which these roles are explicit or implicit depends in the main on the size and structure of the companies, but it also varies between different countries. For example, in Germany there is a greater tendency than in the UK or Greece to establish clear structures and to designate responsibilities formally. Beyond any explicit rules laid down by a company/ organisation, formally defined statuses can also be established by law, such as the role of representatives of employees on works councils in Germany.

Third, there is the category of *work holders*. Employees who can – and in terms of our approach should – play a full part in EMAS are an important category in their own right. Even though they hold interests, stakes, knowledge and status, subsuming them under one of those "holder" categories would hide their specific role. Whether they participate in EMAS and if so whether this happens at the decision-making stage (decision to participate) or during implementation can vary.

Beyond the company/site, there are a number of other holders. The fourth category is *knowledge holders* who in the context of EMAS are the external environmental consultants who play an extremely important role in the introduction of EMAS. Their participation mainly consists of giving advice on the application of EMAS to the specific site conditions and it usually ends after the first EMAS cycle.

Fifth, there are two different kinds of *status holders* outside the company, namely the external environmental verifier who participates at the end of the EMAS cycle in a strictly defined role, and the registration bodies, who participate by co-operating with environmental agencies and by accepting or rejecting registration. Local authorities responsible for the area where the site is located should be included in this category. They represent the "local state".

Sixth, there are *holders of a spatial location*, i.e. neighbours. Neighbours are in principle a very important holder group, potentially suffering heavily from environmentally damaging behaviour.

Seventh, there are *interest holders*, i.e. those persons or organisations who have an awareness of the site's activities and an interest in participation. In the context of EMAS these are in the main (local) environmental groups. They would be likely to participate indirectly by demanding that an environmental management system be established or by criticising the way it was being introduced.

Finally, there are *stake holders*, a category which covers all those who could potentially be affected by what happens at the site, regardless of who and where they are. For example, there are the very important groups of customers and suppliers. Whereas all of the other holders participate by using their "voice", customers have the very powerful instrument of choice. In choosing environmentally friendly products, they can exert strong pressure on companies to participate in EMAS. On the other hand, EMAS sites can put similar pressure on their suppliers.

By using Schmitter's "holder" concept the "question of political design" for participatory governance is answered in a particular way, i.e. by "choosing the apposite criterion according to the substance of the problem that has to be solved or the conflict that has to be resolved" (Schmitter 2002, 63). This seems to be both possible and plausible for EMAS, or for self-regulation in companies in general. But it has to be stressed that this implies that an answer to the "question of political design" is focussed on *effectiveness*, because only those possessing some quality or resource to solve specific problems or to resolve specific conflicts related to EMAS are given an entitlement to participate.[7]

But there are other standards (or norms) than effectiveness by which actors are entitled to participate – namely, *legitimacy* and *ethics*.

The standard (or norm) of legitimacy is introduced by the politically defined and legally required procedures and contents of EMAS. The required procedures and contents are crucial because the scheme and the credibility of its implementation are politically and socially accepted only when they are secured. In this context the role of *status holders* can be seen as crucial. They have their *status* by being "recognised by the authorities ultimately responsible for decision and formally accorded the right to represent a designated social, economic or political category" (Schmitter 2002, 63), and it can be argued that they have been "recognised" and formally accorded a specific right of representation because otherwise the scheme runs the risk of losing or never acquiring legitimacy. In short, this is political and societal acceptance.[8] In this respect, the "granting" of a status does address the effectiveness of a scheme, but not in a mere economic or technical sense but in a purely political way. Furthermore, status holders possess a *specific* quality or resource that entitles them

---

[7] This is the dimension of governing characterised as *first order governing* by Jan Kooiman (2002). Other governing orders – second order governing and meta governing (as Kooiman phrased them) – are characterised by other standards (or norms) than effectiveness – namely *legitimacy* and *ethics*.

[8] On this, see the discussion at EU level in the context of EMAS II about the involvement of "interested parties" and employees (section 3.2.4) and also the discussions in the three member states about the involvement of certain status holders (representatives of collective actors) in the national organisational framework for the implementation of EMAS.

to participate because this quality or resource is "designed" by political decision and can be enforced by the authorities ultimately responsible for that decision.

Maybe the term "ethics" is misleading because of its vagueness. In this context the term can be linked to policy specific "images" or paradigms, underpinned by communicative rationality based on dialogue or, more broadly, deliberation. The development of "images" implies a *linguistic coding of problem definitions and patterns of action* which are binding through a common understanding of what is "good" or "bad" or even "appropriate" or "inadequate" for society as a whole or for a specific sector of society. This has to be established through dialogue and debate. In our case, sustainability in general as well as "good housekeeping" and "corporate governance" represent such "images". Provided such "images" have acquired the status of a "meaning system" (Scott 1994, 70ff) shared by an organisation, they can influence the answer to the "question of political design" for participatory governance just like statuses "granted" to actors by political authorities. This is because they influence the choice of the "apposite criterion" for those who should be participating according to the nature of the problem to be solved.

To sum up, participation in EMAS is mainly determined by measures of the effectiveness of this business-related instrument of self-regulation. Nevertheless, legitimacy plays a crucial role. To guarantee political and societal acceptance of the instrument some actors are assigned a specific political status which permits them to participate. Furthermore, particular "images" or common understandings are needed to agree the "apposite criterion" for identifying those who should participate.

Although these reflections are important for answering the question "Who participates?", another question remains: "How do different holders participate?" or "What does participation under specific circumstances mean?".

In examining participatory governance in the context of EMAS, it is clear that share holders "participate" in a way different from other holders. Based on their property rights, owners can take final decisions.

There are also differences between the multiple status holder category and the other "holders". As mentioned above, status holders possess a specific entitlement to participate which is "designed" by political decision and which can be enforced by the authorities ultimately responsible for making decisions. In other words, status holders possess a bargaining power embedded in the politically based "right to represent a designated social, economic or political category" (Schmitter 2002, 63), be it the EMAS system as such (in the case of verifiers) or the local state or employees with particular status, such as works council or safety representatives. Such status and the bargaining power derived from it can limit the discretion of share holders. This depends on the specific definition of status, which can vary between countries, as in the case of representatives of employees. Bargaining power and different kinds of participation can be acquired in various ways.[9]

Bargaining power can also be acquired by holders through other means than status. For instance, customers have exit options. They can choose the product or the service of another company. But this kind of bargaining power need not be expressed

---

[9]   Schmitter (2002) has elaborated a set of principles which are important for structuring governance arrangements in such a way that one can talk about participatory governance.

by direct communication. Even the perception of an alternative choice (in the market) can be influential. Furthermore, "holders" (especially neighbours as "holders of a spatial location") can acquire bargaining power by threatening to go to court, where there is evidence that they are suffering from the illegal activities of a site.

The other holders (especially knowledge holders, but also work holders when their interests are not covered by status holders) have just the one option of participating through debate or consultation by trying to convince others by putting forward "good reasons". They can rely only on "voice" and not on bargaining power based on status, exit options or threatening to go to court. Nevertheless, the "voice" option should not be underestimated in the context of EMAS, where ignorance and uncertainty are crucial issues and knowledge is a scarce resource.

However, it should be emphasised that participation in a context like EMAS is mainly based on communication. As a mode of interaction, communication is not only crucial for consultation or debate (or arguing) but also for negotiation (or bargaining).[10] However it makes a difference only if

- *consultation* does take place, and those who are "consulted" feel free to scrutinise the reasoning of the "consultant", or
- *negotiation* is appropriate where "good reasons" can get support from some kind of bargaining power.

---

[10] For the debate on arguing and bargaining (or debate and negotiation) see Heinelt et al. 2001, 13.

Chapter 3

The Negotiation and Renegotiation of a European Policy Tool

Annette Elisabeth Töller and Hubert Heinelt

3.1 Introduction

This chapter deals with the development of EMAS as a policy instrument. In addition to providing a full account of decision-making on EMAS at the European Union level we focus on the following specific questions: (i) How are the conceptual roots of EMAS I connected to the sustainability debate? (ii) Which "holders" participated in the decision-making on EMAS? (iii) How far are the participation of "holders" and issues of sustainability addressed in EMAS I and in other (competing) schemes? (iv) What role does the Article 19 Committee (established by Article 19 of the EMAS I regulation) play with regard to collective learning in general and with regard to participation and sustainability issues in particular? (v) How should EMAS II be judged compared with EMAS I and the experience of this scheme, not least in relation to participation and sustainability issues?

The chapter has five sections. Section 3.2 deals with the development and content of EMAS I; section 3.3 examines the relationship between EMAS I and ISO 14001; section 3.4 deals with the role of the Article 19 Committee in respect of the development of EMAS I and II; section 3.5 analyses the negotiations over EMAS II; and section 3.6. deals with the role of the Article 19 Committee in operationalising EMAS II. Section 3.7. offers some concluding comments.

3.2 EMAS I

3.2.1 The Conceptual Origins of EMAS

The first EMAS regulation, adopted on June 29 1993, has three conceptual origins based on different sets of ideas and debates.

The first set of ideas for EMAS was a debate on the aims and instruments of environmental policy, a debate which is still taking place in the EU member states as well as within the EU institutions (see Heinelt et al. 2001). At the beginning of the 1990s it was felt that the traditional "command and control" regulation was no longer adequate given its limited success in reaching defined objectives on the one hand and the threat of increasing environmental damage on the other. A number of criticisms were put forward. First, traditional regulatory law was separated into different elements (water, air, soil) and thus did not focus on the possible cumulative or dangerous interrelationships between the elements. Second, traditional regulatory law was – by its nature – not able to foster proactive and innovative behaviour but rather produced defensiveness on the part of those affected. Third, traditional law tended to concentrate on remedying existing environmental damage and took less account of the need for precautionary action (Heinelt et al. 2001). Finally, regulatory law has suffered from a severe implementation deficit, since those affected have not been motivated to comply and public authorities have lacked the information (knowledge) necessary to enforce the law (see Majone/Wildavsky 1979; Mayntz 1980). As a way of tackling these problems, the EU's Fourth Environmental Action Programme recommended in 1987 integrated policy instruments. At the same time, the 1986 Single European Act (SEA) laid down the precautionary and the "polluter pays" principle (Article 130r, Paragraph 2 EEC Treaty). The Fifth Environmental Action Programme stressed the idea of shared responsibility for environmental protection between the state and different "target groups" in society instead of unilateral statutory intervention. One of the most important of the target groups was manufacturing industry (European Communities 1993, 26-31).

The second set of ideas for EMAS emerged from the dialectical development on the perception of industrial risks and associated regulation that took place during the 1980s throughout Europe, following the earlier debate in the United States (European Commission 1992a, 5). On the one hand, after a number of severe industrial accidents during the 1970s and 1980s, such as Seveso, Flixborough and Basel, there was by the late 1980s strong public pressure for stricter statutory control and regulation of industrial hazards, especially in the chemical industry. On the other hand, the chemical industry itself, having abandoned the "defensive approach" of the seventies (Franke/Wätzold 1996, 178), wanted to reduce the risk of hazards. They wanted to do so for several reasons: first, because ecological damage produced high liability costs; second, because the chemical industry had started to suffer from a bad image as a "dirty industry"; third, because in some extreme cases (such as Boehringer Ingelheim) accidents which damaged the environment and human health led on to site closures (see Franke/Wätzold 1996, 177). However, the industry still wanted to avoid draconian legal regulation. Thus, the chemical industry (beginning in Canada and the USA in the 1970s and following in Europe in the 1980s), started to develop voluntary initiatives – such as environmental reporting and internal audits. These audits had three objectives: they aimed to reduce the risk of hazards internally; they were geared to improve the bad image of the industry with both the general public and shareholders, and they tried to show to the general public as well as to politicians that self-regulation was in hand and thus there was no need for regulation by the state.[1] One development was the "responsible care" initiative which was introduced by the Canadian Chemical Producers' Association in 1984 and taken up more widely by the chemical industry. One difficulty with the responsible care initiative was that it lacked credibility (Franke/Wätzold 1996, 175). In 1988 the International Chamber of Commerce (ICC) presented a paper which

---

[1] One possible kind of draconian regulation was a measure similar to the American "Superfund Amendment and Reauthorization Act", under which certain companies were required to publish data on their emissions (see European Commission 1992a, 6).

tried to define the core elements of eco-audits on industrial sites (ICC 1989). The paper stressed the voluntary nature of audits and was meant for internal debate within industry. Subsequently, national standards bodies, started by the British Standards Institution (BSI), began to develop standards for systematic environmental audits, taking the basic framework from quality management audits. The European Commission, against the background of the debate on suitable instruments for European environmental policy, used the ICC paper around 1990 as a basis for developing an obligatory European policy tool (on the subsequent development, see Malek/Töller 2001).

It was only after the first informal and a first formal EMAS proposal had been presented, that the third set of ideas began to influence discussion. The debate on sustainability entered the arena, after the Rio Conference in 1992 had identified sustainability as a collective policy aim. This notion built in part on an already developed concept and in part provided a general label for a number of ideas. In the context of an eco-management scheme, sustainability focused on increasing the responsibility of industry and on introducing an integrated environmental perspective into the management of industrial sites. This required a five-fold extension to management thinking: (i) content (environmental concern as one aim among others); (ii) responsibility (for external effects); (iii) space (beyond the boundaries of the industrial site); (iv) time (long term perspective beyond the everyday affairs of the company); and (v) third parties (such as future generations or endangered species).

3.2.2 The Long Negotiation Period over EMAS I

There was a strong British influence in the lead Commission department, since, by April 1992, the UK's British Standards Institution had developed a standard for environmental management systems (BS 7750, see Koch 1993, 29). This meant that the British had, on the one hand, the expertise, and on the other, a strong interest in exporting their concept of environmental protection to the European level.[2] Whereas the early informal drafts envisaged compulsory participation by industrial companies (Franke/Wätzold 1996, 175), the first formal draft proposal (European Commission 1992a), presented in March 1992, based on Article 130r of the EEC Treaty referred to voluntary participation, as a result of intense and united lobbying of the Commission by industry. This first official proposal already contained the main elements of the later regulation, namely that companies would have to go through a certain procedure and produce an environmental statement. After external verification the company would be put on an EMAS register. The proposal already included the two fundamental pillars of the scheme, the continuous improvement of environmental performance and compliance with relevant environmental laws. Different from later versions, but similar to "new approach measures,"[3] Article 4 of the proposal envisaged a mandate to the European Standardisation Organisation

---

2   For the concept of regulatory competition, see Héritier et al 1996.
3   The regulation was not a "new approach measure", since these had to be based on Article 100a of the treaty (see Waskow 1997, 59).

requiring it to develop a European standard for environmental management schemes. Accordingly, no comitology procedure was identified at this stage. Once the mandatory nature of the scheme had changed to a voluntary approach, lobbying by industry ceased to be united. The UK and German lobbyists remained active, though representing opposing positions (see below).

The European Parliament, through the consultation procedure, based on an October 1992 report by MEP de la Graete (European Parliament 1992), proposed 75 amendments to the Commission's proposal (OJ No. C 42/44). Both Parliament and the Economic and Social Committee (ECOSOC) stressed the need for specific criteria for the assessment of environmental improvement and argued for greater employee participation (OJ 1993 No. C 42/44).

In the Council, the main reservations were those of the German delegation, since the German Economics Ministry, representing German business and commerce, was totally opposed to the introduction of a European management scheme. Furthermore, they feared competitive disadvantage for German industry, because the high level of environmental regulation in Germany would be much more demanding than in any other member state. Moreover, introducing a management scheme was not familiar to the "engineer driven" German approach, which focused rather on introducing new technologies for improving the environmental performance of companies (Franke/Wätzold 1996, 193). Another reason for opposition by German business was that in Germany standardisation in the field of environmental management had got under way quite recently.[4] Thus, German companies would not gain any advantage, whereas British and Irish companies would. The different interests in Germany, however, did not unanimously share a negative view of EMAS. The Environment Ministry as well as the Bundesrat (Länder chamber) were generally in favour of the scheme, proposing only minor amendments. Environmental NGOs in Germany, although generally in favour of an environmental management scheme, rejected the Commission's draft, as they criticised the lack of objective assessment criteria. They feared that the scheme would "turn into a public relations instrument rather than the internal revision instrument originally intended" (Hildebrandt/Schmidt 1994, 69).

At the end of 1992, a number of minor modifications were made, following the experience of a pilot study undertaken by the Commission. These modifications included a move towards BS 7750, aimed mainly at making the regulation easier to apply at site level, but also to introduce clear separation of the internal auditor from the external, independent verifier (see European Commission 1992a; Franke/ Wätzold 1996, 187).

By the end of 1992 the Council had reached a compromise on the text of the regulation. In achieving this, the UK delegation had played an important role. By this stage, the proposal had gone through major changes (Waskow 1994, 7). For example, the explicit reference to delegation of standardisation functions to CEN (The European Standardisation Committee) was removed. Instead, a procedure,

---

[4]   In October 1992 DIN (Deutsches Institut für Normung/The German Standardisation Institute) and the Environment Ministry set up a committee which did not meet until February 1993.

including a comitology committee composed of member state representatives, was introduced. This compromise was, however, still opposed by the German delegation. Whereas the Council had hitherto required unanimity to pass this kind of regulation, it was the fact of the Maastricht Treaty coming into force in 1993 which finally persuaded the German delegation to "move". They realised that they could be overruled by 11 other member states. It thus seemed more sensible to give up outright opposition and to aim for some amendments which would be of benefit to Germany (Malek/Töller 2001, 45). They did this by putting forward their ideas for clarifying the assessment criteria for environmental improvement. As a consequence, it was written into the regulation that BAT (best available techniques) were to be applied in order to achieve environmental improvement. This approach reflected the German, technology-based approach and improved the competitive position of German companies, since according to German environmental law, they had to introduce best available techniques anyway. The British delegation, however, was successful in adding their concept of "NEEC" (not entailing excessive costs). At the Council meeting on 22-23 March 1993 the final text was agreed. Although the Commission had presented a modified proposal based on the amendments proposed by the EP and presented to the Council on 16 March 1993 (O.J. 1993 No. C 129/3), this was not taken into account by the Council. On 29 June 1993 the regulation was formally adopted.

3.2.3 The Main Features of the Scheme

Under this regulation, industrial sites[5] could, on a voluntary basis, put themselves forward for EMAS registration. If they decided to engage, they had to fulfil a number of obligations. The system had two core elements. They have been described as two "pillars", which the participating site had to commit itself to: one was the continuous improvement of environmental performance and the other one was compliance with existing legal requirements.[6]

The procedure (see Article 3 of the regulation) starts with evidence demonstrating that there is an overall environmental policy which includes the commitment to continuous improvement and to legal compliance.[7] At the same time, an initial review is carried out. This review starts by identifying the environmentally relevant activities of the site, and then analyses and evaluates their environmental impacts. The regulation provides guidance on the kinds of environmental effects that have to

---

[5]   After revision at the end of the 1990s the scheme was extended to all organisations, not just industrial sites.

[6]   It has been argued that the duty to comply with legal obligations merely reflects a duty which already exists outside the scheme (see Lübbe-Wolff 1994, 361). There is, however, a gap between legal obligation on the one hand and real environmental performance on the other. Thus, making legal compliance subject to the scheme i.e. external verification, the publication of data and registration (including the possibility of being removed from the register) gave this obligation new force.

[7]   This company policy has to be adopted at the highest management level and has to be a formal document (Annex I A). It has to be based on good practice (Annex I D).

be considered.[8] Second, there is scrutiny of the relevant legal requirements (Annex I B) to establish how far compliance has already been achieved. By unpacking the broad environmental policy to focus on the environmentally relevant activities of the site and their evaluation, an environmental programme is constructed which (i) establishes clear aims for environmental improvement, including the measures for achieving these aims, (ii) specific deadlines and (iii) identifies the people responsible (Article 2c and 3c, Annex I A, C).[9]

Following this, a management system has to be established to implement the environmental policy and the environmental programme (Article 3c, Annex I B). The management system includes definition of responsibilities, communication arrangements and training of staff on environmental protection issues. Finally, an internal environmental audit has to take place (after the initial implementation of EMAS and regularly from then on) which systematically monitors procedures and policies for environmental protection on the site. Based on the findings of the audit, amendments are made to the environmental objectives and the management scheme. Then an environmental statement is written which aims to present the implementation of EMAS in a way suitable for the general public (Article 5). At this stage, the internal part of the procedure is over.

The environmental verifier is then invited to state that EMAS has been correctly implemented. In addition, the verifier is required to validate the environmental statement, in other words, to check whether the information provided corresponds to reality. When the verifier has concluded these tasks and has validated the site, it can be included on the EMAS register by the competent body and the environmental statement is published (Article 8). The site then receives what is called a participation statement which can be used for PR purposes in general, but not for product advertisement (Article 10). Registration is valid for three years. If the site wants to remain a EMAS registered site, it has to go through the whole procedure again from the beginning (i.e. conduct internal audits, update the environmental programme and statement, bring in a verifier etc.).

Although the legal form of a regulation meant that EMAS was directly valid and applicable (i.e. it did not need to be transposed into national law as directives do), nevertheless it needed specific member state administrative arrangements in order for it to work. Thus the regulation placed a duty on member states first to set up a system by April 1995 for accrediting and supervising environmental verifiers (Article 6) and second, by June 1994 registration bodies ("competent bodies") had to be established in all the member states (Article 18).

---

[8] These include: any kind of emission into water and air (including noise and dust), waste especially hazardous waste, contamination of the soil, the use of soil, water, air and other natural resources and their effects on the environment and the eco-system. These cover both everyday activities as well as specific circumstances (such as accidents) (see for further detail Annex I B.3). Technical procedures should be included as well as the environmental responsiveness of contractors and users (Annex I B4).

[9] The environmental programme should also identify measures for correcting the programme, if necessary (Annex I A5b).

3.2.4 EMAS I – Participation and Sustainability

Although the integrated and voluntary nature of the scheme reflects some elements of sustainability policies, implementation at site level revealed a number of weak points, with regard to both sustainability and participation.

As was pointed out in Chapter 2, sites participating in EMAS are not automatically on the road to sustainability. Rather, this is the case only in particular circumstances. Some of these circumstances are addressed in the regulation (such as the use of endogenous capacity for change and innovation, the need for legal compliance and of continuous improvement in environmental performance), but other aspects are not explicitly or systematically covered in the regulation. These include a range of environmental issues that do need to be addressed. For instance transport, products and indirect environmental effects were not adequately covered. Other missing issues included the life cycle approach, the entire supply chain (which was not possible because of the restriction of the scheme to the industrial sector) and the integration of environmental activities across all departments. The provisions of the regulation for the participation of relevant "holders" (see Chapter 2) are not very explicit. The role of environmental verifiers and competent bodies is defined in a relatively clear way, since they constitute a crucial part of the system. Top management representatives (as status holders) have to be committed to the production of the environmental programme (Article 3e). With regard to "work holders" (employees) there are (in the Annex) some provisions, which stipulate (i) that the environmental management system must ensure that employees are informed about the environmental policy and programme, and (ii) that measures to make them aware of the relevance of the application of the programme and of the environmental impact of their activities are introduced. In addition training in respect of environmental measures was required (see Annex I B 2) but no systematic participation by those employed in a company was expected. Contractors were also mentioned. The management system had to ensure that the procedures for the selection of suppliers were specifically mentioned. Overall, there was no strong emphasis on participation in the scheme, since it was not obligatory to have employees participate in all stages of the scheme nor to include suppliers and customers. Other "holders", such as neighbours or local environmental groups or other persons or groups who could be interested, were not mentioned at all.

3.3   EMAS and ISO 14001

As already mentioned, parallel to the development of EMAS, national standards bodies, such as the British Standards Institution (BSI) were working on national standards for environmental management systems. The British standard was adopted in April 1992 (Koch 1993, 29), followed by others in Ireland, France and Spain. At the international level, in 1992 the Strategic Advisory Group on the Environment (SAGE) at the International Standards Organisation recommended the development of an international standard for environmental management systems. In September 1996, ISO 14001, based in great measure on international standards for quality management (ISO 9000ff), was agreed.

ISO 14001 is of relevance only in so far as there are different kinds of relationships between the EMAS regulation and the international standard, which played an important role in the implementation of EMAS as well as in its revision. The core of the relationship between EMAS and ISO 14001 is intense competition between the governmental and the non-governmental scheme, although there is a strong tendency to deny this, especially by the Commission (see Wätzold/Bültmann 2000). The next sections outline the major differences between the two schemes and describe the different levels of interaction between them.

A common characteristic of both systems was that both of them adhered to the basic principle of pro-active and self-governing action on the environment by companies, and were thus voluntary. However, one fundamental difference is the fact that ISO is applicable world wide whereas EMAS is restricted to the EU. Also, EMAS is an official system which involves the state in setting up arrangements for the accreditation and supervision of verifiers, even though this involvement varies a great deal between different member states. Nevertheless, the impact of this typically hierarchical form of state intervention is reduced by the nature of the regulation ("procedural law") and by the voluntary basis of the system. ISO, on the other hand, is based on an agreement between the members of the international non-governmental standardisation organisation and is, therefore, essentially an instrument of self-government.

The ISO standard is seen to be more systematic and user-friendly than EMAS and major differences can be seen between them (see Dyllick 1995; Benchmark/ EEB 1995; Kottmann 1998). One of the most important differences is the restriction (with some exceptions) of EMAS I to industrial operations, whereas ISO 14001 is open to all sectors. Moreover (and partly as a consequence of this difference), the basic unit of EMAS I was the site, whereas ISO 14001 is based on the organisation as a whole. Focusing on a specific site allows (i) for a detailed environmental review inside the company and (ii) through the public environmental statement for a better identification of responsibilities from outside the company.

Another significant difference is that ISO, unlike EMAS, does not require a systematic initial review. Furthermore, the obligation in the ISO standard to continuously improve and prevent environmental pollution relates to the development of the company's management system and therefore only indirectly to the aim of improving environmental performance. EMAS lists a set of environmental issues and principles which have to be taken into account (Annex I C and D of the regulation), whereas, under ISO 14001, companies have a wider discretion in deciding which environmental impacts should be taken into account. There is no requirement to apply BAT in order to reduce environmental impacts. Furthermore, under ISO 14001, compliance with regulations is not given as much weight as in EMAS. Finally, under ISO 14001, neither the publication of an environmental statement nor formal registration (including the possibility of being removed from the register) are provided for. Therefore, after certification, there are no further direct or indirect control or monitoring mechanisms that can be drawn upon by the public or employed by registration bodies.

Practitioners point out that in practice the two schemes do not differ greatly, provided there is a basic willingness to take seriously the environmental externalities of company action (see Kottmann 1998, 170). However, it should be noted that ISO 14001 contains a number of non-binding "target provisions", such that company concerns are given greater priority than EMAS. In addition, external control has less weight in the ISO standard. All in all, EMAS can therefore be seen as the scheme that has higher environmental requirements compared with the non-governmental ISO standard, based, as it is, on an international compromise.

Three distinct levels of interaction between EMAS and ISO 14001 can be identified. The first level relates to the individual choice of a company to go for either EMAS or ISO. The second level addresses the circumstances under which ISO users can also secure EMAS registration and the related issue of how this is likely to affect the attractiveness of EMAS. The third level focuses on the revision of EMAS I and the negotiation on EMAS II. This third level is examined in more detail below.

3.4 Implementing EMAS: The Article 19 Committee

When analysing the implementation of EU policies, the importance of committee procedures (comitology) tends to be neglected because implementation is likely to be located at national level and the activities of the committees are seen in the main as very technical work by distant officials. If the role of comitology is acknowledged, it is seen as something non-transparent, undemocratic and broadly problematic. We do not subscribe to this view and argue that for EMAS the comitology structure opens up excellent opportunities for (i) participation by various kinds of "holders" and (ii) learning (see Töller/Hofmann 2000). Although the important role of comitology has been recognised in the area of Community environmental policy (see Töller 2002), the extent of this role is exceptionally wide with EMAS and is worth closer analysis.

When analysing the role of comitology in the policy cycle, it is important to distinguish formal and informal functions. Usually there are some formally delegated tasks in the area of implementation, only some of which have been realised in practice, whereas there are also a number of significant practical functions which are not mentioned formally in any document. For the Article 19 Committee (established by Article 19 of the EMAS regulation) there are three kinds of formally delegated tasks. First, there are amendments to the Annexes of the regulation "in the light of the experience gained in the operation of the scheme"(Article 17).[10] Second, there is the creation of guidelines on the frequency of internal audits (Article 4 Paragraph 2). Third, there is the recognition of national, European and international standards for environmental management systems and associated certification procedures (Article 12 Paragraph 1). In addition, on a more informal basis (i.e. without a formally required decision, but mentioned explicitly in

---

[10] Annex I lays down the requirements for the environmental policy, programme and the management system; Annex II defines the subjects of internal audits; Annex III lays down conditions for the accreditation and supervision of verifiers and describes their task at the site. Annex IV focuses on the public statement; and Annex V lays down details of the information that has to be presented to the competent body for registration.

the regulation), the Commission is expected to promote collaboration between member states on the accreditation of environmental verifiers. Of these formal tasks, only one has been put into practice. There have been a number of decisions on the recognition of standards. Comitology, by way of contrast, has taken a wide-ranging role in the implementation of the scheme which reaches beyond implementation into the stages of evaluation and reformulation of the policy. Be the work formal or informal, the secretariat of the committee is provided by the Commission which thereby sets the agenda for the meetings and has a major role to play.

In the following paragraphs the work of the Article 19 Committee is outlined and opportunities for participation by "holders" are identified (3.4.1). Next, there is an overview of the activities of the committee from the beginning of its work in the summer of 1994 until the summer of 1997 when the work of implementing EMAS I was completed and the agenda for developing EMAS II was determined (3.4.2).[11] The third section focuses on where in the work of the Article 19 Committee we can perceive learning about how the scheme was working and how the agenda for revision (EMAS II) was set (3.4.3).

3.4.1 *The Working Style of the Article 19 Committee and Opportunities for Participation*

There are two characteristics that in principle distinguish comitology committees from other committees. First, they comprise only government players (national delegations and Commission officials), and second they can pass legally binding measures, following a specific formal decision-making procedure, laid down in the regulation itself, based on the so-called comitology decision.[12] However, in the Article 19 Committee there are virtually two committees, a formal one and an informal one. The formal comitology committee in principle consists only of government players from the Commission and member states. This "real" comitology committee, comprising the national delegations, formally votes on legal measures proposed by the Commission (this has been the case only rarely, as outlined in section 3.4.2). In the informal committee there are, in addition to the government players, a number of other government players from non-EU member states and non-government players who "meet, quarrel, negotiate or compromise" (van Schendelen 1998, 12). Among these non-government players, there are representatives of interest groups (UNICE, ETUC and EEB), of CEN and ISO, of various national accreditation and registration bodies as well as of verifiers'

---

[11] Further developments in comitology are analysed in section 3.6.

[12] According to this procedure, the Commission presents a formal proposal to the delegates who have to vote on the proposal (votes are counted in the same way as in the Council). According to the "IIIa-procedure" which applies to the Article 19 Committee, the Commission requires a qualified majority in favour of its proposal in order to adopt the proposed measure. If there is no such majority, the Commission has to submit a proposal to the Council which can pass the measure as proposed by the Commission by qualified majority but for modification of the proposal needs unanimity (see Töller 1998, 191). If the Council cannot decide (either by qualified majority or by unanimity), the Commission can adopt the proposed measure.

associations and consultants (see Töller 1998). For the interest groups in particular, the invitation to participate in the committee's meetings is a privileged access to the area of decision-making in the implementation of EMAS, even though they cannot formally participate when it comes to legally binding decisions. This privileged access to the comitology debates by European non-government players explicitly representing specific interests[13] was in the mid 1990s rather unusual but has increased in the environmental sector since that time. In this respect the Article 19 Committee on EMAS made the running for others.

This comitology constellation with both formal and informal elements (including working groups focusing for set periods on a specific issue) represents the core of a wider network of organisations and individuals involved or interested in the implementation of the scheme. There are regular meetings of the Commission with accreditation bodies and competent authorities and there are workshops for verifiers on specific issues. Furthermore, there are regular contacts with the European standards organisation CEN and with ISO as well as with the European Accreditation Organisation (EAC).

If we adopt a multi-level approach, we can identify a typical multi-level constellation including corporate, individual, government and non-government players who can come from subnational, national, European or even international territorial levels. Based on the "holder" idea, we can see that there are only very few "holders" who are not represented and therefore have no access to the centre of implementation at the European level.[14]

It can be argued that those who are not government players, such as the representatives of specific interests on the committee, have only a "voice" to express their views but do not have a "vote" in order to back up their "voice" with power. This is true, but one can also argue that it is appropriate to have legally binding decisions taken by government players, whereas non-government players benefit from the power of the well based argument. Furthermore, non-government players who feel that their arguments ("voice") are not taken seriously in the committee still have the "exit" option. They can make their voice heard outside the comitology structure, be it in the general public arena (e.g. in a newspaper) or by contacting members of the European Parliament.[15] One big problem is that the different interest groups have varying levels of resources which of course influence their chance of making their

---

[13] There is an emphasis on "explicitly representing specific interests" because there are situations where, for example, industrial players as experts (nevertheless implicitly representing specific business interests) form part of national delegations. Furthermore, it is quite usual that organised interests have access to informal circles. However, diffuse interests such as environmentalists have been in a poor position due to lack of resources for lobbying. Thus inviting all of the relevant holder groups to attend comitology debates is a means of improving equal access to decision-making. Of course there are other policies where it would be more difficult to identify all the relevant holders and to select which of them should be given access, than is the case with EMAS (see Schmitter 2002).

[14] Those not represented are mainly neighbours and local groups which have not been included because of their decentralised and diffuse organisational characteristics.

[15] Both these strategies have been applied by the EEB (see Töller 2002).

arguments heard. Typically environmental activists and trade unions have fewer resources – in terms of people, time and challenging arguments – than industry.

Beyond offering access to the decision making circle for various "holder" groups, the Commission has also promoted Europe-wide contact between some "holders", such as certain groups of sites, verifiers, consultants or accreditation and registration bodies. This has provided opportunities for communication as well as to enhance the application and acceptance of the scheme.

3.4.2 The Work of the Article 19 Committee

The Article 19 Committee started its work on 12 December 1993. Between then and the summer of 1997 they met fifteen times, spending nearly 200 hours together discussing EMAS. However, as we can see in Table 3.1, only a small percentage of the meeting time was occupied with taking formal decisions. Less than 5 per cent of the time was used for voting on a proposed measure (Column 5). Long term and short term preparation and discussion of the planned measures (Column 3 and 4) make up only a little over one-third of the debate. Only five formal decisions on two issues were taken. One which occurred at a relatively early stage (in March and September 1995) was the recognition of three national standards (British, Irish and Spanish) as partly meeting the requirements of EMAS. The other was the recognition of ISO 14001 and associated certification procedures in March 1997 (OJ 1997 L 104/35, 37). Details of these formal decisions can be found elsewhere (see Töller 1998, 188). The only point worth emphasising is that after a total failure of the committee procedure at a very early stage (due to an unfortunate constellation of interests and poor preparation by the Commission) the Commission officials in charge rapidly learned how to secure the co-operation of committee members and how to build support among delegates in order to reach consensus beyond the qualified majority needed. No other formal votes were taken, not even on the redrafting of Annexes or on the auditing cycle.

Almost half of the meeting time of the committee (44 per cent) was occupied by issues about the implementation of EMAS (see Columns 1 and 2). Since the EMAS regulation needed almost directive-like *translation* into national law, the adoption of these measures by the member states was a point of central interest in the committee. Systems for the accreditation of environmental verifiers had to be established and competent bodies for the registration of sites had to be designated. Actions to achieve these objectives were regularly reported by each member state. Member states having problems in making their systems work were criticised by the Commission and other delegations. When, at the beginning of 1996, the scheme began to be applied, it was of prime interest to the Commission to collect the latest figures from the members of the committee on the numbers of verifiers accredited and sites registered.

Table 3.1 Items Discussed in the Meetings of the Article 19 Committee, 1994–1997

Meeting number, date	1 Practical aspects of implementation	2 Implementation in member states	3 Longterm discussion on planned measures	4 Concrete discussion on a proposed measure	5 Vote on proposed measure	6 Discussion on evaluation	7 Discussion on revision
4, 6/29/94	2	1	4	·		2	
5, 12/5/94	3	1	2			1	
6, 3/1–2/95	6	1	6	1	1		
7, 6/9/95	2	1	1	1			
8, 9/14–15/95	1	1	2	1	2	2	
9, 1/19/96	2	1	2				
10, 3/14–15/96	4	1	2			2	
11, 6/12–13/96	6	3	2			3	1
12, 9/17–18/96	2	2	3	1		2	1
13, 12/9–10/96	2	2	2	2		1	
14, 3/18–19/97	1	2	1	2	2	1	1
15, 6/30–7/1/97	1	2	2			3	2
	32 (28%)	18 (16%)	29 (25%)	8 (7%)	5 (4.5%)	17 (15%)	5 (4.5%)

Source: Minutes of the 4th–15th meetings of the Article 19 Committee.

About 16 per cent of the items on the agenda refer to these issues (see Column 2). Another feature of implementation was the *interpretation* of the central terms of the regulation. Since at the formulation stage of EMAS I there had been little experience of environmental management systems, the regulation raised more questions than it answered because key features of the scheme were ill defined. Pilot projects launched by the Commission and other initial practical matters reported by delegates or brought up in workshops clarified the nature of the problems. For instance, the meaning of central concepts such as "site", "initial review" or "legal compliance" was anything but clear. Furthermore, these concepts had to be translated into the language of the verifier's practical work. In this respect the development of the *General Guidance Document for Accredited Verifiers*, under Regulation 1836/93 and on the *Verification and Validation Approach* was a crucial part of the activity of the committee in co-operation with the Commission (European Commission 1996a, 4). The Commission officials on the committee had their own way of drafting this kind of document. They would first present a small working document on a particular issue to the delegations. Then the document would be discussed in the committee, and finally a couple of revised working documents would be patchworked together into a larger document. Nearly a third of the items on the agenda refer to this kind of work, translating the regulation into practice (see Table 3.1, Column 1).

3.4.3 Evaluation, Learning and Setting the Agenda for Revision

About 15 per cent of the meeting time of the Article 19 Committee was spent on the evaluation of EMAS I as shown in Table 3.1, Column 6. The initiative for debating the evaluation of EMAS came from both the Commission and the national delegations. The Commission had initiated pilot studies on the development of EMAS, for instance, on the application of the scheme in particular sectors or in SMEs. Some of these studies resulted in clear recommendations for revision. The Commission also distributed questionnaires to explore better ways of implementing EMAS in the member states. Furthermore, the Commission received information about the practical problems of applying EMAS from the workshops they organised for different groups of "holders", such as consultants, verifiers, accreditation organisations or companies. This information ranged from general questions about the application of EMAS to more specific issues on how to include suppliers or issues about deregulation. The results of all these workshops, studies and inquiries were then presented to committee members and evaluated in debate. Member states, too, had initiated evaluation studies on specific aspects of the implementation of EMAS which delegates presented to colleagues on the committee. In June 1998, when discussion on the revision of the regulation had already started, more general evaluative studies on the functioning of EMAS were presented to the committee by various national delegations and by the Commission (see Hillary 1998a and b; Steger 1998).

The results of evaluation were also distributed to relevant "holders". Furthermore, the European Trade Union Confederation used the Article 19 Committee to distribute its guidelines for participation of employees in EMAS (see minutes of the 13th meeting, 9 and 10 December 1996).

Evaluations of this kind were clearly intended to throw light on the functioning of the scheme. But learning took place not only through explicit evaluation of the scheme, but also through the whole process of implementation. In particular, the work of defining the central concepts of the scheme and then operationalising them for practical application meant that the committee members learnt a lot through their discussions. Without anticipating the analysis of the negotiation over EMAS II in section 3.5,[16] it should be noted that there were some occasions when the learning in the Article 19 Committee led on to certain decisions about the EMAS II regulation.[17]

Another significant experience for committee members was the drafting of guidance documents which was not formally stipulated in the EMAS I regulation. What national delegations learnt from this experience can be seen in what member states did when negotiating EMAS II in the Council of Ministers. They separated the more important from the less important issues. While providing a more precise definition for a number of important elements of the regulation (some categories were defined for the first time,[18] and others were defined in a more precise way[19]), others were delegated explicitly to the area of comitology where the Commission was formally entitled to adopt guidance documents.[20]

Other issues also link learning from evaluation to EMAS II. Based on pilot projects at the EU level and experiences in individual member states, there were also recurrent debates in the Article 19 Committee about whether the scheme could be extended to other sectors than industry. The Commission invested in some pilot projects in "new" sectors in order to find out "which elements of the regulation need to be changed in the revision and which sectors it is feasible to include" (Commission in the minutes of the 12th meeting of Article 19 Committee, 17/18 September 1996). The British delegation for example reported its experiences with applying EMAS in local authorities and the Austrian delegates spoke about extending EMAS to transport and banking (see 12th meeting, 17/18 September 1996). Other member states followed with further sectors, such as forestry and building. The "lessons" learnt became the basis for extending the scheme to all possible sectors, although at the beginning of the revision process there had been no overall consensus whether the scheme should be extended to just some more sectors or to all sectors (see European Commission 1997b).

---

[16] It can be argued that the analytical separation between implementation, evaluation and revision is not only artificial, but false, as the involvement of an "implementation committee" in the revision of the regulation shows (see Töller 1998).

[17] There are, of course, methodological problems in stating that there are causal interrelationships between learning in the committee and decisions made in the context of the revision. What we can point to are coincidences and plausible interrelationships. There is, however, some evidence for causal interrelationships as we can find in the minutes situations where specific aspects of evaluation are explicitly discussed with regard to the revision of the regulation (see for example the debate in the 14th meeting on 18/19 March 1997, day 2, "exchange of views on the revision of the regulation"). Furthermore, in the 15th meeting of the committee the Commission distributed a number of working documents with the heading "lessons learnt from the implementation of EMAS in SMEs and in sectors not covered by the regulation up to now", and the UK delegation distributed a paper on "lessons learnt" from implementing the scheme in local government.

[18] Such as "continuous improvement", "environmental performance" and "environmental impact".

[19] Such as "environmental policy" and much of Annex V (on the accreditation of verifiers).

[20] This formal delegation had advantages for member states in relationship to the informal work. Besides enhancing transparency, the Commission was obliged to formally adopt the guidelines, which enabled the member states to have an influence, because the Commission needed their formal vote.

Another example of the link between debates and learning in the committee to modifications made during discussions about EMAS II are the conditions of registration and suspension from the register. There were several discussions on this in the committee raising the problem that in EMAS I there were few common criteria for suspension, even though the Commission tried to give advice on the issue in the committee meetings and competent bodies met in order to debate these issues. EMAS II at least provides greater clarity on these issues and requires the consultation of "appropriate interested parties" (see Article 6).

Furthermore from surveys and other sources, members of the Article 19 Committee learnt lessons about the quality of (early) environmental statements. The June 1997 working paper of the ad hoc working group on the revision of the regulation stated that up to then "the environmental statement was underexploited" (European Commission 1997b). Consequently, the new regulation not only contains a new Annex of two pages which outlines the minimum content for an environmental statement, but the drafting of additional guidance by the Article 19 Committee is also provided for (see Annex II, 3.1.).

The issues listed above show that there was indeed learning in the Article 19 Committee about the functioning of EMAS I which had consequences for the formulation of EMAS II. They also demonstrate that the Article 19 Committee had a major impact on the revision of the scheme. From the summer of 1996 onwards, there were recurrent debates on the revision in the Article 19 Committee, as can be seen in Table 3.1, Column 7. On 2 and 3 June 1997 there was a major meeting of the "ad hoc EMAS review working group", a sub-entity of the Article 19 Committee, in Amsterdam, where the general direction of the revision was discussed and in part agreed. The following ten principles for the revision were identified:

- Continuity for the users – no major surprises;
- Work with ISO 14000 series – complementarity/compatibility;
- Meet stakeholder requirements – address the right audience;
- Regulatory coherence – link with other EU instruments;
- Ensure added value – cost savings, better image, regulatory benefit;
- Work with the market – feasibility, saleability and promotability;
- General applicability – for all sectors and all sizes of business;
- No technical trade barriers – consistent with WTO/TBT rules;
- Clear simple language – avoidance of doubt;
- Environmental performance – clarity about the resulting environmental gain (see European Commission 1997b).

These ten principles, which underpinned the general direction the revision should take, were presented to the 15th meeting of the Article 19 Committee at the end of June 1997. Although positions within the committee varied considerably, in attempts to weight and to elaborate on these principles, there was no significant dispute over the general direction of the revision. On the basis of this debate in the Article 19 Committee and another one in the 16th meeting of the committee in Stockholm, in December 1997 the Commission presented a first draft for a revised regulation. It contained many features of the later EMAS II regulation (closer alignment of EMAS and ISO 14000,[21] the extension of the scheme to all sectors, more clarity of language, as much continuity as possible plus various specific modifications, such as those relating to the environmental statement). Thus it can be stated that when early in 1998 the revision went out for wider consultation (although details of the draft continued to be discussed in the Article 19 Committee throughout 1998), the agenda for revision was already broadly set, as a result of the great influence of the members of the Article 19 Committee.[22] This was not surprising nor challenged. It made sense since it was the national delegates on the committee who had the relevant experience of the implementation of the scheme and who would be in the Council working group anyway, plus most of the "holders" who could participate in the debate directly or indirectly.

Overall the success of the Article 19 Committee derived from the fact that it functioned as a hinge first between the European and the national levels, and second between government players and non-government representatives (the relevant "holders") in the scheme. Whereas the Commission could commission pilot studies and other research, national officials from the ministries involved as well as relevant NGO representatives could provide relevant information and experience to the Commission, which the Commission could not get from elsewhere. On the other hand, these players, governmental, semi-governmental and non-governmental, secured a channel of influence at the European decision-making level. The fact that 95 per cent of the committee's meeting time was used for informal debate of different aspects of implementing, evaluating and revising EMAS shows that, in the vast majority of the meetings, all the participants of the Article 19 committee, including the non-government ones (the "holders"), were able to participate and have their views heard, even though some of them were not entitled to take part in the few formal decisions that were made.

Thus, a comitology constellation, going beyond formally legitimised government players by including almost all the relevant "holder" categories, represented a win-win arrangement for all involved.

3.5   Revising the Regulation: EMAS II

3.5.1 Consultation and the First Proposal

The 1993 regulation stipulated that by the summer of 1998, the Commission had to review the scheme in the light of the experience gained during its operation, and propose any necessary amendments (Article 20). As outlined earlier, the process of revision had already started in the summer of 1997 in the Article 19 Committee, when the Commission presented ten principles for revision to the national delegations and other representatives on the committee. Following the committee session in Brussels

---

[21] The forms of this alignment (EMAS as a building block of ISO 14001) were debated in the Committee a little later in January 1998.

[22] See section 3.6 for a discussion of the committee's work on the revision of the regulation.

and a second one in November 1997 in Stockholm, the Commission presented a first draft in December 1997 (European Commission 1997a).

Based on this draft proposal the Commission initiated a wider consultation process asking all interested parties for their opinion. It was not until October 1998 that the first official proposal was presented (European Commission 1998). In it, the Commission outlined the aims of the revision: improved contribution by EMAS to the overall aim of sustainable development; clarification of the relationship between EMAS and international standards in the area of environmental management; stronger employee involvement; better understanding by the public; increased coherence in the transposition of the system to national level (European Commission 1998, 13). There were several key features of the proposal. It was proposed that EMAS sites could automatically receive the ISO 14001 certificate. This was to be achieved by aligning EMAS closer with ISO 14001 in those areas where there was already a great deal of overlap (mainly the environmental management system, see European Commission 1998, 40; Krisor 1998). The extension of the scheme beyond the industrial sector was to include all organisations and not just some additional sectors so that the entire supply chain could be integrated into EMAS.[23] This necessarily meant abandoning the site-oriented approach and the introduction of the organisation as the basic unit of EMAS, just as in ISO 14001. In order to respond to those critics who feared a devaluation of the scheme if any kind of organisation, regardless of their environmental impact, could aim for EMAS, the Commission proposed to restrict participation to those organisations with "significant environmental impacts". The significance of these impacts has to be evaluated by the organisation itself (Annex VI, 6.4, European Commission 1998, 55).

In suggesting changes to the environmental statement, more flexibility was the aim. Although, companies with several sites would not have to present a statement for each site, those sites with significant environmental effects would have to be included (Annex III 3.4). The information would, however, not have to be contained in a single report. Instead, certain kinds of information could be presented to particular target groups. Even though there had to be "free access" to information, this did not mean that the organisation had, as had been the case hitherto, to print and distribute a document (Annex III, 3.6). Organisations were encouraged to use all available options (such as electronic publishing, libraries etc., see Annex III, European Commission 1998, 46). The environmental statement had to be updated internally and validated externally each year (see Annex III, 3.3).

No specific measures were proposed to address the expressed aim of enhancing employee involvement (see Article 1 Paragraph 2d of the proposal).

The auditing cycle could be decided locally based on the circumstances at the site (Annex II). Article 10 of the proposal stated that member states should check whether EMAS registration could be used to avoid duplication of effort by both

---

[23] In the course of the implementation of EMAS I, a number of member states (Austria, Denmark, Spain, the UK and in 1998 Germany) extended the scheme under Article 14 of the regulation to other sectors, such as transport, banking, tourism and public authorities (European Commission 1998, 7).

inspection organisations and competent bodies, which was a very vague allusion to possible substitution measures.

Finally, on comitology structures, the establishment of additional consultation groups for accreditation and registration bodies was proposed. The aim of these groups was to enhance harmonisation of the application of the scheme in the member states (see European Commission 1998, 7). In addition, the Commission suggested that the regulatory committee procedure be replaced by a management committee procedure which would have the effect of reducing member states' formal procedural power (see Töller 2002).

At the stage of finalising the first formal proposal by the Commission, there was a majority of national delegations and other relevant parties in favour of the general direction the regulation should take. However, there were three important players, the German delegation, the European Environmental Bureau (EEB) and the European trade unions, who were opposed to many of the elements of the planned revision. Their main criticism focussed on the strengthened links between EMAS and ISO 14001 and the proposal to drop BAT. They also argued for a strong commitment to legal compliance and to strengthened participation by employees (this last element was particularly supported by the EEB and the ETUC, see EEB 2000; Malek/Töller/Heinelt 2000). After the Commission's formal proposal had been presented, they changed their strategy. Instead of trying to influence the Commission and (other) national delegations in the Article 19 Committee, they turned their attention to influencing the European Parliament (EP) which had begun its first reading. Bearing this in mind when reading the following sections on the first and second readings in the Parliament, we can see that the three players were quite successful, even though in the second reading the Parliament dropped the BAT concept despite strong advocacy by the Germans and the EEB.

3.5.2 The EP's First Reading and the Views of ECOSOC

In accordance with Article 189c of the old treaty (before the entry into force of the Amsterdam Treaty) the European Parliament engaged in the revision of the EMAS regulation under the co-operation procedure. This was later changed and the co-decision procedure was adopted. The Economic and Social Committee was also consulted. In its first reading, the EP suggested nearly 60 modifications to the Commission proposal (OJ 1999 No. C 219(362, 399). ECOSOC made a modest number of comments and proposed very few modifications (see OJ 1999 No. C 209/43).

The Parliament did not want to see participation in the scheme restricted to those organisations with "significant" environmental impacts. It proposed a clearer definition of the commitment to legal compliance (which was also suggested by ECOSOC) and explicit reference to the possibility of offering substitution measures for EMAS participants. In addition, the EP proposed a greater commitment to involve employees and specified ways of achieving this. ECOSOC, too, suggested greater employee involvement. Both the EP and ECOSOC argued that the Best Available Technique (BAT, without reference to costs) would need to be applied in order to achieve environmental improvement. Furthermore, the level of discretion exercised by an organisation in determining, evaluating and communicating the environmental impact of its activity should be reduced. When using ISO 14001 as a building block for EMAS, organisations should have to demonstrate clearly that what they had done to achieve ISO status fulfilled the requirements of EMAS. The EP wanted to extend the audit cycle to three years (the Commission had proposed a one year cycle). In addition, it demanded more rigorous training for verifiers including a compulsory practical examination, which should take the form of conducting three verifications under the supervision of an experienced verifier (like the British system).

The EP suggested a clarification of the concept of "stakeholders" in the regulation. It defined stakeholders as those bodies which had a specific and legitimate interest in an organisation. Recognised stakeholders would be: employees, neighbours, financial partners (banks, insurance companies, shareholders), business partners and interest groups (consumers, environmental groups, trade unions). In particular, the environmental statement provides the opportunity for stakeholders to express their views. In addition, they would have a voice in the process of registration of an organisation.

The EP also wanted the fact of registration to be used on products and in advertising. Finally, the EP wanted the Commission to draw up guidelines on the design and content of the environmental statement, and did not accept the idea that it should not necessarily have to be printed and presented to the public (this last point was shared by ECOSOC, see OJ 1999 No. C 219/5, 362, 399).

Even though the Parliament's position based on its first reading was not entirely coherent, it nevertheless aimed at a greater level of participation by interested parties and at strengthening the credibility of the scheme. Following the entry into force of the Treaty of Amsterdam, the EP confirmed its first reading position under the co-decision procedure (OJ 1999 No. C 279/253, 274).

3.5.3 The Commission's Amended Proposal

In response to the Parliament's position, the Commission presented an amended proposal in June 1999 (European Commission 1999). However, only some of the Parliament's suggestions were included in the new proposal. 18 out of 59 were totally, partially or in principle accepted. The Commission dropped the requirement of "significant" environmental impacts and accepted stricter requirements for the accreditation of verifiers. It also accepted a clearer definition of stakeholders which, however, was not as clear as the one the EP proposed ("persons or groups, including authorities, which are interested in or affected by the environmental protection activities of an organisation").

The Commission did not, however, accept the BAT argument. It put forward the view that BAT was a concept that had its origins in industry and therefore was not appropriate if EMAS was to be extended to all sectors. (European Commission 1999, 3). The Commission did not take on board the EP's suggestions on enhancing employee involvement, as there were very different traditions on this in the member states. It was suggested that a more suitable way to address this issue would be a best practice guide to be produced by the Commission in co-operation with employee representatives (European Commission 1999, 4). Furthermore, the Commission did not accept the Parliament's proposal to put greater emphasis on legal compliance. It moreover criticised the idea of using the fact of registration on products or for advertising. The Commission held to the idea that the environmental statement did not have to be published but had to be accessible. Overall, the Commission seemed rather unwilling to accept the major amendments proposed by the Parliament.

3.5.4 The Council of Ministers' Common Position

The BAT concept strongly favoured by the German delegation was challenged by other member state delegations. However, a common position was reached at the Council meeting on 24/25 June 1999, although it was not published until February 2000 because it took time to find a new EMAS logo acceptable to all parties.

In its common position, the Council accepted 15 of the 18 amendments proposed by the Parliament and accepted by the Commission. In addition, the Council made a number of further amendments. Among the amendments which the Council accepted from the EP, there was a more explicit definition of stakeholders (though to ensure consistency with the terminology of ISO 14001, the notion "stakeholders" was replaced by "interested parties"). It also stressed the relevance of employee participation by stipulating the "active" involvement of employees (Article 1 Paragraph 2a) and made a number of suggestions on how this could be achieved (Annex I B 4).

Furthermore, the Council accepted the general idea that the EMAS logo could be used for advertising products, activities and services, but only under conditions which would have to be laid down in the comitology procedure (Article 8 Paragraph 2e).

The main amendments, apart from those accepted from the Parliament and mere editorial changes, focused on more detailed definition of the core terms of the regulation in Article 2, such as continuous improvement of environmental performance, prevention of pollution and differentiating between environmental objective and environmental target. The consequence was that by putting more detail into the regulation itself, less discretion was left to the informal comitology procedure (less in relation to what the Commission proposed and compared with the EMAS I regulation). In general, validation of the environmental statement had to be annual, though exceptions would be possible, if the Commission in comitology procedure could elaborate appropriate guidelines (see below). In addition, the Council introduced a completely new paragraph stating that the Community's and the member states' public authorities should examine whether, in defining criteria for public procurement, EMAS participation could be taken into account in order to encourage participation in EMAS (Article 11 Paragraph 2).

The Council also made amendments to the Annexes which, as recent research on comitology has pointed out, often contain the material core of a regulation (see Töller 2002). These amendments mainly served to clarify the requirements for EMAS recognition. The audit cycle should not be longer than three years (Annex II B 2.9). The Commission had left this open. On the question of whether a single printed version of the environmental statement had to be produced, the Council proposed a compromise between the Commission's and the Parliament's ideas. At the first registration and every three years from then on a consolidated and printed version of the environmental statement should be produced (Annex III), and more information had to be presented in the statement (Annex III 3.2). More emphasis was put on the requirement of independence on the part of the verifier, but a practical examination was not included (Annex V, 5.2). Examinations had to take place every 24 months (not every 12 months as proposed by the Commission). It was specifically mentioned that accredited verifiers wanting to work in another member state should not be discriminated against (Annex V, 5.3). Furthermore, the verification process was defined in a more detailed and precise way (Annex V, 5.4). Finally, the Council outlined a number of environmental issues that had to be dealt with more precisely. They included direct effects such as risks possibly resulting in accidents and the impact on biodiversity. Indirect effects included environmental performance and practices of contractors, subcontractors and suppliers (Annex VI, 6.2 and 6.3).

On comitology, it is interesting to note, first at all, that the Council replaced the management committee procedure proposed by the Commission by a regulatory committee procedure.[24] Second (and more unusually), the Council explicitly delegated to the Commission a number of tasks for elaborating guidelines following the committee procedure. This delegation of tasks stemmed mainly from the large number of issues for which no final solution could be found in the Council. At the same time, in delegating numerous tasks to the comitology system, the Council codified an informal (and quite successful) practice that had been developing in recent years (see Töller 2002). This practice placed a lot of detailed work on the content of the regulation at the decision-making level below legislation.[25]

This version of the regulation contained in Annex IV a new EMAS logo which had been developed and agreed on in the Article 19 Committee in the course of 1999.

All in all, the Council made the regulation more precise and delegated a substantial amount of detailed work to the comitology procedure.

The Commission issued a response on 10 March 2000, after the common position had formally been passed. It pointed out where the Parliament's proposed

---

[24] This was the old pattern for environmental policy, although the procedure now follows the new comitology decision with slightly modified procedures (see Töller 2002).

[25] These guidelines which the Commission has to elaborate in co-operation with the committee refer to guidance to the verifier on which entity of an organisation is to be registered (Article 2s Paragraph 1, already proposed by the Commission); exceptions to the ruling that the information contained in the environmental statement has to be updated and validated each year (Article 5 Paragraph 2a and Annex III 3.4, introduced by the Council); technical specifications on the use of the EMAS logo (Article 8 Paragraph 1, amended by the Council, the Commission had proposed to discuss the logo in comitology); the definition of circumstances in which the logo may be used (Article 8 Paragraph 2a, introduced by the Council); the definition of the conditions under which the EMAS logo may be used on advertisements for products, activities and services (Article 8 Paragraph 2e, introduced by Council); the definition of exceptions under which the logo may be used on products or their packaging or in conjunction with comparable claims concerning other products, activities and services (Article 8 Paragraph 3, introduced by the Council); examination of the possibility of disseminating best practice for informing the public on EMAS (Article 12 Paragraph 3, introduced by the Council); design and content of the environmental statement (Annex III 3.1, introduced by the Council); guidance on best practice for employee participation (Annex I 4). In addition, the comitology procedure has the power to: recognise standards and accreditation conditions (Article 9 Paragraph a and b already proposed by the Commission); identify exceptional conditions under which an organisational entity to be registered may be smaller than a site

amendments had directly or indirectly been integrated into the common position. The Commission supported the common position without reservation, since it saw that the amendments introduced by the Council served to clarify the text of the regulation, improve compatibility with ISO 14001 and stress the compulsory elements of the scheme.

3.5.5 The EP's Second Reading

In its second reading, the European Parliament proposed 27 amendments to the common position, some of which were repeated from the first reading and others were new suggestions (partly in reaction to the amended proposal, see European Parliament 2000a). In general, the suggestions made by the Parliament were even less coherent than those from the first reading. Although the idea of inserting the BAT concept into the regulation was passed in the Environmental Committee against the position of the rapporteuse, Mrs. García-Orcoyen Tormo, it did not obtain the necessary majority in the plenary session on 6 July 2000. Although the Socialists and the Greens were in favour of the BAT concept, the EPP was against it, putting forward the argument that the concept was applicable only to industrial sites.

The amendments passed focused on greater emphasis on legal compliance and on strict accreditation conditions with the overall aim of strengthening the credibility of the scheme. The Socialist Group was successful in defending their specific interest in the involvement of employees "and their representatives". The EP argued for greater recognition of the needs of interested parties. When informing the general public about the scheme, member states should co-operate with organisations representing business and consumers, trade unions and local bodies (Article 12 Paragraph 1). The EP did not accept the idea that the environmental statement should be published only every three years (Annex III Paragraph 3.1). It also made the point that the text of ISO 14001 should be inserted into the regulation (instead of just a reference to the fact that the requirements of ISO 14001 were to be covered). On verifier accreditation, the EP was seeking a tougher examination of specific knowledge and a practical examination on three sites.

On comitology, the Parliament demanded that the regulatory committee be replaced by a management committee, and furthermore wanted a number of the exceptions that were to be decided upon through comitology procedure to be dropped. Finally, the Parliament wanted the Commission to present a series of reports to the EP.

Overall, the EP once again defended the greater involvement of interested parties (especially employees and their representatives), stricter conditions for the accreditation of verifiers, a stronger emphasis on legal compliance and a clear

---

(Article 2s Paragraph 2, introduced by the Council); modify the Annexes of the regulation. The Council, however, amended the Commission proposal in that not all Annexes should be subject to modification by the comitology procedure. The development of the EMAS logo is not delegated to the Commission (as was proposed by the Commission), as the logo is already defined in the regulation. Finally Annex V, laying down the conditions for the accreditation and supervision of verifiers, cannot be changed in the comitology procedure.

obligation to present the environmental statement to the public. On comitology, it attacked the regulatory committee procedure and tried to restrict the range of decisions to be taken in comitology.

In response to the EP's second reading, at the end of July 2000 the Commission drafted another set of modified proposals (European Commission 2000a). It accepted three amendments totally or partially, namely tougher training for verifiers, encouragement of member states to offer incentives for participation in EMAS and a demand that registration fees be reasonable. Seven more amendments were accepted in principle, including the protection of SMEs, the integration of the text of ISO 14001 and the use of indicators. The Commission did not, however, accept the other amendments proposed by the Parliament. In respect of the stronger emphasis on legal compliance, the Commission put forward the argument that EMAS, as a voluntary scheme, should not replace environmental inspections in the member states (European Commission 2000a, 6). Furthermore, the EP's amendments on the validation of the environmental review and the publication of the environmental statement were not accepted. On comitology, the Commission supported the Council's position. Finally, the Commission did not accept the reporting duties the Parliament wanted to place on the Commission, since they believed it was more important to invest their resources in implementing EMAS. The EP could, however, address questions to the Commission at any time.

3.5.6 The Conciliation Procedure

Since the Parliament in its second reading proposed a number of amendments which at that stage were not acceptable to the Council (24 July 2000) under the co-decision procedure, the pre-conciliation procedure was initiated. The French Presidency had a major interest in getting the regulation passed before the end of 2000 because they considered the adoption of EMAS II as "their" project which they wanted to bring to a successful conclusion. After a number of negotiation meetings, on November 22 a compromise was reached in informal "trialogue" between Council, Parliament and Commission. The compromise contained a number of the suggestions made by the Parliament and thus can be interpreted as a considerable success for the Parliament (see European Parliament 2000b).

One major success for the Parliament was a stronger emphasis on legal compliance, which is now explicitly mentioned as part of the duties of participating organisations (Article 3 Paragraph 2 a) and in relation to registration (Article 6 Paragraph 1, 4th indent). This last amendment requires registration bodies to undertake a check with the competent enforcement authority before putting an organisation on the register.[26]

A second major success concerns the involvement of interested parties. In Article 1 (2) d, relating to employee involvement in EMAS, more emphasis is put on

---

[26] In terms of implementation the intriguing question will be, when is it the case that there is "non compliance"? Is it when minor breaches of the law take place, or only when a legal case is put forward? The latter seems more likely, although the threshold for de-registration might be higher than that for the denial of or delay in registration.

training and active participation by employees. The disputed issue of employee representatives' involvement is solved by the sentence "Where they request to be, any employee representatives shall also be involved" (the same wording is inserted into Annex I B 4). Furthermore, for the environmental statement, the information needs of "relevant interested parties" are explicitly mentioned (Article 3 Paragraph 2 c, Annex III Paragraph 3.1). In addition, the EP succeeded with its demand that "the environmental statement shall be made accessible to the public" (Annex III, 3.6). In respect of measures for promoting the schemes, co-operation with interested parties is suggested "where appropriate" (Article 12 Paragraph 1, second subparagraph). Explicit mention is made of industrial associations, consumer organisations, environmental organisations, trade unions and local institutions. More transparency is secured by inserting section 4 of ISO 14001 into Annex IA.

On verifiers' accreditation and supervision, stricter rulings were introduced in Annex V which deals with accreditation and supervision conditions. The verifier has to show knowledge of administrative requirements and of the "environmental dimension of sustainable development"(Annex V 5.2.1 b and c).

On the guidelines to be developed in comitology, the Parliament successfully argued for a sentence to be included in recital no. 16 that "in drafting such guidelines, the Commission shall take account of Community policy on the environment and in particular Community legislation as well as international commitments where relevant" (see also Annex VI, 6.1). This explicit mention of Community policy is intended to emphasise that EMAS is a Community policy and not an (international) standard and thus has to follow policy demands and not only the needs of business.

Finally, the Parliament was successful in requiring reports from the Commission on a couple of issues.

Overall the result of conciliation has to be interpreted as a major success by the Parliament in pushing through most of its demands which – although not entirely coherent – tended to strengthen the participation rights of interested parties and to improve the credibility of the system. It should be emphasised that without the "weapon" of co-decision the Parliament would not have been able to secure this kind of outcome.

3.6   Implementing EMAS II: The Article 19 (14) Committee

In contrast to EMAS I where the task of editing guidance documents was not explicitly mentioned (except for those on the audit cycle), EMAS II is much clearer on what is expected from the comitology procedure. As mentioned in section 3.5.4, extensive delegation to the Commission in co-operation with the Article 19 Committee (which after adoption of EMAS II is now the Article 14 Committee) has been provided for in the EMAS II regulation to edit guidelines containing detailed regulation on specific issues. These refer to:

- guidance with regard to the question about which part of an organisation is to be registered (Article 2s Paragraph 1);

- guidance to organisations on best practice with regard to employee participation (Annex I 4);
- guidance on the logo (Article 8);
- guidance on best practice for informing the public on EMAS (Article 12 Paragraph 3); and
- guidance on the design and content of the environmental statement (Annex III 3.1).

Whereas the uninformed external observer would be likely to expect the committee to start with the editing of this kind of "soft law" after the regulation had formally been adopted, in reality the situation was different. Parallel to the negotiation of the regulation, the Commission had in the course of 1999 already started to develop guidance documents on all the major issues referred to in the as yet unadopted regulation in order to have them operational shortly after the final adoption of the regulation (which came about after significant delay). This means that matters which could not be agreed on in the Council (because they were too technical or contested) were regulated in parallel in the comitology arena. As early as the spring of 2000 five draft guidance documents had been prepared:

- guidance on organisational structures suitable for registration to EMAS;
- guidance on the environmental statement;
- guidance on the use of the EMAS logo;
- guidance on employee participation; and
- guidance on the identification of environmental issues and assessment of their significance.[27]

As is not uncommon in the work of the Article 19 Committee, different players took the lead in the various working groups and made a substantial contribution, not least because the Commission did not have the staff capacity to provide all the necessary input. This gives those who do the preparatory work a good opportunity to put forward their ideas, but there is no guarantee that these ideas will find the support of colleagues on the committee. Whereas Sweden led the working group on the environmental statement, the German delegation focussed on elaborating the draft on the identification of environmental issues and assessment of their significance, and Italy led the working group on the unit of the organisation to be registered. ETUC took the lead in the working group on the involvement of employees, whereas UNICE was active in drafting the paper on the use of the EMAS logo. Papers drafted in the working groups were regularly presented to and discussed in plenary sessions of the Article 19 Committee.

---

[27] This last document was edited without any explicit reference to the regulation.

Brief analysis of some of these drafts is provided in the following paragraphs, but only in so far as they are of relevance for the themes of participation and sustainability.[28]

The guidance on *organisational structures suitable for registration* is aimed at organisations and verifiers and reflects the shift from the site principle to the organisation principle and of the inclusion of the service sector, where a predominantly geographical approach for defining the unit to be registered does not make sense. The document, which lists eight different kinds of organisational structure and how to deal with them, basically aims at preventing organisations with several sites from "cherry picking" – from registering only those parts of the organisation with the smallest environmental relevance and thus undermining the purpose of the scheme. It also provides guidance to verifiers to come to grips with the different organisational structures, when checking out an organisation (see European Commission 2000b). Overall, without analysing the document in detail, it is intended to create a credible system for the extension of the scheme to all types of organisation and to all sectors.

The guidance document on *environmental effects* is intended to clarify Annex VI in which long lists of direct and indirect environmental effects as well as reflections on establishing criteria for assessing the significance of the environmental effects have been laid down. All of them have to be taken on board by any organisation participating in EMAS. In doing so, the organisation has to take three steps: first, to identify all environmental effects; second, to define the criteria for significance; and third, to identify those environmental effects which are significant (see European Commission 2000c). Whereas the advice on how to deal with the various indirect environmental effects (such as product related issues, capital investment, choice and composition of services etc.) has yet to be included in the draft (extensive suggestions made by the German delegation have not been accepted), there is some advice on how to assess significance (see European Commission 2000c). Interested parties ("holders") are of relevance here in so far as their views on the significance of environmental aspects constitute one criterion among others, such as the potential to cause environmental harm, the extent and frequency of the harm and the existence of relevant environmental legislation (see Annex VI, 6.4c and European Commission 2000c).

The guidance document on the *environmental statement* can – like Annex III – be seen as a direct result of collective learning in the Article 19 Committee about the weak points in EMAS practice in respect of drafting and using environmental statements. Although there is now more flexibility in how to present the statement, organisations have to respond clearly to the "information needs of interested parties

---

28 Whereas the guidance papers on employee participation and on the environmental statement directly affect the opportunity for participation by "holders" in the scheme (to a lesser extent this is true also for the paper on the logo), the papers on the unit to be registered and on the environmental effects only indirectly affect the issue of sustainability. Their detailed contents determine to a great extent the credibility of the scheme, credibility which is seen to be a necessary but not a sufficient condition for reaching sustainability through EMAS.

[...] Openness, transparency and periodic provision of environmental information are key factors in differentiating EMAS from other schemes. These factors are also important for the organisation in building confidence with interested parties" (European Commission 2000d). On greater flexibility, the document suggests that placing the statement on the web is a good and cost effective way of making the statement available to a large number of people. However, it must be possible to offer a printed version to those interested parties who do not have access to the web. It is important to include the same type of information in subsequent statements in order to present data that are comparable over time so that interested parties can identify significant trends.

Particular emphasis is placed on the kind of information that should be provided for the different types of "holders". In this case, the users of the environmental statement are important "holders". For example, for the local community the health implications of products and emissions might be of great interest, as well as information on the number and kind of complaints received by the organisation and how they were addressed. Customers are an influential group of "holders", but it is more difficult to identify their information needs, since the relationships between an organisation and its suppliers and its customers are likely to be long term and of an indirect nature. In this respect, the document is not very helpful. Employees are mentioned as an important "holder" group in respect of the environmental statement. It is likely that they would be most interested in the link between environmental issues and their working conditions, plans for inhouse training in the environmental field and the implementation of the environmental management scheme. Financial institutions and investors are another important group of "holders" with regard to the environmental statement. They would in the main be interested in the environmental strategy and performance of an organisation at the corporate level, the relationship between environmental and financial information, compliance with legal requirements, the quality of the environmental management system and any risks arising from dangerous processes and substances. The final category mentioned in the document are "other social partners", mainly consumers and their organisations and environmental NGOs. They are mainly interested in the environmental policy and performance of the organisation in a local context and on "topical issues in the political realm or in the media", such as recycling activities, the elimination of toxic substances or the origin of certain raw materials (see European Commission 2000c).

As a measure to operationalise the new Annex III, the guidance document on the environmental statement constitutes an important step towards much better communication between organisations and "holders".

Finally, the guidance document on the *participation of employees* emphasises that the involvement of employees in implementing EMAS is not to be seen as an additional burden to management but as an opportunity to develop a more effective management scheme and to ensure the continuous improvement of environmental performance. In addition, employees' involvement can improve morale and give rise to pride in working for a "non polluting" organisation (see European Commission 2000e). Organisations are encouraged to introduce specific procedures for employee involvement, proof of which can be provided to the verifier. Employee involvement should take place at all stages of EMAS implementation, i.e. policy formulation, review, the establishment of an environmental management system, the development of the environmental programme and the drafting of the environmental statement. Employee involvement should be supplemented by education and training measures on environmental issues which may well go beyond EMAS. Furthermore, the guidance document recommends the introduction of suggestion and reward systems.

These documents have not been adopted formally, first because the legal basis for their adoption has not yet been established, second because after the completion of the conciliation procedure, a check has to be made on whether the draft guidance documents are still in line with the final outcome of conciliation. However, we can conclude that in operationalising EMAS II (before the regulation was formally adopted) the Article 19 Committee did a useful job to make the scheme work, particularly in respect of participation and its overall credibility which, as argued earlier, is a necessary but not sufficient condition for achieving sustainable outcomes through the application of EMAS.[29]

3.7 Conclusions

A number of conclusions can be drawn on the relationship between participation and sustainability in the context of decision-making on EMAS at the European level. It should be noted that these conclusions are not based on the results of our work at the national level nor on our local case studies.

Although the main conceptual origins of EMAS are older than the sustainability debate, EMAS I was considered to be an innovative policy tool at the time of its adoption because it reflected a number of ideas taken from the sustainability debate. Thus, whilst EMAS can be seen as a tool for achieving (or at least working towards) sustainability at company level, that does not imply that all companies registered with EMAS have achieved sustainability objectives.

In the transposition into national law and application at site level, a number of deficiencies became clear in the EMAS scheme, including issues of participation and sustainability. Participation by "holders" (interested parties) was not emphasised and many dimensions of sustainability, such as products, indirect environmental effects and supply chains, were not explicitly or adequately addressed.

In the world of environmental management systems, EMAS has a serious competitor in ISO 14001. There is, however, a major difference between the two standards in that ISO 14001 clearly puts less emphasis on participation and sustainability issues.

The implementation of EMAS I through the committee procedure (comitology) allowed the majority of relevant "holder" groups to participate. Major players (member states, environmental interests, trade unions and industry) were given the

---

[29] That does not mean that there is no criticism that can be made of the details of the guidance documents. The aim of this section was to show the involvement of the Article 19 Committee in operationalising the scheme, especially with regard to the issues of participation and sustainability.

opportunity to engage in implementation, evaluation and reformulation in the Article 19 Committee. A wider net of informal structures, consultations, workshops and seminars offered access to further relevant players. By establishing this open, informal consultation process, the Commission offered opportunities for participation and influence and obtained a wide range of information based on experiences from specific "holder" groups. This has had a major influence on the implementation and revision of EMAS.

During the revision of EMAS, greater involvement by "holders" was already included in the Commission's initial proposal. The European Parliament clearly emphasised participation, be it by employees (and their representatives), special "holder" groups or by the general public. Great stress was put on defining more clearly who the relevant "holders" are and how they could be integrated into the system. In the end, the Parliament – through the co-decision procedure – was relatively successful in achieving its aims. The Council of Ministers, by making definitions and Annexes more precise and by delegating more detailed work to comitology, significantly improved the profile of sustainability in the new regulation, for example, by inserting a detailed list of indirect environmental effects (including biodiversity). The Parliament also profiled sustainability issues by ' pushing for measures to enhance the credibility of the scheme.

Comitology had already done a great deal of work on operationalising participation and other issues that organisations should take into account, before EMAS II was finally adopted.

PART II
COUNTRY STUDIES

Chapter 4

The Eco-Management and Audit Scheme in Britain

Randall Smith and Femke Geerts

4.1 Environmental Policy in Britain: A Brief Overview

Until the later decades of the twentieth century, environmental policy in Britain was characterised as politically uncontentious, focused – in terms of implementation – at the local level and essentially gradualist and pragmatic in its development since the mid-nineteenth century. It was furthermore dependent on technical expertise and based on an approach that emphasised informal regulation through negotiation by interested parties rather than by use of formal standards of control (Lowe/Ward 1998). Even in the first decade of UK membership of the European Community, despite two Environmental Action Programmes, there was little by way of major impact on this domestic style of environmental management. Indeed, policy developments tended to reinforce this distinctive approach.

However, by the mid 1980s, it had become clear that the UK policy stance was not as successful as had been expected and domestic practices were being challenged by both other European governments and European Community environmental programmes. Britain was accused of insularity and labelled "the dirty man of Europe" (Rose 1990), challenges echoed within the country by the Royal Commission on Environmental Pollution and by environmental pressure groups, which turned to the European Community institutions as powerful allies in their campaigns against the inadequacies of British environmental policies and practices.

The UK approach to environmental issues changed in the second half of the 1980s. Margaret Thatcher, Prime Minister throughout the decade, became convinced of the political, and possibly scientific, importance of addressing the global environmental agenda and the issue of sustainable development. The daily broadsheet newspapers began to appoint environmental specialists in the mid 1980s. Macrory (1999) stated that membership of environmental organisations in the early 1990s exceeded the total membership of political parties. Garner (1996) reported that the overall membership of national and local environmental groups in the UK in 1990 was about 4.5 million compared with 2.5-3 million in 1980, which was double that of 1970, which in turn was double that of 1960. Friends of the Earth set up a branch in Britain in 1970 and Greenpeace in 1977. Not only have the numbers of people expressing concern about the environment increased, but the visibility of some environmental groups has been enhanced by direct action and civil disobedience, not least through the animal liberation movement and ALARM UK, the network of some 300 local anti-roads groups (Connelly/Smith 1999; Doherty 1999; Seel/Paterson/Doherty 2000).

The passing of the 1990 Environmental Protection Act (EPA), informed by the principle of integrated pollution control (IPC), was a highly significant point in the development of environmental policy in Britain, but it must not be forgotten that it was preceded by a great deal of legislation relating to planning and protection. Indeed it can be argued that the 1990 EPA brought together in one act ideas and policies contained in legislation, EU directives and local government regulations going back many decades. Nevertheless, the 1990 EPA was important in signalling that environmental policy was now a strategic issue for the nation. This was reinforced in the 1995 Environment Act, which brought the Environment Agency into being to operate the IPC regime (Weale 1997).

4.1.1 The Origins of British Environmental Policies: Pollution Control, Land Use Planning and Energy Conservation

Close inspection reveals a history of concern for regulating the environment dating back to the middle of the nineteenth century. The earliest legislation addressed issues of *pollution control.* The Alkali Inspectorate (the world's oldest national pollution control body) was established in 1863 and in the ensuing years extended its responsibilities to industrial air pollution in general. Other inspectorates came into being, but their mode of operation did not vary hugely. The approach was to negotiate with those responsible for the pollution in order to agree on the level of pollution that could be countenanced, the "best practical option". This consensual approach is attributed to the first chief inspector, Angus Smith, whose view was that improvements would be best achieved by working with rather than against industry (Weale 1996). Legal proceedings were rarely invoked and Vogel (1986) reported that between 1920 and 1966 just two firms were prosecuted for breaches of the Inspectorate's authority. Control of domestic sources of pollution, such as smoke from coal fires, was the responsibility of local authorities, notably under the 1956 Clean Air Act. Codification of existing practice on pollution control was to be found in the 1974 Control of Pollution Act, though it has not been fully implemented.

The standing Royal Commission on Environmental Pollution, set up in 1969,[1] recommended the rationalisation of the range of pollution agencies, but this was not achieved until April 1987 when Her Majesty's Inspectorate of Pollution (HMIP) was created. The philosophy of HMIP was to licence sites through a system of integrated pollution control still based on the "best practical environmental option" approach. This system of environmental pollution control became known as "best available techniques not entailing excessive costs" (BATNEEC) under the 1990 Environmental Protection Act. The development of HMIP departed from its original objectives with the rise to prominence of the EC in the determination of Britain's pollution control practice. Prior to the development of EC pollution control policies

---

[1] The year before the creation of the Department of the Environment, which was the result of amalgamating the Ministry of Housing and Local Government, the Ministry of Transport and the Ministry of Public Buildings and Works.

in the 1980s it had been possible to suppose that Britain could have maintained its flexible and discretionary style of regulation. Such an expectation however, according to Weale (1996, 116), came increasingly under strain as the effect of EC directives began to work themselves out, not simply in the substance of specific policies, but also in the very mode by which policy was organised and conducted.

The privatisation of the water industry resulted in the creation of the National Rivers Authority (NRA) in 1989 with, inter alia, pollution control powers for water. The HMIP, the NRA and the waste regulation function of local authorities were brought together, as a further measure of integrated pollution control, into the Environment Agency, which came into being in 1996 with a sustainable development objective, tempered by economic considerations.

Parallel with the history of pollution control, environmental policy in Britain is also embedded in the system of *land use planning*. In the nineteenth century, local authorities were granted powers to redevelop the slum areas of Britain's industrial cities, but it was not until the early twentieth century that the legal basis for a land use planning system was established. The main legislation is the 1947 Town and Country Planning Act, albeit amended in a number of important ways in the second half of the twentieth century. The core of the legislation is that, with the exception of agriculture, any proposed development of land is subject to planning permission from the local authority based on local development plan policies. There is a system of appeal if a local authority turns down a planning application. In controversial cases a public inquiry can be held, and the final authority lies with central government (the Office of the Deputy Prime Minister, since May 2002, in England).

Environmental considerations have played their part in decisions on planning applications, not least in the context of preserving the countryside. The well organised environmental movement in Britain also had its origins in the nineteenth century with the appearance of the Commons, Open Spaces and Footpaths Preservation Society in 1865 and the National Trust in 1895. The Council for the Preservation (now Protection) of Rural England was set up in 1926 and the Ramblers Association in 1935. Europe's largest single interest group is the Royal Society for the Protection of Birds (RSPB) which was founded as The Fur and Feather Club in 1889. It now has over one million members and is a highly significant pressure group on environmental matters.

As well as pressure groups, official conservation agencies have been established. They are consulted by local authorities for their views on the environmental implications of major planning proposals. Under the 1949 National Parks and Access to the Countryside Act, the Countryside Commission and the National Parks Commission were set up. The latter has a major amenity role in identifying areas for designation as national parks, areas of outstanding natural beauty, heritage coasts and local countryside parks. Its name has changed several times since 1949 and became English Nature under the 1990 Environmental Protection Act. The Countryside Commission plays a conservation role with responsibility for recommending areas to be designated sites of special scientific interest, as well as national and local nature reserves. In April 1999 it came together with the Rural Development Commission to form the Countryside Agency.

Even though the official conservation agencies play no formal part in the planning and decision-making process, they have to be consulted for certain kinds of proposals for development, and their views can be influential. Local planning authorities are encouraged to consult with non-state conservation interests in drawing up their development plans and in coming to a view on specific planning applications. The planning process in Britain is one where, under the 1968 Town and Country Planning Act, there is a requirement for wideranging public consultation. This provides opportunities, on the one hand, for environmentalists and other members of the public to have their say and, on the other hand, for developers to mobilise support in favour of a particular proposal. However, only the latter have the right of appeal against a decision. Since the early 1990s local authorities have been required through their planning responsibilities to pay increased attention to environmental issues (Department of the Environment 1992b and 1994a). However, proposals are being considered in the early years of the 21st century to make some radical changes to planning law in England. For instance, it is being argued that decisions in principle on big projects like power stations, major airports, road and rail schemes should be a matter for Parliament with public inquiries at the local level focusing only on details and not on the principle. Whether this would invite escalation of direct action is a matter for debate.

Fiscal measures have featured strongly in the policy debate about changes to the generation, transmission, supply and *conservation of energy*. Until the 1980s, the supply of energy in the UK was provided by the nationalised industries of gas, coal, electricity and nuclear power. In the 1980s and 1990s under the Conservative administrations led by Margaret Thatcher and John Major, policy on the production and delivery of energy changed, though existing concerns about the diversity and security of energy supply remained high on the agenda. The state monopolies were opened up to competition and subsequently privatised. Since then mergers and take-overs, plus government imposed restructuring, have transformed the energy industry, not least with the creation of multi-utility providers. These changes in the ownership of former publicly owned assets, and the progressive liberalisation of the energy market, have transformed the circumstances in which, and the means by which, Government can implement energy policy (House of Commons Trade and Industry Committee 1998, Paragraph 22). These major changes were taking place at the same time as greatly heightened awareness of the environmental implications of energy production and use.[2] Holdgate (1994) identified two specific charges that were levelled against the energy industries:

- they were squandering irreplaceable natural capital by depleting fossil-fuel reserves too rapidly;
- they were releasing pollutants which were damaging components of the environment, thereby reducing the productivity of crops and forests, eroding biological diversity, corroding materials and buildings, and damaging human health (p 86).

---

[2] In 1992, the Department of Energy was abolished. The supply side went to the Department of Trade and Industry and the demand side to the then Department of the Environment.

He argued that a wide range of measures would be needed to counter environmental damage, including flue-gas desulphurisation and other abatement technologies, as well as energy conservation measures. The latter would include economic incentives (see below), and he referred to the taxation on domestic fuels and the increase in road fuel duties, outlined in the government's programme on climate change (Department of the Environment 1994c). The overall aim was to reduce CO_2 emissions in the UK to 20 per cent below the 1990 level by 2010.

Holdgate also emphasised the provision of advice and information on energy efficiency, the focus of a report by the recently established House of Commons Environmental Audit Committee (1999b). It in turn quoted a report from the previous year by the House of Commons Trade and Industry Committee (1998) on energy policy that "there is a crying need for the integration of environmental priorities with energy policy, rather than the one being a tardy intrusion into the latter" (Paragraph 21(e)). The Environmental Audit Committee demanded an effective energy efficiency strategy, which should be "an integral part of the existing energy policy mantra of diversity, security and sustainability" (Paragraph 6). It advocated a sustainable hierarchy to provide a framework for UK energy policy. "This hierarchy should start with demand management and the promotion of end-use energy efficiency, then energy supply from renewable resources followed by combined heat and power and fossil fuels, in order of efficiency and carbon intensity, and nuclear power" (Paragraph 24). The Government's response to this was said to be unenthusiastic (House of Commons Environmental Audit Committee 2000b, 13).

Following the general election in June 2001, the re-elected Labour Government announced a review of energy needs in the light of the Kyoto agreement and as a result of concern that the UK in the longer term would become a net importer of gas and oil from what have been called "unstable regions of the world". It was argued that renewable sources of energy such as wind, tide and hydro-electric schemes would not produce enough electricity to achieve the target of generating 10 per cent of UK supply from such sources by 2010. The prospect of a return to nuclear power as a renewable source was raised.

4.1.2 The Politics of British Environmental Policy in the Late Twentieth Century

In the 1980s the government, with Margaret Thatcher as Prime Minister, became convinced of the importance of addressing environmental issues, not least through the influence of Sir Crispin Tickell, former UK Ambassador to the United Nations and a leading environmentalist. In 1988, a year after the publication of the Brundtland Report (World Commission on Environment and Development 1987), the UK government announced its conversion to sustainable development and the need to safeguard the environment. In 1989 Thatcher replaced Nicholas Ridley as Environment Secretary by Chris Patten, regarded as a very "green" Cabinet Minister. In 1990 the Environmental Protection Act was passed and the White Paper, This Common Inheritance, published (Department of the Environment 1990). This Paper was the first comprehensive overview of environmental policy in Britain, making a clear statement that the principle of sustainable development should inform the policymaking process, but the general response to it was more than a little sceptical. There were few new commitments. Most of the 350 measures in the

White Paper had already been put into effect or at least announced. Nevertheless, a large number of government departments signed up to the Command Paper, a Cabinet committee for co-ordinating policy was announced, individual departments were expected to nominate ministers to take responsibility for environmental issues and annual progress reports were promised.

The White Paper established a small Panel on Sustainable Development chaired by Crispin Tickell and reporting directly to the Prime Minister. It also set up a larger Round Table on Sustainable Development comprising people from a variety of backgrounds. This wide range of representatives ensured that the argument would not be ignored that sustainable development provided opportunities for economic growth, for the creation of new markets and the development of new products, what environmental protagonists labelled "harnessing greed in the public interest". The emphasis on sustainable development was accompanied by growing openness in the policy process (Carter/Lowe 1998; Knill 1998; Knill/Lenschow 1998b).

Enthusiasm for environmental policy issues waned during the premiership of John Major (1990/97), possibly because "the onset of recession in the early 1990s and various government-led deregulatory initiatives dampened Whitehall's appetite for more radical change" (Jordan 2000, 259). Critical comments were made about the British government's response to the Earth Summit in Rio in 1992. The four 1994 policy documents published in the wake of the summit (Department of the Environment 1994b, 1994c, 1994d, 1994e) were said to be largely rhetorical – "documents containing elegant academic appraisals of environmental problems, full of good intentions, but without any important, new, explicit commitments" (Gray 1995). Gray, however, was not uncritical of these critiques. His conclusion was that whilst there had not been a sea change in British environmental policy, the Conservative government had done more than pay lipservice to a fashionable notion. There had been some response to both scientific and political pressures, the latter coming from the European Union, from the increased – albeit temporary – popularity of green politics, from the privatisation of the utilities and from the unrelenting pressure of the environment lobby.

Labour Prime Minister Tony Blair began his new government in May 1997 with enthusiastic commitment to a green policy, rebuking the US government for its reluctance to address climate change issues and addressing a meeting of the UN in New York on the lack of progress made in implementing the 1992 Rio agreements (Jordan 2000, 257). Domestically, the Labour Government created in November 1997 a new House of Commons Committee of backbench members to address environmental audit issues. In January 1998 the government formally announced that it wanted all local authorities to have adopted Local Agenda 21 plans by the year 2000 (Department of the Environment, Transport and the Regions 1998a).[3] In February of that year the Department of the Environment, Transport and the Regions (1998b) issued a consultation paper on a revised strategy for sustainable

---

[3] In December 2000 the Sustainable Development Unit asked Regional Directors in the Government Offices in England and Wales to find out what had happened. 357 of the 384 local authorities had strategies in place. For a commentary on participation strategies in Local Agenda 21 programmes, see Young (2000).

development, in which it was proposed that the recently established Regional Development Agencies in England should have explicit responsibilities for developing policies on sustainable development. In June 1998 the House of Commons Environmental Audit Committee recommended a more active role for the Cabinet Committee on the Environment and the Green Ministers Committee (House of Commons Environmental Audit Committee 1998b).

A year later, it concluded that the Cabinet Committee was "not driving the government's pursuit of sustainable development nor acting in a positive way to unearth and deal with planning conflicts" (Paragraph 12). It also concluded that the government had not made a full across the board commitment to environmental reporting, environmental appraisal and the introduction of environmental management systems (House of Commons Environmental Audit Committee 1999a).

In July 1998, the government announced its intention to place a new duty on local government to put sustainable development at the heart of council decision-making as part of the process of modernising local government (Department of the Environment, Transport and the Regions 1998d). This announcement was made at the same time as the "Best Value" initiative which placed a duty on local authorities "to deliver services to clear standards – covering both costs and quality – by the most effective, economic and efficient means available" (Paragraph 7.2). Whereas the latter became a formal requirement under the 1999 Local Government Act,[4] the former was changed to a power rather than a duty. Also in July 1998, the government published the long awaited and much watered down White Paper on an integrated transport policy (Department of the Environment, Transport and the Regions 1998e).[5] Towards the end of the year, the government produced a set of sustainability indicators to be placed alongside headline inflation and GDP figures, in order to try to check whether the economy was or was not on a sustainable growth path (Department of the Environment, Transport and the Regions 1998f).[6]

In May 1999, the government published its revised strategy for sustainable development (Department of the Environment, Transport and the Regions 1999). It wished to see sustainable development frameworks in place in all English regions by the end of 2000, and produced guidance on their preparation early in 2000. This was followed up by a central concern for sustainable development in the government's planning policy guidance on regional planning (Department of the Environment,

---

4   The 1999 Act required local authorities to set performance and efficiency targets and to review services over a five year period, including consideration of alternative ways of achieving them.

5   The ten year transport plan published in July 2000 referred to a £180 billion investment, equally divided between rail, road and local transport (Department of the Environment, Transport and the Regions 2000e). Plans for the first five years were published in December 2000, and the inclusion of 40 road schemes was strongly attacked by environmental pressure groups.

6   The 13, later 14, then 15 headline indicators were to be accompanied by another 150 supporting indicators, reflecting individual policy areas. The handbook explaining these indicators emphasises the balance between the economic, social and environmental dimensions of sustainable development (Department of the Environment, Transport and the Regions 2000d).

Transport and the Regions 2000b) and the production of a good practice guide on sustainability appraisal of regional planning guidance (Department of the Environment, Transport and the Regions 2000c).

The Local Government Act 2000 gave local authorities in England a new power to promote and improve the economic, social or environmental well-being of their local communities and placed a duty on them to prepare community strategies to improve well-being and to work towards sustainable development. Statutory guidance on how to prepare a community strategy was published at the end of 2000 (Department of the Environment, Transport and the Regions 2000g) and a consultation paper on the power to promote well-being appeared at the same time (Department of the Environment, Transport and the Regions 2000f). The overlap with the existing Local Agenda 21 initiative is substantial, as it is with the promotion of local strategic partnerships,[7] and for some deprived areas the community strategy takes the form of a neighbourhood renewal strategy (Nock 2001).

These exhortations from central government were reinforced in a December 2001 White Paper on local government (Department of Transport, Local Government and the Regions 2001) called *Strong Local Leadership-Quality Public Services*. It asserted, under the heading of "Delivering Sustainable Development" that strong community leadership "means developing social capital by supporting civic engagement and networks of neighbourhood organisations. It means enhancing environmental quality by reducing waste, energy use and air pollution and improving public space. And it means safeguarding the interests of future members of the community" (Paragraph 2.8).

The government's own definition of sustainable development encompassed (i) social progress which recognises the needs of everyone; (ii) effective protection of the environment; (iii) prudent use of natural resources; and (iv) maintenance of high and stable levels of economic growth and employment (Department of the Environment, Transport and the Regions 1999, 9). The White Paper[8] on sustainable development also announced that the Panel on Sustainable Development and the UK Round Table were to be subsumed into a new Sustainable Development Commission, reporting directly to the Prime Minister. The final report from the Panel highlighted issues of particular importance for the new Commission, including the need to develop better means for determining the real cost of environment policy, to deal more effectively with the disposal of waste and to

---

[7] Local strategic partnerships are single, strategic, non-statutory, non-executive organisations consisting of the public, private, business, community and voluntary sectors. They are intended to ensure different initiatives and services work together for the economic, social and environmental regeneration of the area, improved health and quality of life, and sustainable growth. They are aligned to local authority boundaries and are to be established in all areas of England (Hamer 2001).

[8] Jonathon Porritt, Director of Forum for the Future, welcomed the conceptual leap forward compared with the first White Paper on sustainable development. By his reckoning it was "the first government document of this kind to do justice to the integrated nature of sustainable development, the economy, society, environment and overall quality of life" (Porritt/Levett 1999, 4). He looked forward to seeing this integration operating in practice although he regretted that the publication was almost target free.

ensure the quality and supply of fresh water. In the middle of 2000, the announcement was made that Jonathon Porritt would chair the new Sustainable Development Commission.

The preliminary conclusion on the achievements of the Labour government is that whilst the public commitment remains, "at the start of a new millennium, environmental policy finds itself in the rather paradoxical situation of being both an established part of British politics [...] and marginalized in the sense that few politicians champion it with passion [...] sustainable development has made headway only when political and economic circumstances have permitted" (Jordan 2000, 259, 275). In October 2000, at the launch of the Sustainable Development Commission, Prime Minister Tony Blair made a major speech at a Confederation of British Industry-Green Alliance conference, the first since his early public commitment "to put environment at the heart of government". As well as acknowledging the wide range of intractable environmental problems, he called for partnership between the green campaigners, business and government. He challenged the top 350 companies to report on their environmental performance by the end of 2001. In the same month, there were newspaper reports of possible alliances between NGOs such as Friends of the Earth, Greenpeace and major clients of the advertising industry such as Texaco and Monsanto (Nisse/Jury 2000). The following month saw the publication of the UK programme on climate change in advance of an international conference in The Hague (Secretary of State for the Environment, Transport and the Regions et al. 2000). The broad ranging strategy was aimed at reducing emissions of greenhouse gases to a level that would more than meet the UK's commitments to the Kyoto protocol. The Environment Minister, Michael Meacher, claimed that "this is by far the most far reaching climate change programme produced by any country in the world". Perhaps environmental policy in Britain has begun to have more political champions than hitherto. There is some evidence to suggest that political rhetoric is beginning to be accompanied by a modest level of political action, despite the setbacks at the International Climate Conference in The Hague in November 2000, the highly contentious proposed energy policy of the United States of America, following the election of President Bush and the huge contrast between Prime Minister Blair's speech at a London conference organised by the World Wide Fund for Nature and the Royal Institute for International Affairs on 6 March 2001 where he claimed international environmental leadership and the content of Chancellor of the Exchequer Gordon Brown's budget speech the following day with its paucity of environmental measures and cuts in the cost of motoring.

Certainly, environmental concerns did not feature centrally in the April-June 2001 general election campaign in the UK, despite the problems of a major foot and mouth epidemic, alarm about BSE and CJD and worries over GM crop trials. The Labour Party, alone of all the major parties, did not give environmental issues a separate section in their manifesto. More controversially, in the middle of the election campaign, the Lord Chief Justice (Lord Woolf) called for the establishment of an environmental court or tribunal to rule on environmentally sensitive issues, where the scientific evidence would be in the public domain and policymakers would be required to justify their decisions. The idea of an Environmental Auditor General based in the National Audit Office has also been floated.

Following the re-election of a Labour Government in June 2001, the large Department of the Environment, Transport and the Regions was split into two and the Ministry of Agriculture, Fisheries and Food was abolished, largely because of its poor performance in respect of the foot and mouth epidemic, the GM crop trials and BSE. One of two new ministries covered transport, local government and the regions. The other took on the environment, food and rural affairs. In May 2002, a separate Department for Transport was established and the responsibilities for local government and the regions (including planning and housing) were placed in the new Office of the Deputy Prime Minister.

4.1.3 Environmental Standards

The British approach to environmental policy has traditionally focused on regulation through negotiation and compromise in particular contexts. It has been contrasted with the practice in other countries, not least other member states of the European Union, of the application of uniform, and often rigorous *environmental standards*. The UK government has tended to resist the imposition by the EU, for instance, of uniform emission standards, arguing that emissions are not a problem unless they damage targets. Protection of the environment should be undertaken in a flexible rather than a uniform way. However the practices of the NRA and HMIP in the early 1990s began to take the issue of regulating quality objectives very seriously and to operate more formally at a greater distance from the regulated organisations. Carter (1998) argues that "the establishment of uniform air and water quality standards and the standardisation of pollution control procedures has upset the cosy, gentlemanly British style. Specifically, the EC directives on bathing areas, drinking water and on pollution from large combustion plants [...] have all contributed to [a] changing approach to environmental management". In its report on setting environmental standards the Royal Commission on Environmental Pollution argued that the nature of environmental concerns had changed dramatically over the last 30 years (Royal Commission on Environmental Pollution 1998). In its response (Deputy Prime Minister 2000a) the government agreed that this was "a wide-ranging and evolving area of policy" (Paragraph 4).

More broadly, concern has been expressed about the growth in the number and range of inspection arrangements in many policy fields. The House of Commons Environment, Transport and Regional Affairs Select Committee commented in its report on the Audit Commission (2000, Paragraphs 40-49) on the danger of over-inspection and proliferation of inspectorates, particularly in respect of local authorities. In response, the Government hoped that the activities of the Best Value Inspectorate Forum would help to avoid duplication, not least by the nomination of a lead inspector for each local authority (Deputy Prime Minister 2000b, Paragraph 12).

4.1.4 Fiscal Measures and Economic Instruments

Another way of addressing environmental problems rather than through the regulatory approach is by the use of *economic instruments*, such as environmental taxes, or tradable pollution permits (a device promoted for the international level at the Kyoto climate change conference in December 1997 and in existence in North

America for some years). The idea was attractive to the former Conservative government, which appointed David Pearce, a keen advocate of environmental taxes, as an official adviser in 1989. Some developments along these lines have taken place such as the imposition of value-added tax on domestic supplies of gas and electricity in 1993, a tax on landfill waste in 1996 and differential pricing for lead-free and ultra-low sulphur petrol. Finally, a tax on all non-domestic users of energy, the climate change levy, was introduced in April 2001.[9]

An aggregates levy is planned from April 2002, which is also intended to be "revenue neutral". Part of the revenue raised from the latter is to finance a Sustainability Fund, aimed at delivering local environmental benefits to areas subject to the environmental costs of aggregates extraction. Road pricing has also been advocated. Meanwhile, the motor car industry is planning on the assumption that the internal combustion engine will be overtaken by new technologies, such as hydrogen powered cars, in the next ten to twenty years (Harrison 2000), and the UK government expects to consult on and then produce by the end of 2001 a Powering Future Vehicles Strategy (House of Commons Environmental Audit Committee 2001).

Gordon Brown, the Labour government's Chancellor of the Exchequer was, before his package of 22 reforms in the 1999 budget, under fire from the new House of Commons Environmental Audit Committee for not giving adequate recognition to environmental issues. It called, inter alia, for the establishment of a Green Tax Commission. The House of Commons Public Accounts Committee now covers environmental performance in their deliberations on all government spending. Tindale and Hewett (1998, 52) forecast "a rolling programme of green tax measures, rather than the large packages introduced in the Scandinavian countries". The House of Commons Environmental Audit Committee (2000c) argued that the government should "set out a plan for developing, implementing and evaluating its programme of environmental tax reform" (Paragraph 6) and reiterated its support for a Green Tax Commission, whilst recommending that the new Sustainable Development Commission should include among its responsibilities the provision of advice to the Treasury on sustainable development economic instruments (Paragraph 29). At the end of 2000, the Committee was arguing that it could not carry out its work effectively without the support of an independent environmental audit facility, specifically an Environmental Auditor General within the National Audit Office (2000d, Paragraphs 78 and 146). The government response to this idea was very cautious (Deputy Prime Minister 2001, Paragraphs 71–74).

---

[9] "This new levy is accompanied by a reduction in employers' national insurance contributions, which, together with amounts being allocated to environmental improvement, mean there will be no net gain to the Treasury. The environmental improvement amounts include extra relief for capital expenditure on energy-saving works. This means that, at least in theory, businesses could avoid having to pay higher fuel bills and actually pay less tax overall by investing in projects to increase their energy efficiency. Money spent on the projects attracts tax relief which may well exceed the climate change levy being paid" (Andrew 2001). A survey of local authorities by the Councils for Climate Protection programme calculated that an average council in England and Wales would benefit by £ 52,334 per year (Matthews 2001).

Helm (1998) has argued that it has proved difficult to introduce economic instruments partly because of their income effects – "to introduce an instrument that will have a significant impact on the level of pollution, the tax rate needs to be relatively high" (p 12) – and partly because the institutional arrangements in the UK (the Environment Agency and the Department of the Environment, Food and Rural Affairs) have not been appropriately geared to this approach. He reported five economists among the 9,000 employees of the Environment Agency (p 15) and argued that the very wide range of responsibilities of the main government department (the then Department of the Environment, Transport and the Regions) had meant that it had been unable to focus on core environmental policy activities (p 16).

4.1.5 Environmental Management in Britain

Against this policy background, the development of environmental management systems in Britain can be seen to follow a similar trajectory. Up to the end of the 1980s the development of formal environmental management systems including environmental audit, was modest in both the private and public sectors. In general, the adoption of an environmental management system (EMS) derives from the development of "upstream" approaches to environmental management, which are no longer concerned just with "end of pipe" solutions. "Management systems aim to pull a potentially disparate system into an integrated and organised one. To that end the system covers not only management's responsibilities but the responsibility and tasks of every individual in an organisation" (Welford/Gouldson 1993, 73). The EMS approach is based on earlier quality systems approaches and on the idea of continuous improvement.

The interest in environmental management systems in Britain rose in the 1990s, in large part as a result of the increased role and activities of the British Standards Institution (BSI). The BSI has a long history of providing technical standards and related services to industry and has been in the forefront of environmental standards for over 40 years (Sheldon 1997, 165). In April 1992, the British Standards Institution launched BS 7750, a standard intended to encourage organisations to establish a structured system for measuring, managing and improving their environmental performance. A final version was published in March 1994. It was a development from British Standard 5750 which set standards for products and services. The Institution developed BS 7750 as part of an environmental initiative designed to aid organisations in coping with a relatively new aspect of industrial life where there were few established work practices and processes. BS 7750 covered by-products such as wastes and emissions and required the development of an environmental policy and management programme, including an audit to test whether the programme's objectives were being achieved. The scheme was voluntary and allowed individual organisations to set their own environmental targets. It was company not site based. External verification did not imply public availability of environmental management records. BS 7750 provided the basis for

ISO 14001.[10] In addition, BS 7750 was so influential in the development of EMAS that substantial pieces of the standard's text are quoted in the advisory annexes of the regulation (Sheldon 1997, 169).

Private business welcomed the BSI and ISO standards and linked them to its own practices such as the Responsible Care Programme of the Chemical Industries Association, the International Chamber of Commerce's Business Charter for Sustainable Development, the Confederation of British Industry's Environment Business Forum, the Environment Forum of the Engineering Employers Federation and a number of other sectorally based codes of practice (House of Lords 1993, Memorandum by the CBI, 27).

More recently, the British trade unions have also supported the introduction of environmental management systems. In 1996 the Transport and General Workers Union (T&G) decided to campaign for a statutory Eco-Management and Audit Scheme, to negotiate for EMAS at site level and to educate its 10,000 safety representatives in environmental issues (Dalton 1998, 52). It produced a safety representatives handbook in 1998 which included environmental issues and it has also published environmental reports on eco-auditing and trade unionists (1996), pesticides reduction (1997) and a guide to environmental management systems called *Workplace Pollution Reduction* (Transport and General Workers Union 1999). In introducing this latter report, the Environment Minister, Michael Meacher, stated that if EMSs were to work "it is essential that creative and practical partnerships are formed. Trade unions are one of the key stakeholders in the environment. It is only right, therefore, that they should be consulted on the implementation and operation of workplace environmental management systems. As, indeed, should all employees." Meacher established in 1998 the Trade Union and Sustainable Development Advisory Committee (TUSDAC), jointly chaired by himself and John Edmonds, General Secretary, GMB Union. In a mid 1998 survey by the T&G and De Montfort University of 340 British organisations that were ISO 14001 certified or EMAS verified, 22 per cent responded. Of this total (75), just 19 (about 25 per cent) stated that they involved their trade unions in the accreditation process (Transport and General Workers Union 1999, 19).

Against this backcloth of environmental management EMAS was implemented in Britain.

4.2   Implementing EMAS in Britain

4.2.1 Establishing the EMAS Machinery

In an early explanatory memorandum (Department of the Environment 1992a) released shortly after the March 1992 publication by the European Commission of the formal proposal for an eco-management and audit scheme, the UK government was particularly concerned that duplication of auditing and verification procedures

---

[10]   BS 7750 was withdrawn in 1997. In April 1997 the EU officially recognised ISO 14001 (see Chapter 3 above).

associated with national, European and international standards should be avoided and that the proposed administrative structure should be workable and not be an undue burden, particularly for small companies. A consultation process on the viability of such a scheme was launched in August 1992 and, shortly after the formal adoption of the scheme in June 1993, a further consultation paper focusing on the implementation of EMAS was published the following month (Department of Trade and Industry/Department of the Environment 1993).

One proposal was to extend the remit of the UK Ecolabelling Board and to designate it as the competent body for EMAS. However, in May 1994 the Government announced that, in the first instance, the Department of the Environment should take on this role, partly because the Ecolabelling Board was facing some unexpected difficulties and partly to keep arrangements simple. A consultative body was also established to offer advice to the Department of the Environment in relation to EMAS. Its terms of reference were to offer general advice to the competent body and in particular (i) to advise on the operational aspects of EMAS underlying the relationship between the competent body, the body for accrediting verifiers and the enforcement authorities and (ii) to advise on a strategy for informing industry, local government and other interested parties about EMAS and for encouraging participation. Its first meeting was in September 1994. It was chaired by a Department of the Environment official and its membership included people from industry, local government, the voluntary sector, the trade unions, the utilities, the accrediting body, an academic expert, a verifier, the National Rivers Authority and Her Majesty's Inspectorate of Pollution. Apart from practical matters, the DOE felt that an advisory group was important to counter any concern that the government might not be seen as sufficiently open in carrying out the role of competent body.

The July 1993 consultation paper also suggested that the UK body for accrediting environmental verifiers should be the National Accreditation Council for Certification Bodies (NACCB)[11] and this was confirmed in October of the same year. "Essentially this utilised the existing institutional structure which surrounded other voluntary standards in the UK" (Gouldson/Murphy 1998, 92). In April 1994 the NACCB produced draft criteria for the accreditation of environmental verifiers and in May began to establish an Environmental Assessment Panel to bring together people with relevant experience from industry, the small business sector, local government and the regulators. Following a pilot project involving a number of organisations which were thought to have the capability to become environmental verifiers, the criteria were adopted in January 1995. In July 1995 the NACCB published a supplement called *The Accreditation of Environmental Verifiers* (National Accreditation Council for Certification Bodies 1994, 1995a, 1995b).

---

[11] The NACCB was set up following publication in July 1982 of the White Paper, *Standards, Quality and International Competitiveness* (Department of Trade 1982) to advise the Secretary of State for Trade in carrying out the function of accreditation. The July 1993 consultation paper noted there was at the time no environmental expertise on NACCB's Council. "It would therefore need to be reconstituted to take on sufficient members with a broad understanding of the environmental challenges facing UK industry" (Paragraph 15).

In August 1995, the NACCB was disbanded and the United Kingdom Accreditation Service (UKAS) inherited its roles. Input to the development of UKAS' environment related services has been through what is at present called the EMS Technical Advisory Committee, which has representation and expertise from parties interested in the environment and its protection. UKAS has a section that accredits certification bodies, including those for EMS, and has expert assessors available to it who are knowledgeable about EMAS issues. Its responsibilities include supervision in the UK of accredited environmental verifiers.

EMAS was officially launched by the government in April 1995. In Britain, four of the first five organisations accredited in August 1995 as verification bodies under EMAS were already accredited as certifiers for BS 7750 (Falk/Wilkinson 1996). The first BS 7750 accreditations had been made in March 1995. As already noted, UKAS' central task was to set standards and competence levels for verifiers as well as to grant and review accreditation status. Concerns have however been expressed about the competence of the verifying bodies to conduct EMAS verifications and validate EMAS statements.

The process of verification offered wide discretion to those accredited, such that the benefits of participation in EMAS have depended to a marked extent on the quality of the verifier's work. "The functioning and reputation of EMAS is highly dependent upon the technical qualification and independence of the accredited environmental verifiers" (Töller 1998). This comment may apply particularly to individual verifiers as opposed to those working in a verifying organisation. The latter arrangement has predominated in the United Kingdom. Cross-national research has shown that individual verifiers, not linked to any organisation, spend just over half the time on a site compared with "organisational verifiers", suggesting that the former "do a scant job" compared with the latter (Hillary 1998a; see also Kähler/ Rotheroe 1999). One of the concerns has been that those who have been accredited to certify management systems for product quality have been applying to accredit environmental management systems without really being qualified (Taschner 1998, 229). Verifiers with a background in quality management systems moving into the environmental sphere would be more focused on systems auditing than environmental improvement. This situation can jeopardise the quality and credibility of environmental auditing services. Hillary (1998c) suggests some form of international standardisation to assist in defining auditors' qualifications. Gouldson and Murphy (1998) argued that, to be fully effective, an interdisciplinary approach was required. They went on to argue that assessment of the environmental performance of a company "demands a contribution of scientific, technical, managerial and legal expertise within the organization responsible for verification. Given that these areas of expertise have tended to be polarized in discrete professions and organizations over time, demands for their integration are likely to meet resistance and inertia [...] it is possible that [...] organizations and individuals from particular professional backgrounds will intentionally or unintentionally shape the delivery of EMAS standards to reflect their own background and values" (1998, 93).

Another concern was the dual loyalty of the verifier both to the client footing the bill and to the public at large in signing the environmental statement as an objective and impartial verifier (Taschner 1998, 231). It has been queried whether UKAS has had the resources to ensure that it has been, in practice, effective in maintaining standards (Gouldson/Murphy 1998, 92).

As already noted, it was decided that competent body status, as an interim arrangement, should be invested in central government, the then Department of the Environment "to keep the arrangements simple by limiting the number of key players involved, particularly in the initial stages of setting up EMAS" (Meacher 1998). However, as the scheme moved beyond the developmental stage, it was argued that the Institute of Environmental Assessment (IEA) should take over the official competent body role from May 1998 at least until April 2001. According to Environment Minister Michael Meacher, this shift reflected "the importance which Government attach to motivating businesses to adopt externally verified environmental management systems" (Department of the Environment, Transport and the Regions 1998g). It was in any case linked to the five year review of EMAS, built into the original regulation. The Department of the Environment, Transport and the Regions' own review included the idea of reducing the degree of bureaucracy and changing the administration of the scheme (Department of the Environment, Transport and the Regions 1998c). A memorandum of understanding was drawn up between the Department and the IEA, whereby a marketing strategy for EMAS was developed with the Department paying for the publicity, which was implemented by the Institute. The Institute was not to receive payment for fulfilling the competent body function. The income was to be derived from the fees charged for registration.

The Institute of Environmental Assessment had been formed in 1990 to improve the standards of environmental assessment and auditing. Its initial sponsors included Anglia Water, BP, Commercial Union, European Land, Laporte and Unilever. Membership was over 500 and included companies, local authorities, educational institutions and consultancies. For this last category there were both associate assessors and registered environmental impact assessors. The former were accepted into membership subject to the provision of a satisfactory reference. The latter had to be associate assessors who had satisfactorily met the quality standards of the Council of the Institute. To achieve this, they had to submit an environmental statement or similar report for review. Continued registration depended on maintaining good quality work, judged by subsequent reviews. In September 1999 the IEA merged with the Institute of Environmental Management and the Environmental Auditors' Registration Association to create the Institute of Environmental Management and Assessment. Table 4.1 shows the governance structures for EMAS in Britain and how these have changed over the years.

The EMAS scheme was required to "go live" by 13 April 1995. The first five company sites in the European Union to participate in EMAS were all in the UK and were registered in August 1995 (Hillary 1997, 141). At the end of 1995 the Institute of Environmental Management (IEM) argued that Britain had "taken a pro-active role in the development of formal environmental management systems, being the first to have a national standard, the first to accredit verifiers, the first to register EMAS sites and the first to establish an EMAS scheme for local government" (IEM 1995, 12). However, this profile was soon eclipsed by developments elsewhere, particularly in Germany (see Chapter 8).

Table 4.1 Governance Structures for EMAS in the UK: Changes over Time

	Competent Body	Accreditation Body	Government Department
10/93		National Accreditation Council for Certification Bodies (NACCB)	Department of the Environment
05/94	Department of the Environment		
08/95		United Kingdom Accreditation Service (UKAS)	
06/97			Department of the Environment, Transport and the Regions (restructuring)
05/98	Institute of Environmental Assessment (IEA)		
09/99	Institute of Environmental Management and Assessment (IEMA) (merger)		
06/01			Department of the Environment, Food and Rural Affairs (after June 2001 General Election)

For small and medium sized companies in Britain (fewer than 250 employees), government grants were made available under the Small Company Energy and Environmental Management Assistance Scheme (SCEEMAS). This scheme could cover up to 50 per cent of the costs of using outside consultants to help the development of an environmental management system which could be recognised under the EMAS regulations. A programme of roadshows was conducted to promote EMAS (Department of the Environment 1997, 76). The scheme was operational for almost three and a half years until its closure in March 1999 owing to poor take-up.[12]

4.2.2 EMAS in Local Government

As well as private sector companies, local authorities in Britain were entitled to register under an EMAS pilot scheme. Since the mid 1980s many local authorities in Britain have developed green policies and produced green charters, which have addressed the environmental implications of the activities of the local authority as a whole, emphasising the corporate nature of its responsibilities (see Morphet 1991). Guidance was produced by a working group of the Central and Local Government Environment Forum, so that an adapted EMAS scheme could be applied in the local government sector (Central and Local Government Environment Forum 1993). It came into operation in April 1995, following a pilot programme launched in 1992 involving seven local authorities.[13] In October 1993 a manual was produced on applying EMAS to local government in Britain (Jacobs/Levett 1993). The government argued that it would "help local authorities fulfil their responsibilities under Agenda 21 and to manage their environmental impacts in a systematic and considered way" (Secretary of State for the Environment et al. 1995, 25). A Help Desk, jointly funded by the Department of the Environment and the Local Government Management Board (LGMB), was set up in January 1995 to promote EMAS (or other environmental management systems) to local authorities and to provide practical support. It was intended to operate for two years but in fact ran for over four years. The Help Desk Project Officer was also on the UKAS Advisory Panel for EMAS. Department of the Environment Circular 2/95, *The Voluntary Eco-Management and Audit Scheme for Local Government*, was issued on 21 February

---

[12] The most common source of enquiries to the SCEEMAS office were consultants who made up 45 per cent of the total, followed by SMEs themselves who accounted for 29 per cent of the total. There was a substantial proportion of enquiries categorised as other, 19 per cent. These were made by a wide range of individuals and organisations from students undertaking research projects, a number of enquiries for capital grants and organisations selling or promoting goods and services. The largest number of applications received from a single geographical region came from the West Midlands. The number of applicants at 92 was 27 per cent of the total and almost twice as high as any other region. This probably reflects two things: the concentration of manufacturing business in the West Midlands and the pressure placed on suppliers by the Rover company to obtain ISO 14001 or EMAS registration (NIFES 1999).

NIFES Consulting commented on the generally poor quality of consultant/client proposals, especially early in the scheme. They noted that it was quite clear that many of the consultancy organisations involved in the project had little knowledge of EMAS and its requirements. A substantial proportion of those organisations who applied for and were given a grant ultimately failed to begin or complete the projects. A total of 321 grants were approved of which approximately 23 per cent were not paid out, in most cases because the projects did not proceed.

[13] Cleveland County Council, Leeds City Council, London Borough of Hackney, Bassetlaw District Council, City of Glasgow District Council, North Wiltshire District Council, Ross and Cromarty District Council.

1995, and the LGMB published a guide for local authority environmental co-ordinators the same year. A major conference on EMAS in local government was held in London in March 1996.

The Help Desk Project Officer at the LGMB undertook regular surveys of local authority involvement in environmental management systems (EMS). In May 1996 he reported that the results of a survey in May 1995 revealed that just under 50 per cent of authorities were implementing an EMS (Riglar 1996). A subsequent survey in July-August 1996 of 140 local authority environmental co-ordinators resulted in 79 returns. 15 were seeking registration in the medium to long term (Glasbrenner/Riglar 1997). The biggest obstacle was the lack of senior management support.

In October 1997, Riglar (1997) reported that three local authorities had achieved registration under EMAS (London Borough of Sutton, Hereford City Council and Stratford-on-Avon District Council) and 33 councils had formally committed themselves to achieving external recognition under one of the standards available. A further 72 councils were implementing an EMS but had not decided whether to seek verification/certification. 113 were investigating the feasibility of EMAS by carrying out pilot schemes. Overall, about 46 per cent of councils were involved in EMAS work of some kind.[14]

Three local authorities, as noted above, had achieved registration under EMAS by the end of 1997. The number of units registered totalled nineteen, of which seventeen were in the London Borough of Sutton. In April 1994, the London Borough of Sutton had decided to seek EMAS registration for the whole of the authority, because the scheme was tailor-made for local government. It recognised that "the largest environmental effects local authorities have arise from the way we deliver services – not on a site-by-site basis, but by functions or units" (Lusser 1996). The target was to complete the process by 2000 and to ensure that all contractors and suppliers had established a recognised environmental management system by the end of 1999. The direct costs of EMAS were estimated in April 1998 to be about £52,500 a year, though this did not include the cost of the time taken up by staff going through the accreditation process and by management promoting environmental awareness in their units or departments (House of Commons Environmental Audit Committee 1998a, Paragraph 448).

In April 1999, the LGMB was turned into two agencies, the Improvement and Development Agency (IDeA) and The Employers Organisation. An EMAS Helpdesk (named PuSH, Public Sector Helpdesk) located within LGMB's Sustainable Development Unit (SDU) was to be part of the new Agency, but following internal restructuring the Unit was absorbed into the Agency's Best Practice Unit at the end of March 2000. The argument was that the modernisation

---

[14] Riglar identified a trend towards increased EMAS participation and performance and attributed this to a number of factors: a settling down after local government reorganisation, greater experience of external accreditation in general, the change in central government, the Audit Commission study (1997), *It's A Small World*, the sustained impact of Local Agenda 21, the widening impact of environmental issues on non-environmental staff, greater responsibilities placed on environmental staff and an increasing number of full time EMAS officers.

and Best Value agenda was of primary importance to local government so that EMAS and sustainable development were of second order importance. This move was criticised partly on the grounds that key interests had not been consulted and partly on the grounds that the argument that a "Rolls Royce service" could no longer be justified was crass.[15] By 2001, the Improvement and Development Agency had appointed a sustainable development consultant to the Best Practice team and was promoting the idea of a Local Sustainable Development Unit to come into being in April 2002. Meanwhile, the Local Government Association had incorporated sustainable development into its Future Work programme for local government. It also produced a guide to mainstreaming sustainable development into the Best Value agenda and the decision-making processes of local authorities (Local Government Association 2001).

4.2.3 The Current Profile of EMAS in Britain

In Britain, EMAS enjoys only a modest popularity compared to countries like Germany and Austria. The number of registered sites has however steadily grown over the years. At the end of June 2001 the list of EMAS registrations kept by the Institute of Environmental Management and Assessment (IEMA www.emas.org.uk) contained a total of 131 sites compared to 62 in December 1997. Table 4.2 contrasts the number of registrations for these two years and their sectors (based on NACE codes).

In the three and a half years between December 1997 and June 2001, 45 new sites in the private sector had registered and 11 had decided not to reregister or had been removed from the register. June 1999 saw the first registration of a site in the EU subsidised pilot scheme on the distribution sector, a regional distribution centre of supermarket chain Sainsbury. Over the course of the three and a half years between December 1997 and June 2001, 36 new local authority sites were added to the list of EMAS registrations. Hereford City Council was the first local authority to register all its operations, but was obliged to withdraw from the scheme in 1998 when it ceased to exist as a separate authority. The new unitary council, the County of Herefordshire District Council, intends to register all its operations by 2003. The number of EMAS accredited verifiers in Britain at the end of 2000 was ten – compared with four at the end of 1995 – out of a total of 295 in the European Union and Norway.

The biggest increase in the number of registered sites between the end of 1997 and June 2001 was in the oil and gas extraction sector. There was just one site at the earlier date compared to 20 by June 2001. This included fourteen BP sites (including six North Sea platforms) and three Shell gas plants. Other significant increases were in chemicals, energy supply and local government. The sector with the greatest number of EMAS registrations in December 1997 was local government, followed

---

[15] At an environmental NGOs organised conference on Leadership for Wellbeing: Democracy, Sustainability and Civil Society, Jonathon Porritt challenged the outlook of an organisation which "on finding it owned a Rolls Royce, decides to take the wheels and doors off, smash the windscreen and remove the engine, and then present it as the modernised version" (Tuxworth 2000, 3).

by the chemical industry in second place and the publishing and printing sector together with the vehicle/ trailer sector in a shared third place. In June 2001 local government still had the highest number of EMAS registrations, followed by the oil and gas sector in second place and the chemical industry in third place. The oil and gas extraction industries (BP, BG, Shell and Conoco) are very much aware of their environmental responsibilities, as is the energy supply industry. The high number of EMAS registrations in the chemical industry can be attributed to the importance of its "Responsible Care" programme.

One reason for the modest number of EMAS registrations in Britain might be the earlier availability and consequent popularity of ISO 14001. In addition ISO 14001, unlike EMAS, does not require publication of an environmental statement and is an internationally recognised standard as opposed to the European focus of EMAS. In March 2000 there were 1,014 ISO 14001 sites in the UK compared to 73 EMAS sites (www.iso14000.com). "For companies operating in global markets, certification to ISO 14001 is making more sense than registration to EMAS. Even in the local authority sector, where EMAS has traditionally been the favoured standard, it is reported by the Local Government Management Board that ISO is now more attractive. The most important reason for registering under EMAS is the credibility that its third party verification of the environmental statement is perceived as providing" (Institute of Environmental Management 1998, 3, 6, 11).

To illustrate the preference for ISO 14001, an environmental code of practice launched in April 2000 set a wide range of standards for UK PVC manufacture, plus targets for future improvement. The standards were claimed to be stricter than required by law and the firms involved are required to achieve ISO 14001 certification by 2002 (ENDS Daily 2000a). Again, BT achieved ISO 14001 in January 2000 and the firm's corporate social and environmental manager said that this was the obvious choice partly because EMAS had been exclusively site based and restricted to manufacturing at the time the company decided to proceed two years previously and partly because "ISO14001 is better recognised internationally and we are now a global company" (ENDS Daily 2000b).[16]

In the local government sector there was an increase in the number of registered sites from 19 in December 1997 to 54 in June 2001. The number of authorities involved rose from three to ten during this period. Prominent in this group is the London Borough of Sutton with 17 sites registered in December 1997 and 44 in June 2001. Five councils had adopted a cross-authority approach rather than using a site basis and thus had all council departments registered simultaneously. In a follow up to its 1997 report, the Audit Commission (1999) published a review of progress in environmental stewardship in local authorities. In summary, it found that between 1994-95 and 1997-98 "local authorities have improved on the strategic front in that more of them now have environmental strategies (up to 67 from 20 per cent); those strategies that exist also now score higher on average against a quality checklist" (Kara 1999, 2–4).

---

[16] It could be argued that under the revised regulation, EMAS can be seen as international and not just European in focus.

Table 4.2   Number of UK Registered EMAS Sites, December 1997 and June 2001

Sector NACE Code	Number of Sites (Dec 1997)	Number of Sites (June 2001)
Oil and gas (11)	1	20
Mining and quarrying (14)		3
Textiles (17)	2	2
Paper (21)	2	2
Publishing and printing (22)	4	4
Coke and petroleum products (23)		1
Chemicals (24)	13	16
Rubber and plastics products (25)	2	2
Non-metallic mineral products (26)	1	2
Basic metal manufacture (27)	1	1
Metal products (28)	3	5
Machinery and equipment (29)		2
Office machinery, computers (30)	2	1
Electrical machinery (31)	1	
Medical and optical instruments (33)	1	
Vehicles, trailers (34)	4	4
Furniture, musical instruments and sports equipment (36)	1	
Recycling (37)	2	4
Energy supply (40)	3	7
Distribution sector (63)		1
Local government (75.1)	19	54
TOTAL	62	131

Source: EMAS registration list.

Environmental management systems were also being developed in British central government. All departments had been required to assess the practicality of developing EMSs for greening operational activities according to the 1995 annual report on *This Common Inheritance* (Secretary of State for the Environment et al. 1995). A guide commissioned by the committee of officials supporting the Cabinet Committee on the Environment was intended to help governmental organisations set up EMAS and if they so wished, to gain certification under ISO14001.[17] A review of EMSs in British government departments was undertaken in September 1997 to determine progress. This review showed that there was a strong preference for ISO 14001 over EMAS.[18] The government was sympathetic to the recommendation of the House of Commons Environmental Audit Committee that 75 per cent of departments should have at least one site certified to ISO 14001 by 2001.[19] In a later report the Environmental Audit Committee commented that it assumed the government had sympathy for the idea "because the proposal is dead" (House of Commons Environmental Audit Committee 2000a, Paragraph 261).[20]

The official guide mentioned EMAS but only to list the differences between it and ISO 14001, to point out that the latter can be used to meet many of the requirements for EMAS and to refer to the review of EMAS, through which the scope of the scheme was likely to be extended. The Government's Sustainable Development Unit reported cautiously to the House of Commons Environmental Audit Committee (1999a, Appendix 21, Paragraph 42) that "to date EMAS has allowed the participation of sectors, other than those specified in the regulation, only on an experimental basis. As a result EMSs within Government have been based first on the British standard BS 7750, and then subsequently when superseded by it on the international standard ISO 14001. The revision of EMAS will provide Government with a further option which will be considered when the regulation has been agreed and adopted and the implications for implementing in Central Government can be fully understood and costed".

The UK government's response to the five-year review of EMAS conducted in 1998 by the European Commission took the form of another consultation exercise with a wide range of interests. The reactions were by and large favourable and the government broadly supported the proposed amendments put forward by the Commission, seeing them as reasonable and proportionate in addressing the weaknesses of the original scheme and as helping to enhance its attractiveness to business (Department of the Environment, Transport and the Regions 2000a). The negotiations following the European Parliament's proposals for amending the regulation are detailed in Chapter 3.

---

[17] See www.environment.det.gov.uk//greening/ems/guidelines/.

[18] Review of Environmental Management Systems in UK Government Departments. http://www.environment.detr.gov.uk/greening/ems/review.htm Accessed July 2000.

[19] See House of Commons Environmental Audit Committee 1998b, Paragraph 158 and Deputy Prime Minister 1998, Paragraph 51.

[20] One of the government agencies that has implemented an EMS is the Northern Ireland Housing Executive which chose ISO 14001 rather than EMAS. They argued that the latter was less attractive, as it had not been initially designed for public authorities. "It had to be tweaked to suit". EMAS was seen as more onerous and complex and the report was thought to be unnecessarily complicated. Most important of all, ISO 14001 was seen as a cross-sectoral standard. "We were keen to achieve something which could set an example to others" (Institute of Environmental Management 1998, 34).

4.3   Conclusions

EMAS may to some small extent have captured the corporate imagination and its precepts are widely held in British local government, even though the level of registration is slight. It certainly has not captured the public imagination, but neither has ISO 14001 nor BS 7750. The number of EMAS accredited verifiers in Britain is minute compared with membership of the Environmental Auditors Registration Association. The voluntary nature of the scheme means that its enthusiasts have a hard task to perform, all carrot and no stick, unless the expectation is created that EMAS registration is accompanied by a relaxation of other regular environmental inspections. Gouldson and Murphy (1998, 90) argued that companies in the Environment Agency's Operator Pollution Risk Appraisal (OPRA) initiative which were EMAS registered were more likely to be placed in a lower category of risk than those that were not. The Royal Commission on Environmental Pollution (1998, Paragraph 6.83) cautiously acknowledged the point that companies which had adopted an environmental management system could benefit from a reduced frequency of inspection. However, the view of the Director of the Environment Agency was that EMAS or ISO 14001 should "bring its own reward [...]. I don't think a lightening of the regulatory load automatically follows" (quoted in Dalton 1998, 251).[21]

The range of interests that can be identified as being EMAS relevant – companies, local government, the general public, the accreditation body, professional auditors and environmental activists – are clearly putting their priorities elsewhere. It is not easy to see sustained involvement over time by a range of mutually trusting key interests in agenda setting, policymaking, implementation and evaluation of this initiative. It does not seem to offer a major component in any overall strategy to increase sustainability in practice.

With the implementation of the revision of the EMAS regulation, it will be of interest to see how central government promotes the idea. It is the responsibility of the Energy, Environment, and Waste Division of the Environmental Protection Directorate of the Department of the Environment, Food and Rural Affairs (DEFRA). The cross-departmental Sustainable Development Unit is located in the Environmental Protection Strategy Division of the same Directorate, together with the Sustainable Development Commission secretariat. Where is EMAS placed on the agenda of the new Sustainable Development Commission? Some of the issues EMAS addresses featured in the Commission's review of the end of its first year (Sustainable Development Commission 2001), but there was no mention of EMAS as such.

---

[21] It is worth noting that paragraph 9 of the preamble to the revised regulation states that organisations "may gain added value in terms of regulatory control".

Chapter 5

The British Case Studies

Femke Geerts and Randall Smith

5.1   Biffa Waste Services Ltd, Redhill

5.1.1 Introduction

Biffa Waste Services Ltd (Biffa), founded in 1919, is one of Britain's largest integrated waste management companies. They provide waste management services to industry, commerce and the public in the United Kingdom and Belgium. Biffa's vehicles collect waste from commercial, industrial and municipal premises and offload it at transfer stations, landfill sites, treatment facilities, recycling plants or other disposal sites. They also offer major companies a range of options at their own sites, such as sorting waste. The company operates 38 landfill sites throughout the UK, managing waste for over 36,000 industrial and commercial customers and 30 local authorities. Their collection services cover around 600,000 households in England and Wales. They also operate waste recycling and treatment programmes. In September 2000 Biffa completed its acquisition of UK Waste. As a result they are now the UK's largest single supplier of integrated waste management services with a turnover of almost £500 million annually. Biffa employs 1,995 permanent staff at their UK operating facilities. These are made up of 1,269 drivers and other manual staff, 299 administrative staff, 204 technical and professional staff, 106 managers, 87 sales staff and 30 supervisors.[1]

The waste management market consolidated in the second half of the 1990s and this process is ongoing. The market is moving to larger regional landfill sites as part of this process. As the market consolidates, Biffa is getting a higher proportion of market share. A few years ago, the largest proportion of market share was 5-6 per cent. Now the top five companies in the market own 35 per cent of the waste management industry. Of these top five, Biffa has the highest market share. Recycling is the biggest change in the waste market. The landfill business, although still profitable, has got a lot tougher with the arrival of the landfill tax. 30 per cent of Biffa's operating divisions deal with landfill, 9 per cent deal with special waste, 10 per cent are taken up by municipal contracts and 51 per cent of the operating divisions deal with industrial and commercial waste (of which 33 per cent are major accounts).

Biffa Waste Services Ltd is a wholly owned subsidiary of Severn Trent plc, one of the UK's largest private water utilities. Severn Trent plc has installed a

---

[1]   Employment figures for June 1998, prior to purchase of UK Waste.

management system to assist in the achievement of environmental goals. This system empowers managing directors of the individual businesses, including Biffa, to develop their own environmental initiatives within the framework of the Severn Trent Environmental Policy. Biffa have documented their own environmental policy and environmental management system and have sought independent certification of this against European (EMAS) and international (ISO) environmental standards. Their performance at March 2001 was as follows: 13 operational facilities were certificated to ISO 14001; 22 landfill sites, 5 special waste facilities, 22 fleet workshops, 19 waste collection and municipal depots, 4 support centres and 1 waste transfer station were all certificated to ISO 9002; and one support centre has been certificated to ISO 9001. Biffa's landfill site at Redhill in Surrey was the first landfill site in Europe to have been registered with EMAS. It achieved its EMAS accreditation in August 1996. The landfill site employs 10 people.

The Biffa landfill site at Redhill in Surrey is situated within former quarry workings. The quarry and landfill site comprises approximately 65 hectares of former agricultural land that was granted planning permission for the working of silica sand and fuller's earth in 1954. The working of minerals was carried out until 1992 to leave a void of nearly 10 million cubic metres which was up to 40 metres deep. In 1986 application was made to restore the excavated site using waste arising from Surrey and adjoining counties. Planning permission for infilling and restoration of the site was granted on 1 March 1989. The planning permission imposes conditions designed to protect the interests of the local community, residents and road users, and to ensure progressive restoration of the site to a satisfactory condition. The operation of the site is regulated by means of a waste management licence issued under the 1990 Environmental Protection Act, an operational working plan which is agreed by the Environment Agency (formerly Surrey County Council Waste Regulation Authority) and an extensive range of other environmental, land use planning and pollution control legislation which directly control the operation of landfill facilities in England and Wales. The waste management licence that is required to operate the site includes conditions limiting the types and quantities of waste that may be accepted at the site. It also requires the on-going monitoring and reporting of local environmental conditions to demonstrate the effectiveness of their control systems.

Following the engineering and preparation of the site during the spring of 1990, the first consignment of waste was accepted at Redhill landfill site in July 1990. Since then some 4,200,000 tonnes have been landfilled. This has enabled 13.25 hectares of the old quarry to be restored to date. This restored area of agricultural grassland is being grazed by sheep and managed to the approval of the County Planning Authority and the Farming and Rural Conservation Agency. The landfill site is surrounded predominantly by agricultural land, together with residential and commercial development to the south and north west. To the east there is a former landfill site that is now restored and alongside the western boundary is a disused industrial site. The northern boundary of the site is delineated by the Redhill Brook and further north there are extensive mineral workings. The site receives about 160 lorries per day. In 2000, it was expected that the Redhill landfill site had approximately 15 years left to operate.

5.1.2 Thematic Background

The waste management industry is very heavily regulated and tightly licensed. The 1990 Environmental Protection Act sets the UK's waste management policy framework. Under the Environmental Protection Act, every waste-producing business has a legal "duty of care" to make sure its waste is disposed of safely. The industry is regulated by the Environment Agency, which inherited the power to grant waste management licences from the former waste regulation authorities. Waste operators need permission to open new operational sites from local authority planning departments which impose stringent land use, environmental and aftercare conditions. Waste operators are also answerable to local authority environmental health departments for odour and noise problems and other public nuisances and to the Health and Safety Executive in the event of a major accident. Waste companies are governed by the regulatory concept of best practicable environmental option (BPEO). This obliges operators to install the latest equipment to prevent potential problems when it is environmentally and financially viable.

The European Community Landfill Directive, approved in 1999, has had a major impact on waste producers and operators. Amongst its requirements, Britain is legally bound to reduce the percentage of biodegradable waste going to landfill. This has demanded new waste minimisation practices in both the commercial and municipal sectors. The Landfill Directive also requires an increasing amount of waste to undergo some form of treatment process before the residue is sent to landfill. The landfill tax is also rising steadily, further discouraging the use of landfill as the main disposal option. The UK strategy for reducing national greenhouse gas emissions, published in 1998, has also raised waste industry costs. Waste management policy in the UK is shifting increasingly from cure (i.e. end of pipe disposal) towards prevention. Since 1994, waste management policy in the UK has shifted towards the environmentally driven concept of the "waste hierarchy". This promotes reduction, reuse and recycling of waste with disposal as the last resort. In 1999 the Government published a new sustainable waste management strategy, which put a strong emphasis on waste minimisation and economic incentives. The Government's waste strategy consultation paper, published in June 1998, sets out seven key commitments to drive policy in the next century:

- substantial increases in recycling and energy recovery;
- engagement of the public in increased reuse and recycling of household waste;
- a long term framework with challenging targets underpinned by realistic programmes;
- a strong emphasis on waste minimisation;
- using the waste hierarchy as a guide, not a prescriptive set of rules;
- creative use of economic incentives like the landfill tax and virgin resource input taxes e.g. sand and gravel;
- increased public involvement in decision-making.

On a broader front, the concept of producer responsibility is being considered in many areas of manufacturing beyond packaging. The European Commission has already drawn up new legislation targeting end of life cars, electronics, tyres and batteries which includes the calculated environmental impact of their disposal in the cost of production. In other words, manufacturing companies are made to finance the collection and neutralisation of their products when eventually scrapped. It is intended that minimising the production of wastes and employing as many reusable or recyclable materials as possible will become new priorities for cost-conscious companies. Within this national and European context, the recycling services that Biffa offers are growing. They are also developing innovative waste minimisation programmes for several major British companies. They have signed contracts with Goodyear, BIP Chemicals and Bass Leisure Retail under which savings through waste minimisation and recycling are shared. In 1999 Biffa launched the first national glass collection service under the auspices of the Brewers and Licensed Retailers' Association, called the Bottleback Scheme. Under the scheme, glass is collected from pubs and other licensed outlets and taken to the nearest recycling plant. Biffa tries to work in partnership with its customers. They work together with companies like B&Q, The Body Shop, P&O and Sainsbury's to deliver cost-effective system solutions to reduce landfill and increase recovery. Biffa has also made alliances with companies which reprocess specific materials, such as fibreboard and newspapers. They also work together with food retailers, the waste management industry and compost producers and users on how to establish cost-effective and environmentally sound alternatives to landfill in disposing of organic waste. Biffa has also initiated contacts, through trade associations, with new sectors which are likely to face costly reclamation obligations in the near future. These include the electronics, pharmaceuticals, pesticides, rubber and plastics industries.

A few major players offering increasingly sophisticated services are dominating waste management in the UK. Waste operators are increasingly working towards the development of integrated waste management services. Although most of the waste is still sent to landfill, a growing proportion is now devoted to waste recovery, treatment and minimisation services in response to regulatory pressure and customer demand. The waste management industry is working closely with businesses, regulators, local and central government in the search for technological solutions to the growing waste problem. Biffa, as well as others in the waste management industry, see themselves as having a central role to play in changing the current unsustainable situation.

Historically, most waste generated in the UK has been buried in landfill sites thanks to an active mineral extraction industry. Increased knowledge of the potentially harmful effects of landfill has led, over time, to the requirement for increasingly stringent site specific engineering and management systems to ensure the protection of groundwater. Today roughly 85 per cent of waste from businesses and households is buried at landfill sites. Most of the rest is burned in municipal incinerators. Special (hazardous) wastes are disposed of by industrial high temperature incineration, chemical treatment plants or at appropriately licensed landfill sites. There is an on-going debate about the relative cradle-to-grave environmental merits of landfill, incineration and recycling. Many of Biffa's European partners (with major incineration industries) are in favour of waste-to-energy plants arguing that they minimise greenhouse gas emissions. The UK government, however, has argued that landfill with energy recovery "in many cases

[...] would represent a more sustainable solution than transporting the waste to an incinerator". This is partly a pragmatic response to the UK's situation. The sheer scale of the landfill industry ensures that it will remain the country's principal waste disposal option, at least in the medium term.

Some of Biffa's older landfill sites may potentially be of greater risk in terms of environmental harm due to landfill gas and leachate (contaminated water). Each new Biffa landfill site is designed to very high civil and geotechnical engineering standards. Detailed environmental impact assessments are carried out to ensure that every control needed to minimise potential pollution of the local environment is put in place. Environmental monitoring is undertaken on each site to assess the effectiveness of the site's control systems. All environmental monitoring records are publicly available from the Environment Agency. Landfill sites in the UK generally start as quarries which have to be restored to the original landscape to the benefit of local communities and wildlife. To this end Biffa work closely together with local authorities, the community and special interest groups. Historically, most of their closed landfill sites have been restored to working farmland. Others are converted into urban or community forests while some include specially created wildlife havens. When sites finish operating, they carry out a five year aftercare programme in line with statutory planning provisions. After five years, Biffa usually retains ownership of closed sites and responsibility for dealing with any pollution problems which arise. Most of the sites restored to agricultural land are leased to farmers.

Biffa has pursued a programme of continuous environmental improvement since 1990. Their environmental programme is driven by Severn Trent plc's Environmental Advisory Committee via the Biffa Board. Internally, the Biffa Waste Services group's environmental management system is managed by their environmental team. In 1998 Biffa published its first environmental performance report. This report won Biffa the 1999 ACCA Environmental Reporting Awards for best "first time reporter". In 1997 Biffa asked AEA Technology to carry out an environmental benchmarking exercise, marking Biffa's performance against the measurements used in the Index of Corporate Environmental Engagement run by Business in the Environment. In the 1997 Index, the Severn Trent Group was ranked in the second quintile of FTSE 100 companies. AEA Technology concluded that Biffa also ranked in the second quintile. In addition, Biffa has achieved an AAA rating in the Corporate Environmental Performance 2001 study undertaken by SERM (the Safety and Environmental Risk Management rating agency).

On the negative side, Biffa's growing fleet of 800 heavy goods vehicles contributes to air pollution and is therefore an area of concern in terms of environmental impact. Their vehicles travel large distances to service their national network of customers. Rising productivity will increase Biffa's fuel needs. Cutting back the fleet of trucks and cars was therefore not felt to be a realistic option. Instead, Biffa has concentrated on making their vehicles as non-polluting and fuel-efficient as possible. In addition, by collecting more waste per route they have managed to improve both mileage and fuel efficiency. The introduction of shift working patterns for city collections to reduce rush hour pollution and congestion has also increased operational efficiency. Trials have been undertaken using cleaner vehicle fuels.

Apart from their operational and corporate activities, Biffa undertakes several social and community activities, some of which have an environmental dimension. In 1997 the company launched Biffaward, an environmental grant-giving fund utilising the Landfill Tax Credit Scheme. Landfill tax was first introduced in October 1996 as a direct levy on the disposal of waste to landfill. It was the first truly "green tax" in the UK and its purpose is to reflect the impact of landfill on the environment and encourage more sustainable waste management by raising the cost of disposal. Under the Landfill Tax Credit Scheme, landfill operators can channel 20 per cent of the tax collected into funding environmental projects. However, any approved project can only receive 90 per cent of its desired funding from the landfill tax. The remaining 10 per cent must come direct from the landfill site or from a third party contributor. Biffa agreed to donate its landfill tax credits to the Royal Society of Nature Conservation to administer under the fund name Biffaward. The aim of the scheme is to divert money back into wildlife and environmental research projects. About half of Biffaward funding is spend on local community-based projects located near their operations. The other half is spent on projects which have wider regional or national benefits. Many projects address social as well as environmental needs. The aim is to involve people across the community in improving their surroundings.[2]

Biffa strives to offer training to employees at all levels and in every job. The University of Warwick Business School and Biffa have developed a programme that offers certificate and diploma courses for the company's managers from the UK and Belgium, focusing on customer service, innovation, creativity and best business practice. They also sponsor further education courses including professional qualifications or masters degrees where appropriate, and encourage other staff, at all levels, to work towards practical-based national vocational qualifications (NVQs). Employees also receive training on environmental issues. New employees receive general information designed to raise their awareness of environmental issues. Biffa's specialist environmental and technical audit teams keep senior staff, including all depot managers, up-to-date on key legislation. Key managers are also sent on environmental training courses run by Biffa's parent company Severn Trent plc. Regionally based Biffa Awareness Days are held every three months primarily for new staff. Employees learn about the company's history and geography, development plans and environmental management.

Most of Biffa's landfill sites have local liaison groups. At the Redhill landfill site, a local liaison group was established in 1993. The group involves members of the local parish council, a local residents' association, local conservation groups, the

---

[2] For example, in 1998 Biffaward together with ten water companies in England and Wales funded a project to protect otters. This money was used to restore riverbanks and wetlands where the endangered animals breed. In addition, otter monitoring task forces of volunteers have been set up around the country. In 1997 Biffaward granted £1 million to the government's Going for Green educational initiative. They also gave donations to help restore a village hall, an old water pumping station and a community drop-in centre. Biffaward also funds many waste minimisation, educational and research and development projects.

County Planning Authority and the Environment Agency. The liaison meetings provide a forum for Biffa and the Redhill site to inform the community of their present and intended operations and for local groups to comment and express their views. The aim is to keep neighbours and the local community informed. Other initiatives at Redhill have included support for a group undertaking local pond conservation work. They have also provided site access to representatives of the Surrey Bird Club. The Redhill site supports visits by local schools and colleges and assists students with project work where appropriate. In addition, they liaise with Surrey County Council and local mineral operators to co-ordinate restoration and boundary treatment works. In 1998, the Nutfield Ridge and Marsh Project was established. This is a partnership between Biffa, Surrey County Council and other local mineral operators to enhance the overall environment east of Redhill. The local liaison group has also been active in the work of the Nutfield Ridge and Marsh Project. This project has resulted in the planting of hedgerows locally, the establishment of a permissive footpath link and the planting of an apple orchard with the local community. In future the partnership hopes to renovate flood plain habitats north of the landfill.

5.1.3 EMAS Procedure

Biffa's landfill site at Redhill is one of the company's older sites and is engineered on the principle of "dilute and disperse". Redhill is the last generation of this kind of site. Modern landfill sites operate on the principle of full containment. At Redhill, quarry rejects and similar imported material have been placed at the base of the site. Commercial, industrial and domestic waste is deposited above this on an engineered base, above the groundwater rest level. The rejects/inert zone between the wastes and rest water level acts to minimise the impact of contamination on the groundwater.

Biffa, as a company in a problematic industry dealing with sensitive issues, is very aware of its corporate image. In this respect, establishing waste management contracts, especially with big customers, is important for them because it enhances their environmental credentials and increases the company's visibility. The decision to register one of its sites with EMAS was based on similar considerations. Redhill is a regional centre so it was decided "to push a few initiatives there", as one of the managers put it. They wanted to push Biffa forward, get a name for themselves and enhance their corporate image. Having an environmental management system was also seen to reduce risks and liabilities. In addition, the larger customers expect Biffa to have certain accreditations. It was also important for Biffa to be the first landfill site in Europe to be registered with EMAS because they wanted to stay ahead of their competitors.

The Redhill landfill site had already been registered with ISO 9002 since December 1992 and with BS 7750 since July 1995. EMAS was felt to be a natural progression from these accreditations. This also meant that Biffa already had the in-house expertise to establish an environmental management system and publish an environmental report. As the certifications manager put it: "We were already looking at these issues, so why not make it official and open?" and "EMAS just required a few tweaks here and there and we had to produce a formal report". There were numerous similarities between environmental management systems and quality management systems. They identified some gaps or new procedures that they needed to work on, such as establishing an environmental policy, a register of legislation, environmental effects evaluation and an annual review. The concept of continuous improvement was one of the main new ways of working. There were no particular reasons for choosing EMAS other than "because it's there". A month after its EMAS registration, the Redhill site became accredited to ISO 14001.

The EMAS process took 18 months from start to finish. A small working group was formed composed of the site manager (local knowledge holder), the certifications manager (systems knowledge holder) and the environmental auditor (technical knowledge holder). Establishing an environmental management system was never their main work and was always seen as an add-on. This working group analysed the operational activities at Redhill using information gathered during the planning phase of the landfill development, data from site monitoring and assessment of potential environmental risks associated with the site and its locality. The potential effects of these operational activities were assessed for a number of receptors (emissions to atmosphere, discharges to water, wastes, land contamination, use of natural resources and nuisance) utilising a numerically based evaluation system (effects register). This system takes into account the frequency, likelihood, potential severity and any mitigating control measures to establish a significance rating. These ratings are used to set objectives and targets for environmental improvement.

The site is audited annually for compliance with legislative and corporate environmental requirements. The effects register was reviewed on an annual basis by the external verifier, SGS Yarsley ICS Ltd. The verifiers were very methodical and considered every little detail while Biffa felt that in the overall context of the site some of these aspects were irrelevant. This initially led to some tensions between Biffa and SGS but the problems are now solved. Only the significant effects are now assessed annually. The effects register has changed and the scoring system is no longer used. The aspect of waste initially presented some problems for the verifiers. It was a question of whose waste because, as a site itself, Redhill hardly creates any waste. SGS was originally a quality systems auditor and thus used to following numerical methods, strict procedures and step-by-step approaches. Environmental auditing is more about negotiating and SGS has consequently been on a learning curve.

Disposing of waste can produce a variety of environmental problems such as loss of valuable land space, soil and water pollution and the release of climate-altering greenhouse gases. Waste management companies release large quantities of these greenhouse gases into the atmosphere. The main sources of emissions are methane formed from decomposing waste at landfill sites and exhaust gases from Biffa's collection lorries. The company is making efforts to "green" its nation-wide fleet of trucks and cars. Since March 1998 they have converted 90 per cent of their trucks to run on ultra-low sulphur diesel, cutting emissions of certain pollutants by up to 90 per cent. They are also making progress in capturing methane gas released from landfill sites and converting it into power. Up to two million tonnes of methane are estimated to originate from UK landfill sites every year. In response, Biffa has invested in five landfill gas to electricity conversion facilities, one of them at Redhill. The electricity produced is exported to the national grid. Eighty eight per cent of the methane generated by the breakdown of waste at Redhill is now used to generate electricity. Two additional landfill gas control systems were introduced in 1998 to minimise methane emissions attributable to their activities.[3] Since 1996 the Redhill landfill site has extended its nuisance control measures, with improved litter netting, close monitoring of cover materials, a new odour control system and an enhanced seasonal insect programme. This has been particularly important as the operational area is very close to housing.

The Redhill landfill site has an extensive environmental monitoring programme. In 1997/98 1,526 measurements were taken by technicians operating the site's landfill gas control system. Groundwater quality is also monitored, with 25 different parameters measured from 11 monitoring boreholes around the site. Sophisticated equipment also closely measures surface water quality and levels of particulate dust, asbestos fibres and noise. Biffa has an environmental data system. During 1999/ 2000 a total of approximately 50,000 environmental measurements were made of landfill gas, groundwater quality and leachate quality to evaluate the effectiveness of Biffa's engineering and control systems at landfill sites. A new high-tech internal IT system which enables information from the field and laboratory to be transferred electronically to a central database has been introduced. It allows Biffa to better manage their environmental risks through automatic checks and validation of day-to-day data, trends and reporting. The company plans to incorporate this data into its company intranet and, following further development work, to make it publicly available through the Biffa website.

The details of the environmental management system are laid down in the environmental manual in accordance with the requirements of ISO 14001 and EMAS. The system is documented and administered through their ISO 9002 quality management systems. The EMS is being implemented throughout Biffa's operations. Each facility operating within the environmental management system will, on a regular basis, review the effects of their activities on the environment and the local community, in conjunction with the objectives and targets specified in the Biffa Group action plan. Following each review, objectives and targets are established as part of a process of continuous environmental improvement. To ensure this policy is achieved, Biffa has identified senior personnel with the necessary responsibility, authority and resources to implement the environmental management system. Since 1991, all Biffa's UK and Belgian facilities have been audited by their own internal environmental auditors. This audit programme, which focuses on environmental compliance, has been certified to the ISO 9002 quality management standard. Biffa has identified several aims in seeking to develop quality and environmental management systems. First of all, it helps them to exceed their legal obligations. Second, by seeking independent certification they aim to achieve recognition of their pursuit of these higher standards. Third, they want to set benchmarks for others in the waste management industry and finally to encourage self-regulation to complement statutory controls.

---

[3] Of Biffa's 30 operational landfill sites, 19 are fitted with active gas control systems. At nine UK sites the captured gas is converted into energy.

Going through the EMAS procedure has had several benefits for Biffa as a company and for the Redhill site. For the company as a whole it has helped to fulfil customer requirements. Many of Biffa's larger clients expect the company to have an environmental management system in place as well as an environmental policy. EMAS has also served as a vehicle for corporate procedures and Biffa now puts great emphasis on its corporate environmental reporting. EMAS has also helped to improve Biffa's corporate image. The EMAS registration has given Biffa a certain profile and has generated a lot of interest and visits. It has furthermore served as a compliance check and has provided reassurance to insurers as well as to senior management. Apart from benefits to the company as a whole, EMAS has also had advantages for the Redhill landfill site itself. It has resulted in a culture change with increased staff awareness of environmental issues. But, as one of the managers put it, Biffa staff are reasonably aware of environmental issues and legislation anyway "because that is their job, that is their licence". EMAS at Redhill has also helped to develop a sense of ownership amongst staff, despite the fact that employees at Redhill have not been directly involved in the development and implementation of EMAS, apart from the site manager. The flagship aspect of the EMAS registration was a challenge to staff at the site. A period of stability amongst staff has helped with the maintenance of their EMS and allowed them to improve areas of environmental performance. There have not been any significant financial benefits as a result of EMAS.

Developing EMAS has not been easy and several difficulties were experienced during the process. The fact that the Redhill facility was the first landfill site in Europe to be registered with EMAS meant there was no previous expertise or knowledge (in the company, the industry or from consultants and verifiers) that could be used.

Biffa has also found it difficult that EMAS only applied to a single site as opposed to the whole company. It furthermore proved very difficult to represent continuous improvement. It has been problematic to see changes or real improvements as a direct result of EMAS at the site level. Biffa's aim is to look at the bigger picture and to incorporate corporate targets into site plans. Biffa has also found it difficult to balance the content of the environmental statement. It required some technical data but should also be user friendly. The target audience for the statement is the local residents near the site. The EMAS statement was distributed through various channels. At the corporate level, it was made available on the company website. It was also sent to customers, big clients and suppliers. At site level, it was posted to all local interests, including the local regulators and the local authority. The site manager has described the EMAS registration as signifying an "open door policy" of welcoming public scrutiny.

The positive experience with EMAS at Redhill led Biffa to the decision to produce a corporate environmental report. Other considerations that contributed to this decision were government pressure, pressure from Biffa's big clients and the pressure to stay ahead in the waste management industry. The company has also been expanding over recent years and wanted to be seen to be involved in environmental matters. Hence visibility and branding were important considerations. Biffa published its first environmental report in 1998. The aims of this report were to highlight the company's performance, to show its expertise and to set objectives and targets. The second report, published in 2000, had less of a marketing role than the first report. The second report looked ahead to the future by setting goals for the company. It included not just environmental aspects but also social issues such as health, safety, training and education. The audience for these reports is internal as well as external and they are therefore not technical documents. In their future reports Biffa want to focus more on issues like biodiversity, corporate sustainability, climate change and energy use and production.

5.1.4 Conclusions

Biffa has got clear ideas about the future of the industry and the consequent implications for their operations. In the decades ahead, waste management will shift increasingly from cure towards prevention. Tighter waste minimisation laws coupled with business investment in cleaner production systems will make industry much more eco-efficient. Companies will also be made increasingly responsible for the safe disposal of products at their life's end, prompting a recycling boom. Widespread changes in energy use to combat climate change are likely to fuel these trends. The waste management industry will be involved more and more in providing recovery services for a multitude of products in addition to its traditional services.

Waste and how society should deal with it are growing political issues. These issues are being hotly debated between industry, government, the regulators and environmentalists. The main avenue through which Biffa is seeking to influence and inform the national waste debate is the Biffa Book series. This series is designed to help Biffa's customers respond to the constantly evolving challenge of waste management.[4]

Another avenue through which Biffa is influencing the national waste debate is by feeding into central government's policymaking process through their response to waste strategy consultations and parliamentary enquiries. In 1999 and 2000, they submitted five such documents, to the Department of the Environment, Transport and the Regions, the Commons Select Committee on the Environment and the Commons Select Committee on Environmental Audit. At the local level, Biffa is consulted by local authorities responsible for household waste collection on the development of strategic waste management plans. They also influence the strategic thinking of NGOs and independent organisations. Biffa's Director of External

---

[4] The first book "Waste: Somebody Else's Problem?" explained the waste management industry and the alternative methods for handling and disposing of waste. It highlighted the advantages for companies which took waste seriously and embraced integrated management policies. The second "Waste: A Game of Snakes and Ladders?" conducted the first analysis of how key industries were responding to increased waste regulation and costs. Eight universities with particular industry expertise undertook this research and scored each business sector against an "environmental checklist". The third, "Great Britain plc: The Environmental Balance Sheet", was a joint project with Manchester University's National Centre for Business and Ecology. It provided the first comprehensive record of air, soil and water emissions from the country's resources by region and their impact on the environment together with waste disposal operations across England, Scotland and Wales.

Relations sits on the advisory panel of Waste Watch and the board of the UK Composting Association.

Biffa describes the situation currently faced by the waste management industry as increasingly complex. On the one hand, the industry is undergoing regulatory, customer and public pressure to shift its services away from disposal and towards minimisation and recovery. On the other, most of the nation's waste still ends up in landfill but needs to be disposed of in the most efficient and environmentally friendly manner. Biffa's approach is to embrace this complexity by developing integrated waste management solutions. The UK continues to have the worst record in Europe for recycling glass and steel and is failing to catch up with EU targets on cutting municipal waste, according to figures released in June 2001 (The Guardian, 6 June 2001). Part of the problem in improving the figures is that the government has opted for a policy of burning waste in incinerators to reduce landfill. Critics say that this policy has also reduced the incentive to recycle.

Biffa's second environmental performance report contains a section on "sustainability indicators" setting out fully audited company environmental data, including five year trends where possible. The report addresses sustainable development issues in line with both the UK government's strategy for sustainable development and the Global Reporting Initiative proposals put forward by the Coalition for Environmentally Responsible Economies. It also includes social and economic aspects of the business. The second report states that Biffa's long term objective is to integrate the concept of sustainable development into all its business activities. It also states that the guiding principle behind their strategy for charitable giving through Biffaward is sustainability. However, as is the case with so many other companies, Biffa also undertakes sustainable development activities because it could enhance their corporate image and because "if other companies are pushing sustainability, we want to be there too".

What holders can be identified in the EMAS process at Biffa's landfill site at Redhill and in the company's environmental policies more generally? The *share holders* are the directors of Biffa. They have participated in EMAS in so far as the decision to participate in EMAS was taken at the corporate level in an attempt to enhance Biffa's green credentials and to increase the company's visibility. It was hoped that the EMAS registration at Redhill would create some media interest and influence corporate procedures. A clear *status holder* is the director responsible for environmental affairs. Another status holder is the external verifier. The *work holders* or employees have not been directly or actively involved in the EMAS process. However, Biffa does feel that the level of environmental awareness of staff has increased as a result of EMAS. Staff do not generally participate in policymaking and decision-making but there is an open policy to keep them informed of activities and decisions. In late 1999 Biffa introduced an intranet system for employees to access company information via their computers. There are two key aims to this: to enable all employees to access relevant and useful information, especially environmental data, from Biffa departments; and to reduce the use of paper, especially form filling, across their operations.

The *knowledge holders* are the three participants of the working group that was established to implement EMAS. These were the site manager at Redhill, the certifications manager and the environmental auditor. No outside consultant was used in the process. In some cases Biffa itself can be seen as a knowledge holder in their dealings with customers. They produce a range of mini-reports to guide customers through the myriad of regulations and fiscal changes which impact on their waste production. These include accessible layman's guides to the packaging waste regulations and the landfill tax, including new opportunities for recycling. *Holders of a spatial location* are the residents living close to the landfill site. They have not been involved in the EMAS process but are being kept informed of relevant developments through the six-monthly meetings of the local liaison group. The local residents are an important audience for the environmental statement.

There are several *interest holders*. An important one is the Environment Agency (EA). They have not been involved in EMAS as such but do visit the site about once a week. The landfill site has not received a "lighter touch" from the EA as a result of its EMAS accreditation. Other relevant interest holders are the local and national media. Biffa comments that the waste industry sometimes experiences difficulty with the media "because they only seem to be interested in bad news". However, the media are important to Biffa because of the politics of what they do. It is important to be seen or heard to be involved in current debates and at the forefront of issues. Biffa's big clients are also interest holders since they exert pressure on the company to establish environmental management systems. Biffa's suppliers, mostly truck and container sales and fleet servicing companies, can be identified as *stake holders*. However, it is mainly Biffa that puts pressure on suppliers and not so much the other way round. Biffa undertakes detailed environmental audits of key suppliers and encourages eco-improvements in the supply chain. Speaking very generally, everybody who produces waste could be described as stake holders to the waste management industry. As one of the interviewees commented "the waste industry is left to deal with somebody else's problems".

There are currently discussions about the future of EMAS at the Redhill landfill site. Biffa finds the new requirement for yearly statements "quite onerous" and suspects that annual reporting will present a problem. This might be a reason not to continue to register with EMAS. It was initially hoped that the corporate environmental report would allow them to register with EMAS, but it would be too much to produce a document of this scale and detail on an annual basis. They also find that whilst the waste management industry and its policies are changing so much so quickly, nevertheless important changes do not take place overnight. It would therefore be more appropriate to look at the long term timescale for environmental reporting.

5.2 Bovince Ltd, London

5.2.1 Introduction

Bovince Ltd is a screen process and digital printing company which was established in 1952 and moved to its present business park site in the east of London in 1980. The park is surrounded by a residential area to the east, other industrial units to the north and south and by large water reservoirs to the west. The family-owned business employs just under sixty people. The broad based ethnic community within the Waltham Forest area is reflected in the employee profile of the company with 26 per cent of the staff coming from ethnic minority backgrounds. Thirty per cent of managers and supervisors are from ethnic minority backgrounds. Only 7 per cent of the company's staff are female and 75 per cent of Bovince's employees are in the works and production areas. Many of the staff have been with the company for more than 25 years. The company's early customers were the cinema trade and whilst film posters are still produced, Bovince's main customers are advertising agencies, councils and London Transport for whom it provides large format advertising posters, bus and bus shelter advertising panels and some point of sale display work. Over the years the type of work carried out by the company has changed. Bovince now concentrates on screen process and digital printing to meet the demands of the advertising and product promotion industries. In 1999 Bovince had a turnover of £3.5 million. The company's 1999 EMAS statement lists all its processes and the various forms of waste these create. The main materials used for printing are ink solvents and substrates (paper and other materials). The main waste products are therefore ink/solvent, paper, plastic, rags and empty ink tins. The studio and camera department generate waste from film processor chemicals and recover silver from the films. The stencil and print department generate VOCs (Volatile Organic Compounds) and ozone through the use of solvents and UV technology. Cleaning screens uses acid and solvent-based cleaners. The delivery and sales departments both contribute to greenhouse gases through vehicle emissions. All core operations generate CO_2 through energy use and consume raw materials through substrate, packaging and office paper use. Their key indicators for environmental effects are carbon output and VOC emissions generated during their day-to-day business. Analysis of these indicators has found that since Bovince implemented its EMS, consumption in all areas has been reduced except fuel consumption (see Table 5.1). The rise in fuel consumption is due to sales personnel adopting wider market catchment areas. Bovince has also installed low-energy sodium lighting throughout their shopfloor areas replacing the usage of fluorescent tubes. Electricity, solvents, water and gas use have all been reduced considerably.

5.2.2 Thematic Background

Bovince has been addressing environmental issues impacting on the print industry since the early 1990s. One of the ways in which Bovince is trying to reduce its environmental impact is by the use of new technologies. Experience at Bovince has shown that these new technologies can bring important environmental as well as cost savings, mainly in the form of waste minimisation. The company has, for example, recently purchased a direct to screen projection system. This system has enabled them to bypass several steps of the conventional film method. Another example is the purchase of an automatic screen-washer with a distillation unit. The unit uses solvents more efficiently than the previous manual process and has thus brought significant cost savings. They have also introduced a solvent regenerate that mixes with the distilled solvent so that it can be used again. The benefit of this is a very significant reduction in the amount of waste solvent sent out for disposal. Special waste is sent to a specialist contractor for recovery and disposal under controlled conditions. Another project was to eliminate the company's use of landfill sites. In 1997 they agreed that their solid waste should be sent to a local power station plant for heat recovery rather than to landfill. Bovince also recycles and reuses its printed waste. The off-cuts from printed posters and any old stock is cut down and sent to a charity organisation called the Children's Scrap Project. They then distribute it to 660 schools for use as writing and painting materials.

Table 5.1   1999 Key Performance Indicators: Percentage Reduction or Increase since Implementing EMAS in 1995, Bovince Ltd

Key Performance Indicator	Reduction or Increase since 1995 (%)
Electricity	-7.06
Solvents	-25.8
Water	-3.88
Gas	-19.64
Fuel	+38.0
Energy (gas & electricity) CO_2 generation*	-14.41

Source: Bovince EMAS statement 1999.
* Figures calculated using DETR greenhouse gases formula.

Bovince has also been working very closely with their main ink suppliers on the production of new UV ink systems (solvent free). Effective solvent management is an important part of Bovince's business. By working in partnership with their key suppliers, the company has been able to reduce the amount of solvents needed. The increase in work output by the digital print system has also meant a reduction in the amount of solvent that needs to be purchased. The fall in water consumption can also be attributed to the increase in digital print work, which has eliminated the need to use screens and the washing required afterwards. A programme of water saving measures has furthermore reduced water consumption. The use of new technology has also drastically reduced the amount of substrate required to print a job.

Bovince uses a variety of benchmarking tools to help measure performance and impacts on the environment. To help gauge their exposure to environmental risk, they have adopted the use of the SERM rating system. The Safety and Environmental Risk Management Agency (SERM) helps companies to assess and compare environmental risks in different business sectors. The rating is based on a mathematical model which calculates the total cost (both direct and indirect) of possible incidents and the likely effectiveness of risk management systems in avoiding or lessening the effects of those incidents. Under this system a company that often deals with hazardous situations but has a good system for managing them may get a higher rating than a company whose activities look harmless but which has poor management systems. The rating resembles a credit rating based on a 27 point scale ranging from AAA+ (top rating, lowest risk) to C- (lowest rating, highest risk). Bovince has been classified as an AAA-risk.

Since 1998 the company has also started to systematically calculate and monitor its CO_2 output by adopting the DETR's greenhouse gas indicators which are based on a United Nations Environment Programme/National Pollution Inspectorate approach. This indicator uses carbon as the common currency for measuring the global warming potential of greenhouse gases. Using the DETR standard means that the company is able to quantify the amount of carbon created by its activities and is also able to measure a reduction or increase in this amount. Bovince's carbon output created by energy use has been reduced by more than 10 per cent between 1995 and 1999. Another tool Bovince uses to measure its environmental effects is shadow pricing. Shadow pricing reflects the cost of pollution to society. It is calculated as if there was an actual market supply and demand for the environment. The concept can be used to cost the implementation of preventive measures. Using this concept, the company can calculate what their reduction in CO_2 would have saved in the cost of pollution to society.

The emphasis on environmental issues is intertwined with the overall quality management approach which has included within it strong social elements, internal as well as external to the company's operations. Internally, Bovince tries to keep staff informed of activities by displaying company objectives at various points around the site, along with its environmental, quality and health and safety policies, its mission statement and its statement of values. A company newsletter is issued to keep staff up to speed with the latest improvements to the business. Staff training and development receive a great deal of attention because education of employees is seen by Bovince as a major factor in the success of the firm's management systems. Employees receive training not only to develop their work skills but also to encourage responsibility and awareness of their environment. There is, for example, training covering the installation of a new computerised management information system, an introduction by one of Bovince's major suppliers to the use of recent advances in ink systems, fork lift training, first aid training, personnel management courses, video-based training for health and safety and training on a digital printer for studio staff. Recognising the need for better training throughout the industry, Bovince has established a new form of modular NVQ (National Vocational Qualification) in conjunction with a nearby college. This allows trainees to learn both theory and skills in-house, eliminating much of the need for classroom-based study.

The company received recognition as an Investor in People (IIP) for their efforts, "to promote a culture that was people-based and not just procedural run" (see Hall 2000). One of the requirements for IIP is a formal business plan, which Bovince did not have. This was developed with the help of an outside consultant. Bovince's IIP programme consists of four principles for continuous staff and company development. The first aim is to continue with its commitment from the top to develop all Bovince employees to achieve the company's business goals. Second, the programme aims to regularly review the needs and plans for the training and development of all its employees. Third, the programme intends to train and develop individuals on recruitment and throughout their employment and finally, to evaluate the investment in training and development to access achievement and improve future effectiveness. Achieving the IIP standard and establishing a business plan has been of great benefit to the company. The director comments: "We are now experiencing much greater commitment within the organisation and profitability has increased by 30 per cent" (www.london.businesslink.co.uk).

Bovince's social activities also include a strong involvement with their external community. The company's interaction with the community is through their Business and Community Development Programme (BCDP). Their BCDP objective is to foster social responsibility that is an integral element of Bovince's focus on internal improvements, or, as the company's directors put it, "we value our links and openness with the local community" (Bovince 2000). Bovince is situated in an area which suffers from high unemployment, so they have recognised the importance of interaction with the young generations in local schools to give an insight into work experience. Bovince aids charities such as the Children's Scrap Project (see above) and Shed 22. The latter is a London Dockland's based charity, which initiates regeneration projects linking business and education. The company has also been working with school teachers in East London on The Big Book Project. The aim of this project is to produce a book which can aid the literacy and numeracy skills of local children and teach them something about the print industry and Bovince. In this project Bovince is cooperating with a local printer and is also hoping to involve a local bookbinder. Bovince also aims to introduce youngsters to the world of work through school visits.

At the end of 1999 Bovince became part of the East London Partnership (ELP), an organisation dedicated to improving the environment, education, health and housing in some of London's most deprived areas. In 2000 ELP and several other organisations have worked cooperatively to create the East London Business Alliance (ELBA). There was a need for a single body providing a coherent business voice for the whole of East London and which comprehensively seeks to address both its economic and social requirements currently and prospectively. The purpose of this new body is to channel the influence, skills and resources of business and to provide a catalyst for the regeneration of the area, for the benefit of all who live and work in East London. ELBA hopes to be a catalyst for economic and social regeneration. Other examples of Bovince's community involvement include the commercial director chairing the regional finals of the Midland Bank Young Enterprise Innovators Awards in 1998. Bovince has also been involved in the National Year of Reading by producing free posters for this event. In February 1999, Bovince was presented with a Business Commitment to the Environment award for their Business and Community Development Programme.

Bovince sees working with other businesses and outsiders as vital for a company's long term survival. As such the company maintains close links with its suppliers, particularly on the ink side. Communicating and cooperating with suppliers can provide opportunities for practical improvements. The commitment to the environment based on BATNEEC principles has led to Bovince pushing suppliers to meet reduced solvent level targets before their implementation date as well as reviewing the environmental credentials of suppliers. In 1999 Bovince introduced the Supplier Annual Award for Excellence. Key criteria for the award were, amongst others, environmental awareness, quality awareness and partnership relationship.

Bovince is also a member of the Department of Trade and Industry's Inside UK Enterprise Programme (IUKE) which can help businesses to help themselves by providing access to networks which enable them to share information, exchange good practice and wherever possible develop new opportunities. Bovince sees this programme as a way of supporting the transfer of environmentally sound technology and good quality management procedures. The company is also getting involved with IUKE's European equivalent called Business Corporation Network. Bovince's commercial director cites strategic and logistical reasons for employing the help of outsiders. As an SME, the company itself might lack the appropriate infrastructure to implement certain activities and using outsiders is seen as a solution. For example, the consultant used for the EMAS registration has been and continues to be a key informant. The company also offers placements to students from UK and foreign universities. The company directors believe that acknowledgement of their social responsibilities has led to various benefits. Visits to the factory from local schools and colleges as part of Bovince's education links programme have meant that the staff at Bovince can develop or enhance their skills in presentation and communication. Staff get a better understanding of all aspects of their work as they explain what they are doing and as they are questioned by the visitors. Students on placements can gain a real life input into their subjects and another angle on business rather than merely the theory. The commercial director commented that "getting outsiders in has been a great process because they have a different perspective" and "we're open to working with anyone who can bring added value". In addition, the works, quality and environmental manager noted that "assessment of our processes by judges and auditors can often point up fresh ideas to meet new challenges of the future".

Gemba Kaizen, the Japanese standard for continuous improvement, has recently been introduced into Bovince. The system is helping the company to eliminate any form of waste or wasteful processes and to use the minimal amount of resources to produce the work. As such Kaizen has got an indirect but obvious link to the company's environmental policy. The basis of the Kaizen discipline is the so-called 5S programme: Sort, Store, Shine and Service, Standardise and Stick to it. *Sort* means to get rid of clutter and to recognise that everything needs to be in place. *Store* means the efficient place to locate the items that are definitely needed. *Shine and Service* means monitoring and restoring the items to their original condition. *Standardisation* means putting standards into place, such as having an area map for all the items. *Stick to it* means keeping the practices and standards up to scratch all the time. Nonconformances are noted and actions taken. The key factor in the Kaizen process is the identification of the company's most value-adding activity. At Bovince this is the moment that ink is printed onto paper because this is the product that the customer receives. During the implementation of Kaizen, Bovince received support from a consultant of the Kaizen Institute in the form of a series of workshops.

Bovince has found that Kaizen is a common sense and low cost approach to improvement and so far the system has had a dramatic effect on the company's processes. For example, based on the *Shine and Service* rule Bovince now employs a full time technical engineer who conducts preventive periodical maintenance. Simple repair requirements can now be made on site thus saving time and reducing transport pollution by not requiring alternative repair services. This also reduces machine downtime and improves machine efficiency. The three ground rules of the Gemba Kaizen system that Bovince operates are, first of all, housekeeping. This is an indispensable ingredient of good management. Through good housekeeping people acquire and practice self-discipline. Without it, it is impossible to provide products or services of good quality to the customer. The second rule relates to waste elimination. This means getting rid of any activity that does not add value. The third rule is standardisation. Standards may be defined as the best way to do the job. Maintaining standards is a way of assuring quality of each process and preventing the recurrence of errors. To Bovince, this means not only adhering to current technological, managerial and ISO standards but also to improve current processes. Overall, the difference between the Kaizen system and other standards is that Kaizen enhances shop floor operatives' skills and helps them to instigate workplace improvements. The Kaizen system gives more ownership of the process to shop floor operatives because it is a bottom up procedure or, as a print manager put it, "it unleashes individual abilities".

5.2.3 EMAS Procedure

In 1992 Bovince realised that it needed to be able to show customers its commitment to quality. The company decided that this could be achieved by obtaining the well known quality accreditation BS 5750, which was officially achieved in 1994. They saw this standard as a framework to improve on the quality of products and services, giving the company a structure that could be continually built and improved upon. The development of better documented design and production as well as the introduction of product quality checking procedures helped the company to improve its operational process. Building on ideas of quality management, Bovince extended its thinking to embrace environmental management. As stated earlier, the company has a history of being involved in environmental issues affecting the print industry. Key staff members have for example been involved in examining a variety of issues affecting printers and the environment with PIRA (Printing Industry Research Association). The drivers for developing quality and environmental management systems are the commercial director (a member of the family) and the works, quality and environmental manager (a long term employee of the company). These two people, and in particular the latter, have not just a work related involvement in environmental improvement but are also personally interested in and aware of environmental effects and sustainable development.

The achievement of ISO 9002 gave Bovince the structure to introduce the environmental standard BS 7750. They wanted a way to monitor their environmental effects but did not want this to be separate from their quality system. Because the two BS standards are similar it made sense for Bovince to integrate them. Bovince achieved BS 7750 in October 1995, followed by ISO 14001. Soon after, the analytical work began for EMAS registration. The first company statement on EMAS was formally issued on 27 March 1996 and referred to EMAS registration as "part of a long term strategy to lead the way in screen process printing through improved environmental performance and investment in cleaner technologies". EMAS registration was formally achieved in January 1997. A consultant, who has been used by the company for a number of years, has been a key person in the whole EMAS process. He offered helpful guidance on practical improvements and was able to give useful insights into the company's operations by providing an outsider's perspective. Bovince received financial assistance for their EMAS registration from the government's Small Company Energy and Environmental Management Assistance Scheme (SCEEMAS).

As part of the EMAS process, a small number of environmental and quality auditors were appointed among the managers, advised by the consultant and led by the works, quality and environment manager. Their role is to ensure compliance with both the environmental and quality management systems. One of the key instruments used by the auditors is the environmental effects register, in which each area of work is listed. As a result of the EMAS process, Bovince reviewed its key suppliers and their environmental credentials. The company's main ink supplier, Sericol, part of the world wide Burmah Castrol group, has itself achieved ISO 14001 and has secured a Queen's Award for achievement on environmental grounds. In relation to small suppliers, Bovince does not expect them to have formal environmental accreditation, but does expect a sensible or quality position. As Bovince is reliant on paper, they see it as essential to support forestry management initiatives from their suppliers.

The early EMAS statements issued by the company were quite basic and covered issues of noise, energy use, raw materials and emissions to air, waste and vehicle use. These were all reasonably low key documents described by the commercial director as "bog standard". The company's consultant and the BVQI verifier have suggested there should be improvements in presentation and style, as well as evidence of continuous improvement in practice. Consequently, in the 1999 EMAS statement, the focus was much more on sustainable development. This statement not only lists the company's processes and their measured environmental effects. It also describes Bovince's community and education initiatives. The statement explicitly states that "business is inextricably linked to the question of sustainable development, as an integral and responsible part of the community and society".

Bovince has been rewarded for their efforts towards environmental improvement and sustainable development with a whole host of awards. In 1993 the company received the Waltham Forest Business of the Year Award for its active role in the community. EMAS status contributed to the achievement in October 1997 of the PRISM award for Environmental Improvement Company of the Year. The company featured in the Good Practice Guide on Solvent Management in Practice published in September 1998 on behalf of the Environmental Technology Best Practice Programme (jointly sponsored by the Department of Trade and Industry and the Department of the Environment, Transport and the Regions). The 1998 EMAS statement won Bovince the 1999 ACCA (Association of Certified and Chartered Accountants) best SME environmental reporting award. Bovince was also one of the nominees for the 1999 European Environmental Reporting Awards (EERA). As part of the DTI's Inside UK Enterprise Programme the company has received an award in recognition of its contribution to improving competitiveness in UK industry and commerce. All this recognition creates a high profile for the company and a great deal of positive publicity.

It would be misleading to suggest that the awards, engagement in local affairs and environmentally responsive corporate objectives were the direct result of achieving EMAS recognition. EMAS has added to the rigour of the company's environmental management system at the same time as providing it with a high profile and positive public image, accompanied by real financial savings in various stages of the production process. Now that the company has various other management systems in place, it regards EMAS more as a background process and as embedded in other systems. The company's commercial director describes EMAS as "a vehicle for the measurement of environmental effects". Bovince's quality, works and environmental manager views EMAS purely as an environmental reporting system. He states that "it is a bit like a rule book, we keep getting back to it".

5.2.4 Conclusions

Bovince is a very innovative and progressive company that is devoting a great deal of resources to achieve continuous improvement in environmental and sustainability issues. Their efforts are particularly impressive considering the fact that they are an SME with limited human resources. According to Schmitter (2002), sustainability and innovation have historically been treated as antithetic. In today's world, he goes on to argue, the two are regarded as potentially more compatible but the suspicion persists that too much attention to sustainability dampens innovation and too much innovation threatens the delicate balance of material and social resources that is necessary if a society is to reproduce itself. In the case of Bovince, however, it seems that innovation and creative thinking contribute to achieving sustainable development. The issues of participation, innovation and sustainability are not working against each other in Bovince, but one actually contributes to the achievement of the others. A good example of this synergy is the application of new printing technologies which has resulted in financial savings as well as environmental benefits. Another example is the participation of staff in the various management systems, particularly in the Kaizen system, which has made employees more aware of waste management and general environmental issues while simultaneously improving staff motivation, involvement and empowerment.

Bovince works on a strategy of integrating ISO 9002, ISO 14001, EMAS, Kaizen, Investor in People systems and other initiatives because they believe that "integrated business systems can support and grow sustainable solutions". The reporting of their environmental effects has provided them with further information to implement improvements to their ISO 9002 and ISO 14001 systems. Bovince soon realised that the EMS could be used as a learning tool for the company's improvement and not just as a procedural complaints process. Monitoring of the effects in each department, analysing them in relation to other departments and seeing how they impacted on the company seemed the most logical route. According to the works, quality and environmental manager "By reviewing impacts in this manner we get more staff interaction and ownership of the improvement activity required" (Hall 2000). Furthermore, the transparency of systems and strategies are seen as being of fundamental importance to business sustainability. In addition, they believe that quality of product or service cannot happen unless environmental effects are considered and linked to the process. They conclude that environmental systems and the reporting of them will not mean anything unless people are involved in their improvement.

Table 5.2 The Bovince Tree of Sustainability

Branch 1. Waste	To continually invest in ways to minimise and eliminate all forms of waste from our process.
Branch 2. Benign emissions	To eliminate benign emissions from our printing processes.
Branch 3. Effluent	To eliminate trade effluent being discharged into sewers and reuse within our process.
Branch 4. Energy	Minimise and reuse energy, to continually upgrade with the most energy efficient fittings. To use renewable energy sources.
Branch 5. Cyclic ways	Develop cyclic processes that use renewable materials and chemicals.
Branch 6. Transport	Use the most efficient ways to transport our products and people, by using alternative technology vehicles.
Branch 7. People and Learning	Communicate with our people to promote the culture of a learning organisation. To use our people learning culture for the redesigning of processes and the development of safe sustainable systems.
Branch 8. Business and Society	To achieve business and society needs by continually pursuing and promoting sustainable solutions with our suppliers, customers, local residents and local authorities, schools, universities and government.
Branch 9. Sustainable Growth	To continually grow the tree of sustainability, to upgrade and redesign it as it changes and matures.

Source: Hall 2000.

The company has developed the so-called "Bovince Trilogy" consisting of three interlinked documents mapping out the company's progress in promoting a sustainable future. The first document is the 1999 EMAS statement, outlining their corporate strategy of alignment to environmental performance improvement and social responsibility. The second document is an article entitled *Industrial Harm to an Ecological Dream* written by the works, quality and environmental manager. This article tracks Bovince's development of integrating ISO 9002, ISO 14001, EMAS, Investor in People and Kaizen systems into their day-to-day activities, along with other initiatives to promote sustainability. The article ends with the concept of the "Bovince Tree of Sustainability" and its nine branches for future improvement (Table 5.2). The third document is Bovince's first sustainable report entitled

Bovince ... The Sustainable Journey. This report is based on previous EMAS statements and the "Industrial Harm" article and looks at the company's economic, environmental and social impacts and reports on current progress against the nine branches of the sustainability tree.

In the "Sustainable Journey" document the company identifies all those groups that are affected by Bovince and its activities. The three directors are the *share holders*. Of these three only the commercial director was actively involved in the EMAS process. He is also an important driver in making the company more sustainable. The *status holders* in the company are the five members of the management team. Only the works, quality and environmental manager is actively involved in environmental and sustainability issues. The employees can be described as the *work holders*. Although staff have not directly contributed to EMAS, indirectly their contribution has been of vital importance. Employees' awareness of health, safety and environmental issues – at work and at home – has strongly increased and more responsible attitudes have been developed as a result. Another category of holders identified by Bovince are their suppliers who function as *stake holders*. Some of the company's suppliers can also be described as *knowledge holders* because Bovince sees its suppliers as "enablers" and as an integral part of their delivery, quality, production and service. Some of their suppliers actively cooperate with Bovince in environmental improvement. The consultant used by Bovince during the EMAS registration is another *knowledge holder* who has been and continues to be an active participant. Other Bovince *stake holders* are its customers who have not been actively involved in the EMAS process in that they do not exert pressure on the company to be more environmentally friendly. The community around the Bovince site are the *holders of a spatial location*. Again, the company has not been put under any pressure from the neighbourhood in terms of their environmental policies. All the environmental and sustainability activities undertaken by the company have been their own initiative. Bovince is an active participant in local community regeneration strategies. Finally, the company identifies another category of holders, namely the environment and future generations and calls these "voiceless stakeholders".

The umbrella for all Bovince's activities is the European Foundation for Quality Management (EFQM) excellence model where the self-assessment process is scored against the percentage criteria within the model. The model identifies nine key areas which companies can use as a route to improved performance. It has provided an effective way for Bovince to coordinate its quality, environmental and people initiatives, as well as helping to further integrate their ISO 9002, ISO 14001, EMAS, IIP, NVQ and Kaizen systems. The nine criteria framework is split into two parts: enablers and results. The enabling criteria are leadership, policy and strategy, people, partnerships and resources and processes to cover what a company does. Results criteria cover what a company achieves. These are customer results, people results, society results and key performance results. By using a scoring system the company can continually check its strengths and weaknesses and because of its dynamic nature the model promotes innovation and learning into the culture of the company.

The works, quality and environmental manager together with the commercial director are the drivers for sustainable development within Bovince. As can be concluded from the company's activities and policies, all three elements of sustainability are emphasised: the environmental, the social and the economic. In relation to the first element, the company is very aware of its environmental effects because it applies several tools to quantify and measure these effects. EMAS was considered to be useful in that it provided the company with a statement of measurables. However, the works, quality and environmental manager was of the opinion that EMAS has "not really been powerful on the people front". By this he meant that there had not been much input into the company's community development programme as a result of the EMAS process. When it comes to the second element of sustainability, the company seems to take its corporate social responsibilities very seriously. It maintains a strong relationship with local communities and charities in the East London area. As stated in their 1999 EMAS statement "Our philosophy is one of openness and active stakeholder involvement, ranging from developing effective supplier relationships to broader community issues like mentoring in schools and working with volunteer organisations". In addition to this, economic considerations continue to be important. Bovince has realised that it pays to be a sustainable company because their economic growth is steadily increasing. They have also used EMAS to achieve competitive advantage with the commercial director stating that "by achieving this award Bovince have set the screen printing industry a very tough act to follow" (www.bovince.com/page_2htm Accessed April 2000).

Bovince is of the opinion that businesses cannot produce sustainable systems on their own. They need support from government as well. They view the role of government as one where there is a combination of policy initiatives. Government can provide a varied framework of best practice information, regulation to improve company standards, reforms to the tax system, encouragement and support for greater innovation and encouragement to the consumer to buy cleaner technology by making the pricing structure more attractive. Bovince advocates that "future success in the business world will depend on a move away from linear, wasteful processes which are toxic, to cyclic processes which conserve resources and are healthy". In summary, looking at a company like Bovince seems to underline the argument that well balanced sustainability policies enable a synergy between innovation, participation and environmental concerns.

5.3   Huntsman Polyurethanes, Shepton Mallet

5.3.1 Introduction

Huntsman Corporation is the biggest privately owned chemical company in the world. It is a family-run business and has its headquarters in Salt Lake City, USA. Huntsman companies today have revenues of approximately $7 billion, employ more than 16,000 staff and have facilities in 43 countries. Huntsman has grown into a world wide supplier of chemicals, polymers and packaging mainly through acquisitions, joint ventures and internal expansion. Their operating companies manufacture basic products for a wide range of industries such as chemicals, plastics, automotive, construction, high-tech, health care, textiles, detergent, personal care and packaging. The Huntsman Corporation manufactures many of the chemicals that are used to build products for these industries. Huntsman Polyurethanes, the polyurethane side of the business, supplies the chemicals and systems that are used to manufacture polyurethane constituents used in everyday products, such as car seats, furniture, refrigerators (insulation foam), shoes (soles), buildings (insulation boards) and binders (resins to bind wood particles together to make high quality boards used in household applications). The site in the small Somerset town of Shepton Mallet is part of the company's polyurethanes business.

The site in Shepton Mallet was established in 1968 by C & J Clark, the well known footwear manufacturer, to manufacture and market shoe-soling chemicals. ICI acquired the Avalon Chemical Company – as it was then known – some 21 years later. In August 1996, under ICI's ownership, the site became the twentieth in the UK to become registered to EMAS. Huntsman's involvement with Shepton Mallet began in July 1999 when they acquired the polyurethanes business from ICI. The site is Huntsman Polyurethanes' commercial headquarters for the UK and Ireland, with supporting facilities of administration offices, a customer service centre, a machine hall for developing and testing new products, stores and two laboratories. Huntsman Polyurethanes at Shepton Mallet employs 81 staff. About half of the employees on the site are involved in manufacturing and a quarter in each of the commercial/ technical and the administrative/customer service/purchasing sections. The 11 acre site in Shepton Mallet is surrounded on three sites by residential housing.

There are three production plants on the site. The polyester plant produces polyols, which in turn form one of the raw materials supplied to the TPU (thermoplastic polyurethanes) plant and the formulation or blending plant. The latter two plants produce TPUs and formulated polyol blends respectively. Polyol manufacturing forms the basis of all production at the site and is used in blended formulations for elastomers and thermoplastic polyurethanes. Elastomers are the rubbery materials used to make shoe soles as well as automotive and furniture applications. TPU is a completely reacted polyurethane, supplied as pellets and ready for injection moulding or extrusion. TPUs are very versatile and can be formed into many products. Tailor-made formulations are created for customers. The footwear and automotive industries are the main market focus at Shepton Mallet. Typical applications for their product are shoe soles and seats, noise insulation and steering wheels for the automotive industry. The main impacts of the production processes on the environment are gaseous emissions, non-hazardous chemical waste and aqueous (liquid) effluent.

At the end of 1997 the TPU plant was taken over by the American company BF Goodrich. This company is a leader in the global TPU business and is therefore an important customer for Shepton Mallet's materials and in turn supplies Huntsman with TPUs for the footwear business. 15 employees transferred with the takeover. Although located on the same site, the BF Goodrich plant is not EMAS accredited. However, Huntsman Polyurethanes and BF Goodrich work closely together on all issues regarding safety, health and the environment for the site.

5.3.2 Thematic Background

The Huntsman Corporation has a commitment to the "Responsible Care" programme of the global chemical industries. This international initiative calls for chemical companies to demonstrate their commitment to improving all aspects of their own and their customers' safety, health and environmental activities. Huntsman's corporate report states that the company is "committed to ever greater levels of achievement in growth, employee safety, product quality, environmental excellence and in bettering the human condition". Integral to Huntsman Polyurethanes' business strategy is a world wide safety, health and environment policy which seeks to ensure that all of its sites are injury-free, protect and promote good health for every employee, and reduce the environmental impact of its manufacturing processes and products.[5]

Huntsman's site in Shepton Mallet, as a chemical company located in a small town with housing nearby, has to be sensitive to environmental concerns. In 1996 the site won the first UK Responsible Care award given by the Chemical Industries Association in conjunction with the Institute of Chemical Engineers. The award was given in recognition of the significant improvements made in their safety, health and environmental performance as well as demonstrating openness in communication about their activities and achievements to their employees and the local community. The 1995 environmental performance report states that it is the policy of the Shepton Mallet site to manage its activities so as to give benefit to society to ensure that as a minimum standard, these activities conform to all applicable laws, regulations and codes of practice as detailed in their register of legislation. The policy furthermore states that these activities should be acceptable to the Shepton Mallet community and that their environmental impact is reduced to a practical minimum. As a site, they have set targets for environmental improvements. These targets include emissions and waste reduction and energy and resource efficiency. Every year details of the site's activities are published including its environmental performance.

A change of major significance for the Shepton Mallet site has been the acquisition by the American Huntsman Corporation of ICI's Polyurethanes business in 1999. Apart from changes to the plant and the operations, this takeover has had implications for its environmental, health and safety policies. Safety is of particular importance to the Huntsman Corporation and the culture of good safety is underpinned by the philosophy that "nothing we do is worth getting hurt for", a motto which is displayed at the entrance of the Shepton Mallet site. The strict safety procedures can perhaps be explained by the American culture of litigation. The Shepton Mallet site is considered to be a safe site with relatively low environmental, health and safety risks. Environmental policies and procedures have not changed significantly since the takeover. ICI did have an environmental management system in place and the Shepton Mallet site has continued with this by incorporating it into Huntsman policies.

For the Huntsman Corporation as a whole, the acquisition of ICI Polyurethanes has been a driver for change. According to Shepton Mallet's SHE (Safety, Health

---

[5] Despite Huntsman's formal commitment to reduce the environmental impact of its activities, the practice of the chemical industry can sometimes tell a different story. The *Independent on Sunday* (1 April 2001) reports how the west Texas city of Odessa has been polluted by one of Huntsman's plants located in the city. Residents have been complaining of health problems and have recently been paid an out-of-court settlement by Huntsman.

and Environment) coordinator, ICI was notoriously bureaucratic with numerous procedures to comply with. Huntsman had grown very quickly globally but did not have many corporate procedures in place. Hence, Huntsman has learned a lot from ICI in terms of environment, health and safety policies. Environmental management systems are in place in all plants owned by Huntsman Polyurethanes. The company has an Environmental Coordination Group (ECG) which is the integration of SHE issues and business strategy on a global basis. The Environmental Coordination Group is a cross-functional committee made up of senior executives from across the company, chaired by the operations director. This group develops strategy, sets priorities and monitors health, safety and environmental performance. Huntsman Polyurethanes furthermore has four departments to develop or advise the business on environmental policies. These are international environmental affairs; environmental technology; safety, health and environment; and the international product SHE group. In addition, there are specialist managers at headquarters and in the separate businesses who deal with specific environmental, health and safety issues.

Huntsman Polyurethanes believe that there are certain responsibilities associated with – as they put it – "being a leading company and major employer in Shepton Mallet". They see this as going above and beyond the management of safety, health and environmental issues. According to the company, it also embraces the development of good relations with the local community based on open and active two-way communication. They aim to play a full and positive role within the Shepton Mallet community and contribute, where possible, to the well-being of the area. "We aim to be a good neighbour." Developing good external relations has obvious benefits for the site and also for the local community.

An important method of working with the community has been via community liaison meetings. Community liaison groups were set up in the early 1990s by the site's communications manager. Since 1993 there has been an annual meeting in which the company explains its activities and provides a site tour. Invitations are sent to up to 100 households nearby. Attendance is usually in the order of 25, including two town councillors. The company is keen to learn of problems, typically of noise and smell, of which they may not have been aware. These meetings give neighbours the opportunity to raise any concerns they might have. Sometimes Huntsman holds open days for visitors, neighbours and families of employees. These events are not just symbolic, undertaken for image improvement purposes, but also to inform the public. The SHE coordinator comments that open days "allay fear and increase awareness and familiarity". In 1989 when ownership of the site transferred from Clarks (shoes) to ICI (chemicals) there was a lot of opposition from the local residents. A full-time community liaison officer was appointed who worked with schools and local residents. Although Huntsman at Shepton Mallet no longer has such an officer, two other employees at the site now share this responsibility.

The site has several links with schools in the area through its schools liaison activities. In Shepton Mallet itself, they have established a business/education link with a local school, creating a partnership in which Huntsman employees can work closely with staff and pupils on special projects or activities designed to create further understanding of the chemical industry. In May 1998 the Shepton Mallet site was awarded a certificate of recognition for its support of the Schools Compact Programme (a system that provides mentors for year 10 and 11 students from among employees of Huntsman Polyurethanes and other local companies). The site also arranges work experience projects.

Huntsman also conducted an environmental project with a local school. The project aimed to link industry with young people to harness their imagination. Working with students from the Cathedral School in the nearby City of Wells, a Zero Emissions Research Initiative (ZERI) was developed. The project was the first of its kind in the UK and its primary aim was to find a new application for the distillate by-product from the polyester plant. It was a reflection of the philosophy of using waste so that there would be no emissions. This project received widespread media interest and recognition, winning a Somerset Business and Education Partnership award, sponsored by British Aerospace and a national "My Place, Our Place" award in 1998, sponsored by the insurance company, Legal and General. This latter award is given in recognition of schools and businesses working together to promote awareness of ecological issues and to find practical solutions to local issues.

Despite the Huntsman Corporation's global operations with over 16,000 employees, the fact that the company is family-owned and family-run is still noticeable in its policies and procedures. A strong sense of ownership and pride in the business has been transferred to the employees at Shepton Mallet. The interviewees at the site have commented that they feel that "Huntsman cares for its employees". In November 1999, just after the takeover of ICI, a member of the Huntsman family came to visit the Shepton Mallet site. Besides its commercial activities, the Huntsman family has a humanitarian mission. For example, at the Shepton Mallet site there is now a health screening programme in place. The site also gives money to local charities.

5.3.3 EMAS Procedure

It was essentially a business decision from company headquarters (ICI Polyurethanes) in Brussels to go down the EMAS route. Because the blending plants, such as the one at Shepton Mallet, are close to customers, formal recognition of an environmental management system would add to selling potential, give a competitive edge, as well as being broadly beneficial. There was no resistance to the idea at plant level. Indeed, the site had been publishing annual statements of safety, health and environmental performance for some time, though in less detail than required for EMAS. The argument of competitive advantage through EMAS status was understood in a company close to its customers, and the future target was to integrate its quality and environmental management systems into an integrated site management system. At the outset in 1995, there were meetings between staff of the three blending plants of ICI in Europe and the verifier, Det Norske Veritas Quality Assurance (DNVQA).[6] The idea of learning from each other did not work out well, as each plant was engaged in different kinds of production processes and they were located on different kind of sites. So the three plants went their separate ways.

---

[6] DNVQA were selected as verifiers for EMAS because they were already the auditors for
    ISO 9001 at ICI and because they were well known in the chemical industry.

The Shepton Mallet site first achieved the BS 7750 standard in May 1996, EMAS in August and IS0 14001 in October. There are two other sites within Huntsman registered with EMAS. There are more sites with an ISO 14001 accreditation. The quality standard QS 9000 was awarded to the Shepton Mallet site in 1998. The QS 9000 standard was created by car manufacturers Chrysler, Ford and General Motors and is aimed at suppliers to the automotive industry.

Implementation of EMAS on the Shepton Mallet site took a total of fifteen months, covering the review and preparation of documentation, the identification of significant effects, the setting of objectives and targets, audits and training and the production of the environmental statement. The work built on the environmental performance reports of previous years, but needed to show the detail laid down by the EMAS requirements. The cost of achieving EMAS status was not insignificant. In staffing terms, it was calculated at £30,000 and the external auditors cost £10,000. The cost of producing the initial statement was also £10,000, with the review statements of 1996 and 1997 costing between £2-3,000.

The significance of environmental effects, both direct and indirect, has been assessed using a procedure designed to reflect the potential seriousness of the effect. This is based on how often it has occurred, effectiveness of controlling it, severity of consequences and extent of effect. Based on this assessment the following significant environmental effects were identified: air emissions, aqueous effluent to sewer, aqueous effluent tankered off-site, land contamination, control of spillages, and energy and resource use. The use of contractors as well as the reduction and reuse of waste were also identified as important environmental issues.

Huntsman is granted consent to discharge their effluent to the local sewer by Wessex Water, who are the local sewerage undertakers. This consent limits the site discharge to ensure that the local sewerage works does not become overloaded and can deal effectively with Huntsman's effluent. This in turn prevents pollution of the local river. All water drained from the production areas on-site is collected in two tanks. They are tested on-site for pH and COD (Chemical Oxygen Demand[7]) prior to discharge to sewer and not released unless they meet with consent requirements given by Wessex Water.

Condensation from the polyester process leads to the production of water which contains some of the process chemicals. It is above the allowed COD concentration and is therefore not suitable for discharge to sewer and is tankered off-site for disposal. Tankered effluent increased by 21 per cent between 1995 and 1999 owing to the increased production of polyester. Target production increases for the near future may result in more effluent being tankered off-site. However, they are trying to reduce the COD in this condensation water so that there might be a possibility to pump it to sewer. Huntsman is talking with the Environment Agency about the best way to deal with this. Discharging the water to sewer would present significant cost savings since at present half of the disposal costs are transport. As well as aqueous (liquid) effluents, other site waste comprises chemicals from laboratory use, some glycol-based waste from the polyester process and a polyol-based waste from the

---

[7]   COD means the amount of oxygen removed by the effluent that would otherwise support fish and other aquatic life.

blending process. These materials are transported and disposed of under the relevant legal requirements. Non-hazardous waste includes packaging, office and domestic waste, which is compressed before being sent to a landfill site. There is a continuous drive to reduce the amount of waste generated by the site in order to reduce the overall impact that the site has on the environment. For instance, old stock is reblended into new material whenever possible. This in turn reduces the amount of material that is disposed of in landfill space.

In November 1999 a £1.3 million site project was sanctioned to handle adipic acid, the major raw material in the polyester manufacturing process, in bulk instead of one tonne bags. This is a major milestone in the site's development and in addition to economic benefits will bring significant improvements in safety and environmental performance. Environmental benefits of this project include reduction in deliveries through larger payloads and the use of rail/road links, reduced noise from relocation of raw material distribution pumps and modified tanker discharge arrangements, reduction in fork lift truck movements and reduced waste streams.

Environmental awareness training on site was developed as part of implementing BS 7750 rather than being EMAS specific, and a booklet was provided for employees. This training was designed to give greater understanding of an individual's impact on the environment and to encourage employees to consider their own environmental impact within their jobs and motivate them to look for ways and means of reducing their impact. One idea that came from the training offered in 1996 (and from an episode of serious non-conformance) was colour coding of the site drains to ensure that employees and contractors understand how the drainage system operates. Red indicated the drains that were monitored, blue dealt with surface water and black covered the usual domestic effluent to sewer. Other training included familiarisation with reorganised waste disposal procedures. All this means that the site is now much cleaner than it used to be. It was found that after the first EMAS registration some issues were still unclear to staff. So they received training on waste issues and disposal. The training was developed ad hoc and several members of staff were heavily involved.

During 1997 – whilst ICI still owned the site – a new system called "audit walks" was introduced to improve safety and housekeeping around the site. Initially all plant operators were trained in how to use the system, and this was extended to all other areas and employees in 1998. The system consists of a total of 22 safety audit walks, designed specifically for each area on-site, that are carried out on a monthly basis aiming to identify safety and housekeeping improvements. The audits are undertaken by pairs of staff who themselves set the standards and therefore have a high degree of ownership of the process. Early in 1999 the site was chosen as the winner of the Polyurethanes Chief Executive Officer's safety award for this system of safety audit walks.

Interviewees at Huntsman in Shepton Mallet described the first EMAS registration as "tortuous". The whole process involved a lot more work and was a lot more thorough than was anticipated. Three people from the United Kingdom Accreditation Service (UKAS) came to the site to audit Huntsman's internal auditor. The verifier from DNV has been very positive and has given a lot of feedback in the EMAS process. The verifier is not just interested in procedures but, more importantly, also in what actually happens. However, the quality systems coordinator notes that one has to be careful that the verifier does not become a consultant. One can negotiate with the verifier but there is nevertheless a fine line between a verifier and a consultant. Staff at Shepton Mallet have also been surprised by the precision that was required in the wording of the EMAS statement. They have found the verifier to be very strict with this, but they also commented that these details focus the mind and improve the operations. The re-registration with EMAS was found to be significantly less difficult.

Staff at Shepton Mallet see EMAS as a learning experience. They claim not to put a show on for the verifiers and that there is an all-year round culture of continuous improvement. Identifying initial improvements was relatively easy but now that EMAS has been in place for a number of years, identifying areas for improvement is proving more difficult. During the first phases of EMAS they have mainly concentrated on waste management issues. In the near future they intend to focus on energy use and reduction. The site is also catching up with the legacy of the building and the operations. The site has grown very quickly and as a result they are now left with three different plants of which some old elements are redundant. They are now cleaning up the site, partly with help from the so-called 5 S programme.[8]

Huntsman Polyurethanes at Shepton Mallet has experienced several advantages as a result of its EMAS accreditation. One of the possible paybacks of having EMAS status is good relations with the Environment Agency (EA) and perhaps less likelihood of disruptive and possibly costly inspections. Because Huntsman goes beyond legal compliance with EMAS, the SHE coordinator commented that "we've got credits in the bank, so perhaps we can be more honest with the EA". It is a reciprocal relationship, with the EA listening to Huntsman's problems and together looking for solutions. EMAS has also provided cost savings, although these are not quantified in any of the environmental statements. Another benefit of EMAS is that environmental improvement has now become part of the site's culture. They feel that EMAS and ISO 14001 have given them something extra. The openness and communication that EMAS requires has been another advantage for Huntsman. As a chemical company in a small rural town it is important "to let people out there know what we're doing, to let them know we are not polluting". As such the main audience for the EMAS statement is the local community. Huntsman Polyurethanes' 1999 Environmental Progress Report states that "reporting on our performance is an integral part of our contribution to sustainable development. It is one of the ways to demonstrate our openness and determination to communicate with our stakeholders, especially customers, employees and legislators".

---

[8]   See section 5.2 for more information on this Japanese continuous improvement scheme. The five S principles constitute a step process which involves the whole engineering team sharing goals and values. *Sort* removes unneeded items from the workplace; *Set* in order stores required items where they can be traced; *Shine* means cleaning and inspection; *Standardise* is the use of visual control as the norm; and *Sustain* uses self-discipline to maintain the standards.

5.3.4 Conclusions

Innovation processes at Huntsman mainly take place in the operational and technical side of the business. It is an innovation-driven company, with about a quarter of employees researching new methods, materials and systems. Shopfloor employees are quite proactive in suggesting solutions to improve operations. It is noteworthy that the majority of environmental improvements are practical solutions that simply seem to make good sense. Often they have the added advantage of savings costs.

Sustainable development strategies are developed by the company as a whole and not so much by the individual sites. Huntsman employs a broad definition of sustainability and refers in its 1999 Environmental Progress Report to the three interrelated business spheres of environmental protection, social responsibility and economic development. It reports on economic performance, health and safety and environmental issues. They are in the process of developing indicators that reflect the social aspects of sustainable development for the next report. Huntsman Polyurethanes is working with the World Business Council for Sustainable Development, leading companies, acknowledged experts and other interested parties to understand the implications and incorporate the concept of sustainable development into their corporate strategy. Huntsman feel that being involved in sustainable development is the right choice morally, but above all they feel that it makes good business sense. Sustainability is viewed as presenting opportunity and potential to provide better service and innovative products to customers. By balancing the demands of economic, environmental and social criteria, they hope to remain both profitable and viable now and in the future. According to Huntsman sustainability strategies and environmental progress can help to demonstrate clear business benefits by improving the stakeholder value of a business. Clear links can be demonstrated, especially through improvements in reputation, increases in innovative capacity, savings through efficiency gains and better awareness of stakeholder and customers' perceptions and needs.

Huntsman Polyurethanes has identified four corporate responsibility objectives. First of all, they want to demonstrate their commitment to sustainable development and be an environmentally and socially responsible member of the industry and society. Second, they aim to earn the trust and respect of key stakeholders and thereby protect their "privilege to operate". Third, they want to encourage the development of an internal culture and mindset that strives to perform ahead of compliance. Finally, they aim to apply the "precautionary principle" to new products and processes, with respect to possible human health risks and environmental impacts resulting from their activities.

In its 1999 Environmental Progress Report Huntsman Polyurethanes identified several stakeholders to its business, namely the media, local communities, environmental NGOs, competitors, schools/universities, customers, legislators and employees. To what extent have these groups of holders been involved in the EMAS process? Internal to the company are the *share holders* of the Huntsman Corporation and of Huntsman Polyurethanes. The share holders of the corporation have only been indirectly involved in EMAS in as much as they determine corporate visions and aims. The decision to register with EMAS was taken at head office (then ICI Polyurethanes). It is hoped that openly communicating environmental performance will increase shareholder values. *Status holders* in the EMAS process have been the safety, health and environmental coordinator (technical knowledge) and the quality systems coordinator (systems knowledge) at the Shepton Mallet site. The *work holders* or employees have not directly participated in the EMAS accreditation, although they have had to adapt to new more strenuous procedures. As stated earlier, employees are actively involved in improving the technical and operational aspects of the work. Trade unions have not been involved in EMAS because they are mainly concerned with health and safety issues. Employees are also an important audience for the environmental statement. No outside *knowledge holders* were used to help with EMAS. However, the verifier and the Environment Agency have contributed more informal advice and feedback. The verifier is also an external *status holder*.

The neighbours of the site, or the *holders of a spatial location*, are an important category. They are kept informed through annual community liaison meetings and file complaints of noise and smell as and when they occur. The immediate neighbours are the most important holders for the Shepton Mallet site. The Environment Agency can be described as an *interest holder* with which they have a good working relationship, based on trust and openness. Relationships with the local authority (Mendip District Council) have also been positive over the years. Finally, the category of *stake holders* covers all those that could potentially be affected by what happens at the site. This category includes the consumers, contractors and suppliers of Huntsman. Contracts are negotiated corporately so the site itself cannot have any influence in choosing suppliers. Suppliers are expected to have environmental statements. They should have at least a quality system in place and perhaps be working towards an environmental management system. There is a general supplier appraisal process and this has not been changed by the takeover from ICI to Huntsman Polyurethanes. Contractors on the site have to comply with Huntsman regulations, but these are more to do with health and safety than with environmental issues. The environmental improvement programme includes a plan to ensure that the contractors are fully aware of the company's environmental management system and, where appropriate, are included in training programmes. Other stakeholders are local businesses in Shepton Mallet with which there are only occasional links. The SHE coordinator has not been able to develop these links further because of the limited time available. There is some communication between the various Huntsman sites in Europe, who can also be described as stakeholders. They share information and best practice, but there is not a lot of commonality in operations.

For the future, Huntsman Polyurethanes has identified several actions. First of all, they intend to better define their role as a good global corporate citizen, to be fair and just to their employees and to play a full and positive role in the communities in which they operate. Second, they want to continue to develop and promote the many environmental benefits of their products which will help others reduce their environmental impact. Third, they aim to include sustainable development issues in their business risk management process and training programmes. Fourth, they want to ensure that their employees are aware of the benefits and responsibilities of pursuing sustainable development. Fifth, they aim to intensify their commitment to product stewardship. Sixth, the intention is to link their business with emerging global values such as caring, respecting and partnering and finally, they intend to be transparent and open in their actions and report on their progress.

Apart from environmental improvement objectives, the Shepton Mallet site has identified several areas for future improvement by means of a survey. Areas for improvement are: more communication; developing a Shepton web page; more information relating to the business results; encourage more direct communication and less use of e-mail; employee empowerment; and more feedback, both positive/ negative, coaching etc. The site will continue to register with EMAS in the future because they do not see it as a high burden and because it has shown certain benefits. Annual reporting is not envisaged to be a problem.

5.4   Sainsbury's Regional Distribution Centre, Basingstoke

5.4.1 Introduction

J Sainsbury plc is one of the UK's leading retailers, operating one supermarket chain in the UK (Sainsbury's Supermarkets Ltd, including Savacentre), a supermarket chain in the USA and a bank in the UK. Sainsbury's Supermarkets Ltd is the largest of the retailing businesses within the Sainsbury Group, serving over ten million customers a week from over 440 stores throughout the UK, and represents about 75 per cent of total Group sales. Sainsbury's Supermarkets employs over 138,000 people (including Savacentre). Of these, 70 per cent are part time and 30 per cent full time. 58 per cent of employees are women.

The Regional Distribution Centre (RDC) is situated in Basingstoke, North Hampshire and is the company's oldest distribution centre. It was built in 1963 on a greenfield site and covers an area of approximately 12 hectares. It is located on an industrial estate off the ring road in Basingstoke. To the south of the site is a railway line, to the north west light industrial and warehouse units and to the north east a modern manufacturing plant. The River Loddon, which flows through Basingstoke rises about half a kilometre from the site, on the other side of the railway line. There are no residential areas close to the site.

The Basingstoke RDC is one of 19 RDCs in the Sainsbury's Supermarkets Ltd Supply Chain Division and is one of the largest within the company. Over the last five years alone, the volume of products delivered by the RDC has increased by over 12 per cent. As part of an EU subsidised pilot project, with assistance from West London TEC (Training and Enterprise Council), the Basingstoke site was part of a trial for EMAS to be made applicable to non-manufacturing sectors. The RDC in Basingstoke has become the first of its type in the UK to achieve registration under EMAS within the "wholesale and retail sector". It was officially registered in June 1999.

The main activity on site is the assembly of individual store orders of products received from suppliers. The RDC employs around 750 people and delivers to about 79 stores in the south of England covering an area bounded by Poole, Swindon, Oxford, North West and Central London, and Brighton. Over 180,000 cases of commodities are delivered from Basingstoke RDC to stores every day, resulting in around 230 journeys. It is a 24 hour operation, with the majority of work taking place overnight. Due to the varying nature of the commodities stored on site, there are several different warehouses with differing temperature controls, necessary for maintaining the correct temperatures that are required for a number of lines. To complement the operation, there is an extensive support function on site covering engineering, cleaning, catering and waste disposal. The site also has its own fuel station and lorry wash. Some of the activities such as catering and hygiene are contracted out.

5.4.2 Thematic Background

J Sainsbury plc is working hard to portray itself as an environmentally aware food retailer, undertaking various activities towards continuous environmental improvement. In March 2000 J Sainsbury plc was named as the leading food retailer in that year's Index of Corporate Environmental Engagement and Performance published by Business in the Environment, ranking third of the 76 FTSE 100 companies surveyed. In 1996 Sainsbury's published its first Environmental Report, or at least it was the first time that this was published as a complete document and presented in a stand alone format. Jonathan Porritt, former Director of Friends of the Earth and currently chair of the Sustainable Development Commission, wrote a Foreword for this report. From 1990 until 1994 he had been a policy advisor for Sainsbury's. The environmental report provided a detailed picture of the environmental performance of the company and provided an agenda for action within the company to minimise the environmental impact of all its activities. It won the company an ACCA First Time Reporter award. The report was written partly for their internal management with responsibility for implementing environmental objectives. It was also written for those contractors and suppliers who could help in this process and for others outside the company who were interested in its performance and approach, such as central and local government, trade unions, academics and journalists. At the time of writing, Sainsbury's Supermarkets is the only food retailer in the UK which produces a full environment report.

There is a director responsible for environmental issues at Group Board level. At senior management level in the supermarkets division, environmental policies are reviewed by the environment committee, on which various departments of the company are represented, such as building services, supply chain and retail, as well as the environmental management department. Its role is to ensure the company's strategic environmental goals are met. The company has produced an internal annual report on environmental performance for the Board since 1995. There is furthermore an environmental management department which is leading the development of environmental management within the UK retail businesses of the company. The environmental affairs department became the environmental management department in the mid 1990s to demonstrate a proactive rather than reactive approach. The company's 1998 environment report set three priorities and six goals with 25 specific targets, designed to contribute to meeting these goals. Environmental management systems and the Basingstoke Regional Distribution Centre are clearly identified in the first of these targets: "By March 1999, formalise responsibilities and procedures for environmental management, incorporate targets into appropriate business plans, and pilot a certified Environmental Management System at a Regional Distribution Centre". The aim was that each operating company in J Sainsbury plc should develop an environmental management system. Sainsbury's distribution depot in Charlton in South East London achieved ISO

14001 registration in April 2000, building on its existing ISO 9000 standard. Sainsbury's commitment to environmental management has saved the company at least £8.5 million since 1992 through more efficient use of resources.[9]

The company has identified three environmental priorities, namely reducing the environmental impact of products, reducing CO_2 emissions and reducing waste. Their overall aim is to reduce the impact of the organisation through a programme of continuous improvement. In this respect their environmental policy is to:

- quantify and monitor all environmental impacts of the business, including new projects and set specific targets;
- comply with current legislation and, where practical, seek to meet future legislative requirements ahead of relevant deadlines;
- integrate environmental objectives into relevant business decisions in a cost-efficient manner;
- require all colleagues to address environmental responsibilities within the framework of normal operating procedures;
- minimise waste, seek to recover as much as is economically practical and ensure the remainder is disposed of responsibly;
- develop appropriate emergency response plans for major incidents in order to minimise their environmental impact;
- influence suppliers of services and own brand goods to reduce their impact on the environment;
- enhance awareness of relevant issues among customers, colleagues and others who have an interest in our business;
- publish information on environmental performance.

This environment policy was first issued in January 1990 and has been revised several times since. Various activities stemming from this environmental policy are testimony to Sainsbury's interest in and commitment to environmental improvement. A series of best practice guides has been developed which are used as an internal tool for ensuring consistent policy and helping to implement environmental procedures. These guides cover a whole range of environmental issues such as refrigerants in refrigeration and air conditioning, timber and forest products, transport efficiency, water conservation, packaging, ground and surface water protection, offices and energy efficiency. The best practice guides are designed to raise awareness, to provide information and identify contacts for further help, and to ensure that consistent action is taken throughout Sainsbury's to minimise environmental impacts wherever possible. Besides being used internally, the guides also help to communicate their policies and practices to contractors and suppliers.

The company has recently opened an environmentally responsible supermarket on London's Greenwich peninsula. This new concept in supermarket architecture has several energy saving features and uses the latest technology. It is the only store in the UK to achieve full marks (31 out of 31) in an independent environmental

---

[9] Sainsbury's News Release (1999) 'Sainsbury's achieve registration to European Eco-Management and Audit Scheme', 10 June.

assessment by Building Research Establishment's Environmental Assessment Method (BREEAM). Sainsbury's is also to buy wind-generated electricity for one of its distribution depots near Glasgow in Britain's first commercial, non-subsidised project of its kind. "It makes good business and environmental sense to try a technology that reduces our CO_2 emissions", according to the company's environmental director.[10] National planning policy is also having an influence on Sainsbury's site search and development activities. There has been a clear shift in store location strategy since the mid 1990s. A greater proportion of new stores are being developed in town centre and edge of town centre locations. The location of these stores offers the prospect of greater accessibility by modes other than the private car and provide the opportunity for linked trips with other activities in the centre, consistent with the government's transport policy. Furthermore, it can also enhance the attractiveness of the centre as a whole, consistent with the government's policy for retail development.

Sainsbury's was one of the first major UK supermarkets to source non-genetically modified ingredients for its own brand products. It has set up a GM help line for customers to seek information and express opinions about GM. Sainsbury's also founded an international consortium of food retailers and industry experts to establish validated sources of non-GM crops, products and derivatives. The consortium encourages long term commitment to farmers and the commodity industry, ensuring a large market for non-GM raw materials both now and in the future. The company is also involved in socially responsible trading, by playing a role in several initiatives designed to ensure basic human and labour rights. They are working with the suppliers of Sainsbury's own brand goods to implement their code of practice for socially responsible trading. Covering issues like health and safety, equal opportunities and the protection of children, the code aims to make sure that basic employment conditions, based on internationally agreed International Labour Organisation standards, are in place. Sainsbury's supermarkets also stock Fairtrade products. Aside from their internal activities, the company is also a founding member of the "Ethical Trading Initiative" (ETI), an association which brings together retailers, trade unions and international charities like Oxfam and Christian Aid. Its purpose is to agree standards for third world suppliers and methods for monitoring performance.

In addition to environmental considerations, sustainability issues play an increasingly large role in the company's policies and practices. In 1996 they began to examine what sustainable development could mean for the retail business. In the 1996 environment report the company's chairman stated that "we hope that by working to improve the social, economic and environmental performance of the Company and embracing the concept of sustainability, this may act as a positive attraction to customers, government and investors". From this statement it is clear that the company wants to operate in a sustainable yet profitable manner and that there might be competitive advantage offered by embracing the concept of sustainability. In the 2000 environment report, the Group chief executive said "I look forward to [...] ever greater progress in the integration of more sustainable principles into our environmental management system".

---

[10] *Independent on Sunday* (1 October 2000).

The company has an environmental management system in place, integrated into regular management. The management system is based on ISO 14001 and is thus transferable to other operational units carrying out similar work. Since 1998 the environmental report has been externally verified by PriceWaterhouseCoopers.

5.4.3 EMAS Procedure

The decision to have the Basingstoke Regional Distribution Centre registered with EMAS was taken at the head office of Sainsbury's and was thus not initiated by the site itself. As stated earlier, it was part of the company's commitment to environmental management as part of its wider environmental strategies. The Basingstoke RDC was selected because it was the most difficult site. The building itself presented difficulties, being the oldest in the distribution network, not originally designed as a distribution centre and with three different temperature regimes. There was, furthermore, a programme of improving industrial relations with the organised workforce on site. EMAS was seen as having a non-political base so that everybody could get involved in its implementation.

As stated before, the Basingstoke site was part of an EU subsidised pilot project, assisted by West London TEC, to make EMAS applicable to non-manufacturing sectors. West London TEC had knowledge of EU funding opportunities and thus served as "the bridge into Europe", as one of the managers put it. They have experienced the involvement of the TEC as very helpful. Sainsbury's were helped in this project by Business Eco Network, a specialist environmental consultancy.[11] The choice of Business Eco Network (BEN) as a consultant for the EMAS procedure was made by the environmental management team at Sainsbury's headquarters.[12] BEN was used as a consultant only in the initial phases of the procedure because staff at the RDC quite quickly developed the necessary expertise themselves.

The environment consultant from Business Eco Network toured the site and conducted an initial environmental review. She pointed out the environmental issues and the identified effects were grouped into the following areas: local environment, noise[13] and emissions; waste; energy and refrigerant use; water consumption and discharge; transport.[14] Volunteers from across the site formed project teams to cover these areas and, with technical guidance, determined the significance of the effects. The teams looked at how resources were being used, suggested areas of improvement, and proposed solutions to the on-site EMAS steering committee

---

[11] This consultancy originated from an organisation called Business Ecologic. For more information on these two organisations, see section 5.6.2 below.

[12] Business Eco Network contacted Sainsbury's to ask whether there was interest in an EMS, as there was funding from Europe. One of BEN's employees was working as a consultant for West London TEC, which had access to the European funding.

[13] A noise survey was conducted to look at noise levels so that its significance could be established. It was found that because of the geographical location of the site and the nature of the neighbours, noise from the RDC did not have a significant impact.

[14] The transport which operates from the site is controlled by the national transport manager at the company's head office. The RDC does not have direct control over the fleet but does exercise a certain degree of influence.

chaired by the distribution centre manager. The intention of this method of participation was to give staff a chance to show that everyone can make a difference to their environment, be it at a local or global level. Each team prepared a report on the nature of the problem and how to address it. Once the teams had put forward their recommendations, an environmental programme was put in place to address the significant issues.

The environmental programme has resulted in several actions being taken. For example, occupation-sensing lighting is being installed in part of the warehouse and offices so that lights automatically turn off when no one is around. Doors between refrigerated and non-refrigerated areas have been replaced with new fast-acting ones which maintain the temperature difference between the warehouses. A water softener has been added to the refrigeration system. This saves water by increasing the amount of time it can be used before it needs to be flushed out and replaced, to avoid limescale build up. Systems have been improved to enable packaging materials – cardboard and shrink wrap – to be separated easily for recycling. In addition, staff awareness on several relevant issues has been raised. General awareness throughout the site has been raised to turn lights and equipment off when not being used. All warehouse staff received a "manual handling" training course to educate them on the safest methods of handling products. This reduces overall waste and encourages recycling. Systems for donation of products and organic waste have been improved, although at present the majority of the waste that the site produces is still buried at landfill sites.

A new system of picking orders has been introduced at the site which is bringing both quality and environmental benefits. Previously, every store order was printed out for the warehouse operative to use. This is being replaced by "finger scanning". This is an electronic system which displays the order on a screen worn on the wrist. As the operative picks individual commodities, a bar code is scanned to ensure that it is the correct one. This will have two benefits: a reduction in product waste from greater picking accuracy and the elimination of paper based orders. Along with this, delivery documentation for stores is now transmitted electronically. In total this will reduce the RDC's paper use by an estimated 2.6 million A4 size sheets every year. Warehouse operators were educated as trainers for this new technology, so that they could train their colleagues. This form of participation has resulted in a transformation in behaviour, performance, productivity and attitude to the job.

Some actions towards environmental improvement had already been taken before EMAS was put in place. For example, the lorry wash operates on a closed system which means that it cleans and is designed to recycle at least 95 per cent of the water that it uses. Other actions were initiated at the company's head office but have affected operations at the RDC. To reduce emissions further, Basingstoke RDC has trialled the use of a solar powered refrigerated trailer. This dispenses with the need for a diesel powered refrigeration unit. A new vehicle that runs on natural gas has been trialled in the distribution division. The advantage of this different type of fuel is that it greatly reduces emissions of particulate matter and oxides of sulphur, and the vehicle is far quieter. To minimise the "empty running" of lorries a new principle called "backhauling" has been introduced. This means that lorries pick up from suppliers on the way back from store deliveries. Backhauling reduces the mileage travelled when moving food around the country, leading to a reduction in CO_2

emissions. Also, drivers are undergoing the "5 star Driver Development Course" which aims to reduce accidents, reduce DETR prohibition orders and improve driver fuel efficiency. Individual tests on drivers have shown an improvement in fuel efficiency of between 10 and 20 per cent (1999 EMAS Statement).

A local environment group was set up with various people in different functions serving on this. The group acted as the management review team when the EMAS system was being set up. The terms of reference of this group were to act as the controlling body, to ensure that other groups (the project teams mentioned above) were working and to ensure that deliverables were achievable and relevant. This group does not function currently, but there is a management review team. Also, the project teams are no longer in existence. At present the RDC does not have a separate environmental manager. The shift manager at the RDC is also responsible for the environment and for implementing EMAS. When an environmental manager is appointed in the future, it is expected that he/she will work closely together with staff responsible for other audits. One internal auditor was trained for EMAS. He received two days of training from a trained auditor (ISO 9000) from another one of Sainsbury's RDCs. The internal auditor is somebody from the shopfloor and not somebody from management. The shift manager comments that this is important "because now the auditor is not seen as somebody who is snooping around". The auditor himself comments that it has been interesting getting to know other parts and staff of the business and seeing what is involved in the activities of the RDC. Basingstoke RDC receives support for its environmental management system from a number of company departments, which offer a wide range of technical expertise. The resources they can draw upon from around the company are the Group energy team, Group legal services, Group waste management department, national transport department, distribution, environmental services and the environmental management department.

The achievement of EMAS registration at the Basingstoke RDC is considered to be important for several reasons. First of all, they see it as a public demonstration of their dedication to improve environmental performance, particularly to their local community. Closer examination reveals that by local community they mean the workforce of the RDC. The EMAS statement is written for the lay person, so that anybody external to the company can read and understand it. Consultants were not involved in the writing up of the statement. It is distributed to the local authority and to libraries, and is also available on the Sainsbury's website. They do, however, get more requests from universities and local authorities than from the general public. Second, EMAS registration is considered to be important because it sets out clear targets to address the RDC's significant environmental impacts. Third, EMAS registration has had financial benefits. The EMAS process was underpinned at the start by ensuring that an adequate number of issues were addressed which would provide cost savings and thus ensure the process had a definable payback. It is estimated that the environmental improvements on lighting, recycling, temperature control and water consumption had resulted in potential cost savings of £45,000 per annum.[15] Others estimate that the financial benefits of EMAS were more modest,

---

[15] Sainsbury's News Release (2000). 'Minister welcomes Sainsbury's EMAS registration', 13 April.

perhaps £20,000 a year return from waste reduction.[16] The fourth benefit of the EMAS registration was that staff at all levels of the RDC (and the company) had been able to get involved in the process. Various staff members pointed this out as the most important outcome of the EMAS registration. EMAS has served as an excellent morale booster for staff at the centre by encouraging greater employee involvement at all levels. Sainsbury's see communication, training and awareness amongst all staff as one of the keys to the success of any EMS (1998 EMAS statement).

As stated before, the decision to have the Basingstoke Regional Distribution Centre accredited with EMAS was taken at Sainsbury's head office. In fact, most environmental policies are initiated at corporate level and very few originate locally, as is often the case in a large company operating various sites. Despite this, the Basingstoke site has fully participated in the EMAS process, albeit with strong support from head office. There was a steep learning curve and, according to two of the managers, the first audit "was like having your teeth pulled". EMAS has resulted in tangible environmental improvements and raised environmental awareness at the site. Monitoring and measurement of key impacts have provided further information for improvements. For example, there is now more focus on water loss which would not have been known, if it had not been monitored. The verifier described the rigour of the data gathering procedures as one of the strong aspects of the EMS. He also described the environmental statement as a strong feature of the system. However, the profile of environmental issues had not been raised as much as had been hoped. After initial enthusiasm, commitment at Basingstoke began to decrease. As one of the managers put it "because there is an awful lot going on at the site, environmental impacts can go out of the window". In these circumstances, the environment is still seen as something extra and not a strategic requirement.

All this means that the Basingstoke site is now experiencing problems with continuous environmental improvement. This is related to several underlying and interlinked problems. First of all, the building itself is presenting difficulties. The RDC was originally developed for distribution, meat and cheese production, and grocery packaging. All production lines ceased in 1989. The warehouses at Basingstoke have been extended a number of times to serve increasing demand. It is an old building to which various structures and functions have been added over the years and this has had environmental effects. The building was not designed to be environmentally friendly or energy efficient. Hence, in terms of achieving environmental improvements, they are struggling against the constraints of the building. As a consequence there has recently been discussion about the future of the Basingstoke site. When they were just about to start to implement the EMAS process, it was announced that the site was going to close in two and a half years' time, although head office has now gone back on this decision because it proved difficult to find a new site. It is however clear that the Basingstoke site does not have a future in the medium/longer term. Understandably, all this has affected morale on the site, not least decisions on whether to make environmental improvements e.g. installation of low energy lighting. If the payback period was more than one and a half years then it was not considered to be an option. Continuous improvement at the

---

[16] Interview with environmental services manager, supply chain division, 10 January 2001.

RDC is also difficult because what is done at the site is fixed. The operation and day-to-day activities of the centre had changed very little since the initial implementation of EMAS. The shift manager commented that "you come to a halt on what you can do". Electricity use has remained fairly constant over the last ten years, despite the increase in productivity. Carbon dioxide emissions, which are calculated using the (former) Department of the Environment, Transport and the Regions' guidelines, have also not been significantly reduced. Certain issues that could perhaps be improved are the responsibility of head office, such as influencing suppliers.

One of the most important aspects of the EMAS process at the RDC has been the involvement of staff. It was deemed important to get their opinions and to talk to management. It proved quite difficult to overcome initial scepticism and suspicion and it was not always easy to get people motivated. In order to get staff interested they had to be convinced that the environment was not something remote, but something that affected them all at site level. The two key people from head office with responsibility for EMAS realised that they needed to bring it down to the local level and to show the local relevance of environmental issues.

The environmental services manager commented that people did not want to hear him talk about the hole in the ozone layer. They wanted to know how he was going to fix the hole in one of the roads on-site. Another example of a local impact was the River Loddon which flows nearby through the town of Basingstoke. The surface water of the RDC goes in the direction of the river and children paddle in it in the town. It was therefore not only important to sell the financial benefits of EMAS, but also to show people the local environmental impacts of the RDC's activities. The two key headquarters people also wanted to get across the idea that every little helps and that they would not have to go for perfection. The assistant manager of distribution environmental services, who coordinated the process on site, said that "EMAS has made everyone very conscious of how they could contribute and showed how individual actions can bring about environmental improvements" (News Release, 10 June 1999).

EMAS, coincidentally, ran simultaneously with Investors in People (IIP) at the RDC. IIP was done through local initiative and there are some other depots and sites in the company which have also been accredited. Sainsbury's is now going for corporate IIP. Basingstoke soon discovered that there was considerable overlap and synergy between IIP and EMAS, particularly because one of the main objectives of EMAS has been to get staff on board. Through the Investors in People standard, the site has identified and trained their own EMS internal audit team. The twice-yearly staff performance and development review is another example of the links between the two standards. It is an opportunity to discuss individual performance and training needs. Staff whose jobs result in key environmental impacts have been identified and further training has been provided if required. Staff are also asked about their understanding of EMAS and their environmental awareness. In addition, the EMAS statement is part of the induction process for new staff.[17]

---

[17] One of the managers was quick to point out that there is a lot of induction in a short time and most people when starting a new job concentrate on doing the job right and forget about the environmental issues.

5.4.4 Conclusions

In a company of Sainsbury's size with a wide variety of operational units, it is difficult to assess the environmental status of one of these units separately, in this case the Regional Distribution Centre at Basingstoke. Environmental improvements at the distribution centre must be viewed in the context of environmental strategies of the company as a whole. The majority of environmental policies have been initiated at Sainsbury's head office and the same goes for EMAS. In addition, the support structures for environmental management in general and EMAS in particular are all located in the company's head office. Despite this somewhat top down approach, staff at Basingstoke do not seem to have experienced this as negative and they have fully participated in the EMAS process. The RDC has had full responsibility and autonomy for their site-based targets and environmental programme. The environmental services manager of the supply chain division, one of two key people who helped to set up EMAS at the distribution centre, is located at the centre in Basingstoke and perhaps this proximity has helped participation and motivation of staff on the site.

The environmental credentials of J Sainsbury plc are well known by others in the industry and by the customers. The company is devoting substantial resources to communicating these credentials to outsiders in an attempt to establish a more green image for itself. While the company undoubtedly wants to create this image to achieve a competitive advantage over others, their attempts towards environmental improvement also seem to go beyond the usual "green glossy" and appear to be quite genuine. Environmental management in the retail sector presents some difficulties. Many of the impacts associated with retailing occur well removed from retail-owned sites. Influencing the many suppliers is also difficult, let alone customers. It is therefore commendable that the Sainsbury's Regional Distribution Centre in Basingstoke has succeeded in gaining EMAS accreditation. On achieving its registration, the Institute of Environmental Management and Assessment commented that "being a retail household name they are raising the profile of the Eco-Management and Audit Scheme. The success Sainsbury's has had lays the foundations for other companies to follow in their footsteps towards EMAS registration".[18] Despite the company's positive actions towards more responsible practices in the food retailing industry, it has to be kept in mind that these kind of policies are not unique to Sainsbury's. A quick look at other retailers' websites reveals that they, to a greater or lesser extent, also operate their fleet on ultra low sulphur diesel, recycle waste, do composting and have various community and social involvement schemes.

Sainsbury's environmental activities are part of wider sustainability strategies, although these policies are not actually labelled as such. As was the case with EMAS and environmental activities, corporate social responsibilities are initiated from the company's head office and are therefore not dealt with by the individual sites. Some of the interviewees described J Sainsbury plc as quite a paternalistic

---

[18] Sainsbury's News Release (1999). 'Sainsbury's achieve registration to European Eco-Management and Audit Scheme', 10 June.

company in that it looks after its employees. There are various schemes in place, such as maternity and paternity leave, dependency leave and sick pay arrangements. The RDC in Basingstoke has an occupational health advisor and an open learning centre with computer-based courses on various subjects such as languages and NVQ for warehousing distribution. Another example of the company's community engagement is Sainsbury's involvement in town centre management since 1995. They actively support and promote initiatives which bring together local authorities, retailers and all those with an interest in protecting and enhancing town centres where they have a presence. Apart from these overall corporate activities, the Basingstoke RDC also has an involvement with its locality. There is a separate budget from which local activities are sponsored. Examples of such activities include managers who run business games in local schools. The RDC furthermore has its own football team and a social club. Sainsbury's thus seems to support a wider definition of sustainability, with its environmental and social activities taking place against a background of economic growth.

In what way and to what extent have the potential participants been involved in the implementation of EMAS at site level and in the company as a whole or, in other words, what "holders" can be identified? The difficulty here is, again, whether to differentiate between participants in the company as a whole and participants in Basingstoke. Within the company, the most obvious group is the *share holders*, which in this case are the directors at Group Board level. The director with responsibility for environmental matters is the only one of these share holders who has been involved, albeit indirectly, in the EMAS process. At the level of the RDC, all employees can be identified as share holders. This is because the site operates a bonus scheme which rewards employees financially if the site performs according to the economic and environmental goals set for it. A second category of participants are *status holders* of which several can be identified. The top level status holder is the director with responsibility for environmental matters. Status holders with more direct involvement in EMAS are the key persons from the environmental management department at head office together with the environmental services manager in the supply chain division. The shift manager at the RDC who currently also functions as the environmental manager for the site is another status holder. All these people have been actively involved in the EMAS process, in particular in training and motivating staff.

The third category of *work holders* are all the employees of the RDC. Their role in EMAS has been important. They were involved in various stages of the process, such as in the initial review by helping to identify significant environmental effects. Audits for EMAS (and for other areas, such as food safety and health and safety) are conducted by internal teams of workfloor staff. There was some support from the trade unions for EMAS. The environmental services manager said this was because everyone was involved in the process from the start and because EMAS was accepted as being non-political or, to put it more simply, "there was no distrust of EMAS because it did not affect pay." The managers at the distribution centre employ various methods of communication from simple team briefings, through structured communications to individual staff development reviews, which allow staff to express interest in the environment and make suggestions for improvement. *Knowledge holders* can be identified within as well as external to the company and the site. Internal knowledge holders are the staff in the environmental management department who developed the idea of participating in EMAS and who have played a role in training the employees at the RDC. Influential knowledge holders outside Sainsbury's are the consultant from Business Eco Network and the West London Training and Enterprise Council. The verifier from DNV can be defined as a *status holder*. He was not actively involved in the EMAS process, but was sensitive to the particular issues and problems of environmental management systems in the retailing business. *Holders of a spatial location* are not applicable in the case of the RDC as there are no residential areas nearby. The distribution centre does have a good working relationship with other companies located on the industrial estate.

The category of *interest holders* are those persons or organisations who have an awareness of the site's activities and, consequently, an interest in participation. Again, because there are no residential areas close to the RDC, this category of interest holders is not prominent. The RDC does send its EMAS statement to the local authority. Finally, *stake holders* are all those who could potentially be affected by what happens at the site. Consumers are the main group in this category. Customer concerns (on issues such as GM products, organic farming) and subsequent pressure are an important driver for change in practices. Competition in the food retailing sector is fierce and it is thus vital that supermarkets keep track of customers' needs and wishes. Interestingly, Sainsbury's itself has named a group of 60 employees *target holders*. When realising that not all staff were going to be "environmental champions", they identified a community of target holders who are responsible for achieving the environmental targets laid down and for motivating and training their colleagues.

What have been the benefits of EMAS and what has it helped to achieve, particularly in the light of participation, sustainability and innovation? This is difficult to assess because EMAS is embedded in the wider environmental policies of Sainsbury's and because it is part of a move towards increased environmental awareness of staff. However, policy processes and procedures have been changed. Managers now get more involved with staff and vice versa. Staff receive training on various subjects (not just environmental issues) and they in turn train their colleagues, so there is an opportunity for employees to bring in local knowledge. In terms of technological production processes there has also been innovation, as described above. The EMAS process at Basingstoke has highlighted that monitoring and measuring of environmental effects needs to be improved. This could help in achieving continuous improvement. The company is already working towards defining product miles in the distribution network in order to reduce mileage and increase vehicle fill.

The most important benefit of EMAS will be when the lessons learned at the Basingstoke RDC are passed on to other operational units in Sainsbury's. A strategic review of the supply chain division is currently taking place. There are plans to replace the nineteen existing regional distribution centres by nine so-called "fulfilment factories" of 600,000 square feet each. This project is in the initial start-up phase. The EMAS registration of the Basingstoke RDC could have spin-off effects to the whole supply chain of Sainsbury's. There is major potential for environmental benefit in the new distribution centres and there could be more benefit than there has been at Basingstoke. As one of the environmental managers notes: "Valuable lessons can be learnt and we must make sure that information does not go to waste." Future plans for EMSs are being reviewed.

Looking at the annual reports of the company over the years, it is clear that the profile of environmental and sustainability issues has been raised. The 1998 annual report has a "Community and Environment" section in which the emphasis is more on the company's community involvement than on its environmental initiatives. The 1999 annual report mentions only briefly the maintenance of high environmental standards as one of its objectives but does not include any detail on implementation. The 2000 annual report devotes two pages to "Environment and Community" and describes the company's support for environmentally friendly and socially responsible production processes such as the Marine and Forest Stewardship Councils. It also gives details of Sainsbury's community involvement, nationally and locally.

Looking to the future, the company has been struggling financially. Since 1998, its profits have dropped nearly £200 million to £541.5 million in 2000. In March 2000 a new chief executive was appointed who developed a radical seven year reorganisation plan that needs to be completed in three years (consequently known as "the seven in three plan"). This plan includes new store layouts and a reorganisation of the supply chain. The whole business has been shaken up and there have been significant changes. The organisation is becoming more centralised. The new chief executive emphasises business effectiveness and efficiency and is keen on environment issues providing there is a business case for the initiatives.

There is a possible tension between sustainable development strategies and fighting to survive as a company. The current ethos is very short term in contrast to the values of the Sainsbury family, who now no longer have an active role in the company. In light of all the above issues, the future of environmental policies at Sainsbury's holds two challenges. First of all, it will be interesting to see what place the environment will take in future business plans considering the current reorganisation and financial difficulties the company is facing. Environmental activities will probably have to take a back seat in comparison to economic considerations. However, environmental and sustainability issues will not disappear off the agenda because the company realises it can gain a competitive edge by establishing itself as a "green" retailer. The second challenge for Sainsbury's is the role of its employees in environmental policies. In a company of Sainsbury's nature – with a large number of employees, with various operational units and with environmental initiatives coming from headquarters – participation and motivation of staff is vital to achieve sustainable and innovative outcomes.

5.5   Stroud District Council

5.5.1 Introduction

Stroud District Council is one of six lower tier local authorities in the County of Gloucestershire and lies between the cities of Bristol and Gloucester in the West of England. It has a population of a little over 100,000 and two main centres, Dursley and Stroud. Both town centres are in need of regeneration. Half of the area of the Council is designated as an Area of Outstanding Natural Beauty. Although the district is predominantly rural, it has a solid industrial heritage, built round the wool industry that thrived between the sixteenth and nineteenth centuries. Many of the mills are still standing, though very few are now used for their original purpose. Artists, writers, craftsmen and women are attracted to the mixture of rural peace and urban awareness and perhaps have been encouraged by the fact that Stroud was the home of the Arts and Crafts Movement at the end of the nineteenth century.

In June 2001, the Council's 55 members were distributed across five political groups: Conservatives (24); Labour (18); Liberal Democrat (4); Green (4); Independent (4); (one vacancy). Stroud District Council is a hung council[19] which had a Labour/Liberal Democrat/Green alliance in 2000 and more recently a Conservative/Liberal Democrat/Independent administration. The Council had five standing committees: Environment, Economic Development and Leisure, Planning, Housing Services and Resources, until it reorganised itself in May 2000 into the Cabinet system of seven members supported by six policy panels of eight members each, plus a Scrutiny Committee and a Review Committee, each comprising twelve members. In terms of share holders, the Green Party former chair of the Environment Committee became the Cabinet member responsible for cross-cutting issues (including the environmental management system developed in the 1990s) until the change of administration. An Independent councillor then became the Cabinet member responsible for cross-cutting issues. Also in 2000, at officer level, the Chief Executive took on responsibility for the effective implementation of the environmental management system from the housing and environmental services directorate, as the policy was recognised as being of central relevance to all parts of the Council.

5.5.2 Thematic Background

It was in 1986 that the first Green Party politician was elected to a local council in the UK, and that council was Stroud. In 1989 there was a major campaign in Stroud to preserve trees that were threatened by a supermarket development. The campaign included direct action, so achieved national coverage. The trees were not cut down and the Green Party subsequently added to its numbers on the Council. One of the political consequences was that the Council changed from a Conservative/ Independent majority to one that by 1993 was completely hung, with a Green Party member chairing the Environment Committee.

In 1991, following a campaign by the local branch of the Friends of the Earth, the Council signed up to a draft environmental charter produced by an environmental policy working party. However, because of the campaign to achieve unitary status for Stroud District Council in the early years of the Local Government Review, the charter had never been formally put forward for adoption and had not therefore been formally published. It was not until 1993 that the Council returned to the draft charter. The Policy Co-ordination Committee noted at its meeting on 27 May 1993 that "progress has been slower than anticipated" and referred the issue to a meeting of the leaders of the political groups. It was agreed that the adoption of the charter

---

[19] A "hung" council is one where there is no overall control by one political party.

and the development of an environment strategy was a political priority. At its meeting on 9 September 1993 the Council's Environment Committee adopted both the environmental charter and endorsed the plans for the development of an environment strategy. The report by the Director of Housing and Environmental Services stated that the Council did carry out many activities in the spirit of the charter, but they needed "to be brought together with ideas, information and policies from many sources to form a coherent framework within which the Council's own policies and programmes can be understood and developed". Such a framework would comprise an environment strategy which would embody a corporate view across all the Council's committees and directorates, and would be consonant with achieving Investors in People status (achieved in June 1999). This corporate initiative required a working group of key officers from all directorates (The Environmental Strategy Group), whose initial task was to develop a timetable for producing the strategy and to estimate budgetary requirements, primarily through the changed use of existing resources. Extensive public consultation was undertaken in 1994 with meetings with a wide variety of interests throughout the District.

Stroud's environmental strategy was launched in the summer of 1995, incorporating the charter, plus goals, objectives and programmes of action. This scene-setting document for environmental improvement was seen to require a system of monitoring and management to achieve the Council's goals across the whole organisation, so in May 1996 Stroud committed itself to EMAS registration and produced an environmental policy statement. In a report to the Council's Environment Committee on 13 June 1996, the Director of Housing and Environmental Services stated that "there is a need to take the Environmental Strategy forward in a form which will enable the Council to address the complex and confusing environmental issues in a systematic and manageable way in order that environmental awareness becomes second nature throughout the workings of the authority. In addition, the Council wishes to demonstrate and involve the public in their concern for the environment and the practical action being taken by their local elected body".

5.5.3 EMAS Procedure

Having decided that it was necessary to lead by example, to operate across the board to address the effects on the environment of all of the Council's activities (including those of its contractors) and to respond to the demands from external funding bodies such as the European Union for environmental appraisals to be included in project bids, Stroud District Council looked for competitive advantage in developing an environmental management system. It concluded that EMAS was appropriate because it required the publication and validation of an environmental statement, which meant that it could "clearly identify to the electorate the actions and achievements which have been accomplished within the District" (Environment Committee report, 13 June 1996, Paragraph 3.4).

In looking at its existing practices, the Council recognised that, despite the best efforts of the Environmental Strategy Group, a major problem was a lack of a well documented and quantified baseline of data from which to measure improvements in practices and services. This was to be addressed by the appointment of an external consultant to work intensively with the Environment Strategy Group members to advise on the reshaping of the environmental strategy to fit EMAS requirements, to provide guidance on the collection of baseline data and to identify significant effects.

Charles Thwaites came to the attention of the Council as a result of a senior officer being at the first award ceremony in the UK for EMAS. Thwaites had spoken at the ceremony about the advice he had been able to give to the City of Hereford which was the first UK local authority to obtain across the board EMAS registration in March 1996. He also had local connections through training and environmental review work for companies supported by Gloucestershire Business Link. Following discussions with the Local Government Management Board and the City of Hereford, Charles Thwaites was appointed by Stroud to undertake preparatory work with the Environmental Strategy Group members in the summer and autumn of 1996.

One of the practices developed in the City of Hereford that was adopted by Stroud District Council was the use of the internal audit section to provide expert, impartial judgements on the activities being audited. Auditing staff had to be trained for their new responsibilities. Hereford had calculated the external costs of EMAS registration as amounting to £9,000 per year, plus the cost of appointing an environmental co-ordinator. Such a post had already been established in Stroud District Council, which was seeking ways of controlling the level of additional expenditure resulting from going down the EMAS road. The stated objective was to achieve full registration of all services by no later than 1999.

The environmental policy statement referred to earlier was transformed initially into a set of 32 environmental objectives under eight themes which formed the core of the first environmental statement, which covered the year April 1997-March 1998. The management system to ensure that each year's programme of tasks was achieved in line with the environmental objectives comprised a number of status holders:

- the Council's Environment Committee which took overall responsibility for the authority's environmental performance;
- the Management Board composed of the Chief Executive and the three Directors of the Council's services who were charged with both reviewing the environmental policy and endorsing the annual programme;
- the Director of Housing and Environmental Services, as a member of the Management Board, who was responsible for implementation of the programme;
- the Environmental Co-ordinator who reported to the Director of Housing and Environmental Services and who was responsible for the day-to-day implementation and monitoring of the programme, supported for purposes of liaison, identification of new issues and achievement by a group of staff from across the Council, known as the EMAS Monitoring Group[20] (previously the Environment Strategy Group);

---

[20] The membership of the EMAS Monitoring Group included: Head of Environmental Health, Environmental Contracts Manager, Principal Estates Surveyor, Planning Strategy Manager, Director of Housing and Environmental Services, Special Projects Officer, Stroud Housing Services Manager, Audit Manager, Council Solicitor, as well as the Environmental Co-ordinator.

- specific service managers who were responsible for ensuring that what they were required to do was achieved, plus checking officers whose task was to ensure that this did happen; and
- an internal audit team responsible for ensuring that the programme was completed in the ways laid down.

The first 1997/98 environmental statement thus listed the 32 objectives and what the local authority was doing for its population. The accredited external verifier was Bureau Veritas Quality International. The statement reported on the various aspects of pollution, comparing where possible with earlier years, noting that there were no legal powers at that time to address light pollution, though relevant planning applications did include a condition about lighting. In the section on the built environment, the report referred to the preparation for publication in 1999 of a draft local plan that would take environmental factors into account. Like the section on pollution, the references to health compared matters such as food safety and speed in approving housing renovation grants with previous years where the information was available. Under nature conservation and leisure, the local authority commented on the production of a District tourism strategy which encouraged sustainable tourism and the establishment of community-based tourism activities, such as local walks.

On energy and water, the focus was in part on housekeeping issues "what we are doing ourselves" and in part on "what we are doing for you". On the former, an action plan was developed with the aim of reducing the Council's use of energy and water by 10 per cent over the following five years. Comparisons were provided between 1996/97 and 1997/98, but there were inadequacies in the monitoring system. The Council started buying electricity from renewable sources for its main offices in July 1996 and from April 1997 it was bought from a Stroud based company that was generating electricity from burning methane gas given off at local landfill sites in the County.[21] By mid 2001 95 per cent of all electricity purchased by the local authority came from renewable sources. Another target was to reduce CO_2 emissions from heat and power by 30 per cent by 2002 using 1995/96 figures as a baseline. This target was exceeded by mid 2001. An energy savings measures budget was introduced to fund initiatives such as timers on water heaters and flushing controls on urinals.

On developments for the population in the District, the Council planned to have insulated and installed central heating in all of the 5,600 dwellings it managed by 2008, and in November 1996 it published its home energy efficiency strategy which was aimed to reduce domestic energy consumption by 30 per cent over the ensuing decade. It encouraged the promotion of renewable energy generation by providing support for the Gloucestershire Hydro Action Group and it also hoped to develop a sustainable housing project as part of the millennium celebrations.

In respect of waste, most of it was sent for landfill, though it was planned to recycle at least 10 per cent of domestic waste in 1998/99, in part by encouraging composting of biodegradable waste in the home. A guide on waste management and

---

[21] It was assumed that the emission of CO_2 gas was nil when burning landfill gas due to the benefit of reducing the more damaging "greenhouse gas", methane.

minimisation was also published in 1997/98 and distributed to local businesses in the hotel, business service and manufacturing sectors. Environmental education and awareness was another theme of the 1997/98 environmental statement, where it was reported that the Council had secured several environmental awards over recent years, the most recent being the 1997 Green Apple Award[22] and grant aided Vision 21, the county wide Local Agenda 21 group for Gloucestershire. Finally, the report commented on the Council's adoption of an environmental purchasing policy in February 1997, which included the redesign of order books to encourage thinking about environmental matters in the purchasing process.[23]

The second environmental statement, covering the period April 1998-March 1999, was published only a short time after the first statement and was produced as an update of progress on the eight themes and 32 objectives of the corporate environmental policy, but including the EMAS logo. Official registration was on 31 March 1999. Of the 238 tasks for 1988/99, 194 were implemented. 14 of the remaining 44 tasks were subsequently completed, deemed as no longer relevant or were transferred into the 1999/2000 programme. On pollution, progress was mixed including, on the one hand, a slight increase in the amount of car travel by staff on Council business and, on the other hand, the conversion of one of the Council's vans to run on LPG (liquefied petroleum gas) and the opening of three kilometres of cycle way to add to the existing 28 kilometres in the District. On the built environment, delays in completing the county structure plan led to delay in the publication of the promised local plan. On health, inspections for food safety were down because of staffing problems and a promise was made to include incidences of public disorder in future years as a measure of the level of crime in Stroud. On nature conservation and leisure, relevant staff took sustainable tourism courses and the Milton Keynes Journal bestowed its Millennium Award for Tourism on Stroud.

Much activity was reported in respect of energy and water, but there was no information about performance in comparison with previous years. The introduction of the management system reported in the 1997/98 environmental statement did not appear to have produced the detailed monitoring by which to make accurate comparisons. The District did exceed its target of recycling at least 10 per cent of domestic waste. The reported figure was 12.7 per cent and the increase was attributed to the introduction of a kerbside collection of recyclable materials serving about 30,000 dwellings in the District and increased composting of domestic waste. However, the figure was about half of the new target for 1999/2000 viz 25 per cent.

On the housekeeping side, spending on stationery had gone down 5 per cent compared with the previous year. There was much more activity reported under the environmental education and awareness section, but no information that showed

---

[22] Stroud District Council won this award from the Green Organisation for the work it had done to promote and use renewable energy.

[23] A December 1997 letter to leading contractors/suppliers stated that "it is likely that issues such as energy use, purchasing, pollution control and waste management will be areas where contractors and suppliers will need to confirm that they are working in accordance with the objectives of the Environmental Policy". Stroud's Economic Development Manager encouraged local companies to address green issues, such as waste and energy.

improvement. This was also true of purchasing, though a survey of the Council's top 200 suppliers or contractors revealed that a third of the expenditure was spent on companies, groups and individuals based in the District, and guidance was produced for staff to encourage then to comply with the Council's environmental purchasing policy. The general tone of this second environmental statement was one of aspiration with a large number of actions initiated, others promised, but very few measures demonstrating improvement in comparison with past years.

Halfway through the April 1998 – March 1999 year, a review of the environmental management system described above resulted in no change to the environmental policy statement but it was felt that the list of 32 environmental objectives made annual programmes for action difficult to manage, and so they were condensed into a more manageable total of 15 for 1999/2000 and beyond, though the eight themes remained unchanged. In the course of 1998/99, the title of the Environmental Co-ordinator was changed to that of Sustainability Co-ordinator, reflecting the broader agenda of sustainable development compared with purely environmental issues.

The 1999/2000 environmental statement was a key element in the process of re-registration under the EMAS scheme by July 2000. It emphasised the broad remit of the environmental management system which was justified "because sustainability development affects human health and well being". One hundred and forty four tasks were identified of which 115 were completed within the year. Three were subsequently completed; five were carried out but did not meet the performance targets laid down; another five were no longer relevant; and the rest were to be considered for inclusion in the programme for 2000/01. The report also anticipated changes in the environmental management system during 2000. A surveillance audit was held in February 2000 as a result of which future audits were to be carried out on a service by service basis and monitoring of the programme's progress was to take place quarterly rather than monthly.

As a statement of environmental performance under the eight themes, the report provided a commentary, a statement of what had been achieved in recent years, what was intended for the following year 2000/01, and a list of performance indicators for six of the eight themes (environmental education and awareness and use of resources (purchasing) did not have such indicators). The performance indicators themselves varied greatly in the kind of measures used. Some of them were input measures such as payments of environmental grants to voluntary bodies, the amount of pesticide purchased, the number of noise complaints received, or food premises inspected, whilst others were output measures such as the amount of herbicides used, the distance travelled by councillors and staff on Council business or the consumption of electricity, gas and water.

Overall the pollution, built environment, health and nature conservation and leisure indicators could not give a clear picture of continuous improvement, though the narrative provided increasing evidence of serious investment in activities to achieve the nine objectives under these four themes. In June 1999, the Council had agreed a draft local plan (including an environmental appraisal) which was out for consultation,[24] and development briefs were promised that would incorporate

---

[24] Over 2,500 representations had been received by March 2000.

sustainable development principles. More Council vehicles were using LPG. A high proportion of new social housing (83 per cent) was on brownfield sites in 1999, though the proportion of new development as a whole on brownfield land was 52 per cent, rather less than the 70 per cent for 1998. The target of building at least 50 per cent of new social housing to lifetime home standards was achieved in 1999/2000, as was the target figure of helping at least 215 households through home repair assistance. The plan to support a sustainable housing project referred to in the first environmental statement led to funding in 1999/2000 of £167,000 to develop with a housing association eight environmentally sustainable housing units. "These homes included a range of sustainable design elements including solar water heating, photovoltaic panels and grey water recycling system". Keys to five of the eight properties were handed over in January 2000, but the completion of the remaining three properties was delayed, because a nearby disused building was found to be home to nesting birds (Stroud District Council News, August 2000, 11).

The indicators for energy and water demonstrated a clear reduction of CO_2 emissions from the Council's own buildings, attributed in the main to the level of 35 per cent of electricity used that came from renewable sources at that time. Whereas the consumption of electricity had fallen by 14.6 per cent between 1995/96 and 1999/2000, the consumption of gas had increased by 2.3 per cent and that of water by 16.8 per cent. The amount of domestic waste recycled had risen from 9.1 per cent in 1997/98 to 15.6 per cent in 1999/2000 (The target of 25 per cent for 1999/2000 reported in the second Environmental Statement had been scaled down to 17 per cent, but this lower target was not achieved. The 25 per cent target was retimetabled for 2002).

For the two themes without performance indicators (environmental education and awareness and use of resources), there were descriptions of achievements or continuing activities, such as the inclusion of a copy of the Council's environmental policy with every contract document, talks on environmental and energy issues to schools and local groups, information leaflets on energy efficiency for social housing tenants, the development of a crime and disorder strategy and the introduction of local producers' markets in Stroud and Dursley.

The EMAS statements are placed in libraries and sent to neighbouring local authorities and EMAS registered local authorities. Copies are also provided on request, mainly to students.

5.5.4 Conclusions

Stroud's environmental statements reflect a whole service approach to EMAS rather than a site by site activity by activity approach that would be typical of a chemical plant or a textiles factory. Whilst specific performance indicators are important for measuring continuous improvement, they tend to focus on the "housekeeping" issues of energy and water and the legal obligations to monitor pollution levels. One of the important roles of a local authority is to keep the well-being of the whole of its area and its population in mind (spatial holders), so that its activities are as much educational, information providing and persuasion as taking direct action to improve environmental sustainability. To be effective in this role, it must show that it is itself acting in accord with the messages it is sending to others.

Stroud District Council is working hard on both these fronts, despite the demise of the EMAS Monitoring Group. The current strategy is to integrate EMAS completely into the Best Value (BV) process to have just one management system. "When service managers do their planning for BV, they will be asked for EMAS information." Environmental issues are incorporated into induction training for new staff (work holders) and there have been specific training events on sustainability and biodiversity. However, the internal strategy for raising and sustaining awareness among staff is as much through networking – using the Intranet, offering advice, regular meeting with service managers – as through training. In part, the regular meetings with the 24–25 service managers (status holders) replace the more formal but apparently not very effective EMAS Monitoring Group, and in respect of sustainability, the focus is on a small number of issues relevant for a particular service manager. The aim is to get service managers to think positively about sustainability as part of a corporate programme, and not focus their attention on what is seen as the bureaucracy of the system. Effective implementation of EMAS can change an administrative culture.

On the range of holders, it is important to stress for Stroud District Council that the *share holders*, in the form of the members of the Council and its Cabinet, had been exposed to green policy ideas in a sustained way over the last fifteen years, following the election of the first Green Party politician to a local council in England in 1986. At times, members of the Green Party had been influential in the running of the Council and had succeeded in infusing green thinking into the policies of their Labour and Liberal Democrat counterparts.

It is of no surprise that this initiative by *share holders* at the political level was reflected in the activities of the paid officers of the Council, particularly the *status holders*. The organisation of the work of the *status holders* began to change towards the end of the last century with the demise of the EMAS Monitoring Group and the relocation of the role of Sustainability Co-ordinator into the Chief Executive's unit. This reflects a combination of administrative centralisation and reinforced decentralisation. Whilst environmental policy and management was formally seen as an across the board role requiring a central capability, the engagement of all rather than about half of the 24–25 service managers through a networking strategy on the part of the Sustainability Co-ordinator bypassing the intermediaries of the former EMAS Monitoring Group could be seen as broadening awareness of and responsibility for effective implementation of the policy. Whether, with the backing of new technologies, this flattened system of environmental management increases the sense of ownership and participation on the part of service managers remained to be tested at the time of the study. The intention was to piggyback on planning for Best Value.

Real efforts were also made to engage with the workforce as a whole, the *work holders*, through the inclusion of environmental issues in induction training, in messages on the Intranet and in inviting ideas at meetings of staff teams, not least on housekeeping issues. The local trade union structure had not been seen as a relevant route for encouraging participation, let alone innovation.

The *knowledge holder* in the form of the external consultant, Charles Thwaites, played a vital role in helping the council to achieve EMAS registration, but his contribution was confined to the early months of the development of the scheme.

The *external status holder*, the independent verifier BVQI, was chosen through competitive tender because of its appreciation of the across the board functions of local authorities as opposed to the typical profile of a particular site of a specific company in the manufacturing sector. The role of a local authority means that it has particular responsibilities in respect of the residents, the businesses and the community groups within its boundaries, the *spatial holders* and the *interest holders*. It has both to provide environmentally sound services and encourage people and organisations to act on their own behalf to safeguard the environment. There were strategies to bring about reduction in the consumption of domestic energy, including a pilot project on eco-housing and to further waste minimisation on the part of local businesses. Composting of biodegradable waste in the home was encouraged. This kind of development is far from confined to local authorities registered under EMAS, but the scheme helps to focus the attention of both the Council on what it and the population as a whole are capable of achieving.

Among the *interest holders* were the local environmental groups. Some of them operated at the level of the upper tier of local government, the County of Gloucestershire including Vision 21, the Local Agenda 21 group. Although supported by Stroud and other local councils, this had experienced mixed fortunes, being a somewhat cumbersome arrangement whereby talking seemed to be more central than action. A more successful relationship existed between Stroud District Council and the Stroud Valleys Project, a voluntary group (*interest holder*) set up in 1988 initially to stop the District Council from knocking down mediaeval buildings to construct a ring road. Council membership of the steering group was reduced in order to demonstrate the independence of the project, even though the core of its funding came from the local authority. Whilst the Council officers responsible for environmental matters were, within limits, working well with the project on such issues as new uses for redundant industrial buildings, there were difficulties in respect of the level of annual grant, which was not determined until very shortly before the beginning of the financial year, and sometimes in terms of the conditions laid down.

One recent condition was the development of a farmers' market, which the project had not included in its own plans. The producers/farmers themselves can be seen, alongside customers and suppliers, as *stake holders*, as can the providers of renewable energy for the District Council's main offices. An environmental purchasing policy was intended to persuade suppliers to think about environmental matters and the Council's own environmental policy was included with every contract document. However, the Council recognised that many of its suppliers, particularly those in the locality, were small businesses which would have difficulty in following the precepts of a very stringent set of environmental requirements. This approach encapsulates the style adopted by Stroud District Council across the board. The aim was to reach achievable and sustainable environmental targets and EMAS registration was a device to help them reach and keep to these objectives.

Stroud District Council has undoubtedly had green issues centrally on its policy agenda in recent years. It has developed an environmental policy, identified environmental objectives and published statements of its environmental performance. Environmental concerns have become more widely owned across the Council at both the political and officer level. In this sense participation on the part of councillors and employees (*shareholders* and *work holders*) has increased. Some key staff, such as service managers and auditors (*status holders*) have seen substantially increased involvement/participation in deliberations on environmental matters. Contractors and suppliers (*stake holders*) have been made aware of the importance of their green credentials and therefore may have undertaken policy changes that enhanced their involvement with/participation in environmental issues.

Some interest groups, such as the Stroud Valleys Project have been able to initiate and develop environmental projects, not least because of annually renewed financial support from the Council. Evidence of increased participation on the part of the local population (*spatial holders*) as a whole can be no more than indirect. The impact of environmental education in schools is unknown. The effect of changes in recycling or waste collection on householders in Stroud did not result in any clear picture of changes in behaviour, and, in any case, it would be stretching the meaning of the term "participation" too far to include externally determined shifts in patterns of household behaviour.

So the evidence for enhanced participation is patchy, though EMAS registration would have played a part in any changes in outlook and action, as would the achievement of Investors in People status. But the question remains whether any changes in policy or practice as a result of participation did result in innovation and sustainability. Some evidence can be provided to demonstrate elements of innovation and sustainability. For example, the willingness to purchase electricity for Council offices from renewable sources enhances both innovation and sustainability.

In the Council, the use of informal networking between officers rather than formal groups to implement changes in practice can be construed as innovative and a more effective use of scarce human resources. Support for interest groups with innovative ideas for the use of redundant industrial buildings is support for innovation. These illustrations show that participation can lead to innovation and sustainability, but it has to remain a weak thesis, on the evidence from Stroud, that there is a strong link between participation and innovation/sustainability, let alone that EMAS is a powerful vehicle for such policy change.

5.6   London Borough of Sutton

5.6.1 Introduction

The London Borough of Sutton is located on the southern side of Greater London. It is primarily a residential suburb, with a large commuting population to central London and a low level of unemployment, though there is light industry and commerce in the area. The population is about 170,000, which makes it one of the smaller of the 32 London boroughs. However, it has more trees than any other London borough. Following the abolition of the Greater London Council in the mid 1980s, Sutton became an all purpose authority responsible for the full range of local government services. The recent creation of the Greater London Authority has yet to have a major impact on individual London boroughs. The political composition of the Council of the Borough has in the last fifteen years been strongly Liberal Democrat. Of the 56 councillors, 46 are Liberal Democrat, 5 Conservatives and 5

Labour. Employees (excluding teachers) total about 3,500. Compared with other London Boroughs, and many local authorities in the UK, Sutton gained the reputation of being a "green" Borough.

5.6.2 Thematic Background

After the local elections in 1986 the Council worked with the local Friends of the Earth group (which become the Centre for Environmental Initiatives) to produce a wide ranging environmental statement, which has served, with amendments, to underpin a corporate approach ever since. The original statement included a commitment to produce an annual report on the state of the environment. Early attention was paid to recycling. The CEI secured funding to research practical recycling and the Council then embarked on a scheme for local community adoption of bottle (and later other materials) banks. These early environmental initiatives resulted from a combination of a strong political drive from the majority Liberal Democrat councillors coupled with pressure and practical action from a few committed and active local people, and the presence of a few champions amongst the officers of the Council. The Leader of the Council stated in evidence to the House of Commons Environmental Audit Committee: "we wanted something that would make an otherwise anonymous London suburb become a little more recognised and this was the issue that we chose" (Tope 1998, 171). The intention was to "make a contribution to a sustainable and healthy balanced environment", to "establish a proper balance between short-term economic requirements and the longer term ecological needs of our community". The policies covered working with the local community as well as affecting the Council's own activities. These early projects established how community involvement was to become a key element to future environmental work.

Just prior to the publication of the 1986 environmental statement, a nature conservation post was established. This became the core of a green team which developed in the late 1980s and early 1990s and was located in what was called the environmental services department. In addition, an Environmental Statement Steering Group was set up in 1988. It comprised officer representation from every department and its terms of reference were to operationalise developments in environment policy, to monitor implementation and to review progress. It was required to report to the team of chief officers and to the Policy Sub-Committee of the Council, which was responsible for the Borough's corporate policies.

A number of developments symbolised the commitment of the Council to environmental improvement in the period between the publication of the environmental statement in 1986 and the decision to go down the EMAS route in March 1994. They included the setting up of an Ecology Centre in 1989,[25] the introduction of a cycling allowance and the provision of cycle sheds in the same

---

[25] In May 2000 the Ecology Centre was awarded the Millennium Marque. This is awarded to organisations that have taken on a project and displayed long term environmental excellence over a five year period. Since its establishment, the Ecology Centre has been visited by 25,000 children.

year, the creation of an energy conservation fund in 1990, support for home composting from 1992 and training on environmental issues from 1989, particularly to underpin a programme of environmental audits and to ensure that induction packages for new staff emphasised that concern for the environment was a core value of the Council.

After the elections in 1990, a number of working groups were established to scrutinise the Council's strategic priorities. One of them concentrated on "Sustaining a Healthy Environment". It was concluded that the language of the policies had stood the test of time, but the commitment to their implementation was too narrowly based. "Champions are not enough to embed any policy initiative" (Tope 1998, 170). To address this the training programme alluded to above was strengthened and extended to a wider group of Council staff, thereby increasing the number of work holders.

External relations with other agencies were enhanced by the adoption in 1996 of a Local Agenda 21 Vision, building on the Rio Summit in 1992. The Vision was developed through local focus groups, a questionnaire and a "Future Search" conference in an effort to include all community sectors in the visioning process. It was organised by the Centre for Environmental Initiatives in partnership with the Council. The vision looked toward "a time with more caring, healthier and safer communities with a stronger community spirit and local economy. A future less dominated by the car, less polluted and lifestyles that have less impact on the local and global environment." It was supported by the Borough and over 85 local community groups and businesses, which can be seen as interest holders.

In order to offer leadership by example to business and community organisations, the Council extended its thinking from one off environmental project work to the development of an internal environmental management system. EMAS was argued to be the "best option" because (i) it required all important environmental impacts to be addressed; (ii) it embraced the idea of continuous improvement; (iii) an externally verified annual environmental statement was required; and (iv) in the UK, there was a local government version of EMAS with a comprehensive guide to help local authorities participate in the scheme. The knowledge holder proponents of a proactive green policy in the Council wanted to ensure that environmental management was embedded in all departments and activities of the Borough, and that any externally imposed initiatives, such as the introduction of compulsory competitive tendering or later Best Value, would not undermine corporate commitment to "sustaining a healthy environment", but would rather enhance the achievement of such an objective.

The combination of the application of EMAS within the Council and the development of a three-way partnership between the community, the business sector and the Council through a Local Agenda 21 Forum was intended to create a "sustainable Sutton".

By early 1995 the Environmental Statement Steering Group had transformed into an Environmental Co-ordination Group which was chaired by the Director of Environmental Services and was responsible to the team of chief officers (the Directors' Team) for co-ordinating all environmental activities across the Council. At the strategic level, it was decided not to take a whole authority approach to EMAS, but to adopt a unit by unit approach with the aim of registering all units by the beginning of the year 2000. "We chose EMAS [...] because it [...] recognised that the largest environmental effects local authorities have arise from the way we deliver services – not on a site-by-site basis, but by functions or units" (Lusser 1996, 29).[26] The first eight units were formally registered on 21 February 1996 and the task was completed by the end of 2000. Sutton was the first local authority to be formally registered under the EMAS system in the UK.

Whereas the commitment to EMAS was essentially to enable the Council to put its own house in order, the strategy for increasing the environmental awareness of the general population of the Borough was through the Local Agenda 21 structure of a forum and working groups, public meetings, advertising and links with a wide range of community groups, the interest holders. The response of schools in Sutton to environmental issues had been good in parts, but patchy overall. "If you go to some schools, it is part of daily life, and if you go to other schools where we have failed or we have not tried, you cannot tell the difference" (Hughes 1998, 175).[27]

The development of Local Agenda 21 in Sutton began in February 1994 with a weekend "Partnership for Change" seminar attended by over a hundred people from the local community, the Council and local businesses. Ground rules were laid down on how to take Local Agenda 21 forward. At the end of 1994, the Local Agenda 21 forum was established and topic working groups were set up in 1995. The forum was chaired by a senior councillor and had senior representatives from the local Council staff, voluntary and business sectors. In September 1995 following the major "Future Search" conference, Sutton's Local Agenda 21 Vision for the year 2010 was born and, after consultation, launched at a fair in Sutton High Street in 1996.

Co-ordination of Local Agenda 21 was undertaken by a locally based charity, the Centre for Environmental Initiatives, core funded by the Council. This voluntary organisation was born from a Friends of the Earth group in 1987 as the Centre for Environmental Information and subsequently changed its name. Its overall aim is "to achieve and sustain strong vibrant communities within a healthy environment". On the information front, it aims to be "a positive and creative place where local people can find out what's going on and how to get involved". On the initiatives front, it "seeks to find solutions to environmental problems in the broadest sense by trying to find common ground and agreement over issues and helping people and organisations plan and implement a variety of relevant projects". CEI can be seen as a knowledge holder and an interest holder.

Six Local Agenda 21 topic working groups were established, three serviced by the CEI and three by the Council. They covered sustainable land use and nature conservation (SLUNC), transport, community well-being, energy, sensible consumerism (renamed consumerism and waste) and local economy.

Whilst the working groups were thought to be working well and to be of mutual benefit to both the Council and the community organisations and individuals involved, the Local Agenda 21 forum itself was facing difficulties with a lack of direction, clarity about its roles, and problems with partnerships. A review was

---

[26] Helmut Lusser was in the mid 1990s Purchasing and Policy Co-ordinator in Environmental Services, London Borough of Sutton.

[27] Patricia Hughes was the Chief Executive of the London Borough of Sutton between the mid 1990s and early 2001.

initiated in 1998. Whilst this review was being conducted, Local Agenda 21 in Sutton had a busy and successful year in 1999, which included several awards – a "Crystal" Green Apple Award for the small grants scheme, and a European Union Sustainable Cities award for "outstanding efforts to achieve a more sustainable future for the Borough". Partnership initiatives developed during 1999 included (among many others) the launching of a farmers' market, the creation of a community orchard using Planning for Real,[28] the publication of a cycle network map and guide, the establishment of a Health and Environment Group to raise awareness of environmental impacts on health and a recycled fashion show to engage young people in sustainability issues. For its part, the Council undertook for the first time an environmental appraisal of its unitary development plan, published a sustainability guide to the draft revised UDP and received the VALPAK award for recycling.[29]

The review of Local Agenda 21 in Sutton took an interesting form. It was decided that a social audit[30] should be undertaken by the CEI and the review subgroup of the forum with advice from the New Economics Foundation (NEF). The exercise was financed by a grant from the Bridge House Estates Trust Fund and the Borough. It was a pilot project in the Foundation's *Social Auditing in the Voluntary Sector* scheme and was believed to be the first social audit of a network, "a complex partnership", rather than an organisation. The three themes of the audit were "remit, structure and inclusivity". The NEF also verified the social audit process using a different member of staff from the advisor. The audit was managed by the CEI working with the members of the review subgroup of the forum and a team of volunteer auditors, with support from the NEF, and the report itself was drafted by the Director of CEI. It produced a revised definition of Local Agenda 21 in Sutton, viz "getting the idea of sustainable development out into the whole community and acted upon". On the organisational side, it recommended the disbandment of the existing forum and its replacement by a small core group of "champions" from differing backgrounds, voted in by a much larger network of stakeholders. This was agreed at the annual meeting of the Local Agenda 21 forum in February 2000. The core group was to comprise between six and eight people (knowledge and interest holders), most of them directly elected by the network but with two nominated members, one from the Council and one from the CEI. This new structure was to be supported by a secretariat based at the CEI. In order to use resources more efficiently the CEI took on responsibility for servicing all the Local Agenda 21 topic working groups, managing the database, organising production of newsletters, reports and Local Agenda 21 events. The network comprised all Local Agenda 21 supporters, topic working groups and action group members, stakeholder representatives and former Local Agenda 21 forum members – about 800 people in

---

[28] Planning for Real is a technique of community participation by which people can identify what needs to be done to improve their neighbourhood.

[29] VALPAK is a collective scheme to capture, recycle and handle packaging waste.

[30] A social audit puts particular emphasis on opening up dialogue with interested parties (stakeholders), identifying both successes and failures and using this information to build measurable improvement in the future. Independent auditing ensures that the process is thorough and that all views and proposed actions are openly reported to stakeholders.

all. In October 2000, a new identity for the Local Agenda 21 forum was adopted: *Sutton Future Network-Living Now, Conserving the Future.*

The Council was fully supportive, in principle, of these developments and addressed its own responsibilities in improving its performance in respect of sustainability by producing an Local Agenda 21 strategy as required by the Government, for the years 2000/2002. Its approach is summarised in Figure 5.1. One of its corporate strategies was from April 2000 to buy electricity for the Council's civic offices, a secondary school and a leisure centre from Ecotricity, Europe's first and largest supplier of green electricity. This was expected to cut the Council's CO_2 emissions by about 2,450 tonnes a year. In May 2000 focus groups were set up as part of the Best Value review of environmental sustainability. It was reported that the Council was poor at communicating ideas and information on environmental sustainability, involving people in developing action and helping them to understand how they could make a difference. These messages reinforced the Council's support for the outcomes of the review of Local Agenda 21 in Sutton described above, that a community based Local Agenda 21 network was better placed to engage local people than the Council on its own. The Local Agenda 21 Strategy and the Best Value review both identified a need to take sustainability forward in the Council by integrating sustainable development principles into corporate management processes (such as Best Value and Asset Management). The corporate responsibility for monitoring progress on sustainability lies with the Council's Resources and Corporate Services Performance Committee.

Since the Local Agenda 21 process was now co-ordinated by the Local Agenda 21 core group at the Centre for Environmental Initiatives, the Council was able to redeploy its Local Agenda 21 co-ordinator into internal sustainable development work. The management processes developed are to become part of the corporate EMAS system. The Council was also able to redeploy its Local Agenda 21 administrative officer to support internal Council environmental strategy work.

The Council is required under the Local Government Act 2000 to produce, with key partners, a community strategy. One of these partners is the core group of the Sutton Future Network which in theory could ensure that this strategy builds on Local Agenda 21 and takes account of community views. The London Borough of Sutton needs to build on the experience of producing its annual community plan which for 2000 was developed in partnership with the Borough Forum, which brings together the Council, the major public agencies, the voluntary sector and business in the Borough, the major stakeholders in the locality.

The Council's links with the business community had included grant aid for the establishment of Business Ecologic, whose function was to provide environmental advice, particularly to small local companies. It soon became self-financing and changed its name to Business Eco Network. By early 1998 about 140 local businesses had attended environmental courses run by the Council and Business EcoNetwork (Everett 1998, 178).[31] In respect of suppliers, the Borough's initial

---

[31] Tim Everett was, at the time of giving evidence to the House of Commons Environmental Audit Committee, the Borough Public Protection Officer. His current designation, at the time of writing, was Executive Head of Public Protection.

strategy was to persuade all suppliers of goods and services to have a verified environmental management system by 1999. It soon became apparent that this was not reasonable, especially for small businesses. A new approach was decided in 1999. Suppliers of goods worth over £5,000 a year were requested to have an environmental policy, and suppliers of goods worth over £100,000 a year were asked to have a verified environmental management system (EMS) or to write, with the Council's help, an environmental improvement programme agreed with the Council. Voluntary sector contractors were also required to have an environmental policy.

One of the flagship projects in Sutton is the development of a mixed tenure "green village" on the four acre site of a former sewage works. The Borough was given permission by central government to sell the land at a lower price to the Peabody Trust instead of having to accept a higher bid from a volume housebuilder. Over 80 zero-energy dwellings are planned by the Trust in partnership with the Bio Regional Development Group, a locally based charity, together with a wide range of offices and community and sports facilities. Energy is to be drawn from renewable sources such as solar energy and a combined heat and power unit which uses renewable tree cuttings to generate energy. It was hoped that the first residents would move in during the autumn of 2001.

5.6.3 EMAS Procedure

As noted in the previous section, the London Borough of Sutton decided in March 1994 to engage with the EU's Eco-Management and Audit Scheme as part of its broad environmental strategy, which itself was one of a small number of corporate goals adopted by the Council. The approach taken was to put forward each year a number of operational units for registration, with the corporate aim of complete coverage of the Borough's responsibilities by 2000. The first eight units were registered in February 1996, nine more in May 1997, a further nine in June 1998, five in 1999, six in September 2000 and the remaining nine were submitted for registration in December 2000. The aim had been to complete registration by the beginning of 2000, but reorganisation within the Council put this back a year. In the later years of initial registration (and early re-registration) the operational units were replaced by larger service units, such that by December 2000 when the whole of the Council was covered, there was a total of 23 service units.

At the outset, in February 1995, the Council decided that its Policy Sub-Committee should take overall responsibility for EMAS backed by the team of chief officers. Within that team, the Director of Environmental Services (later called Director of Environment and Leisure) was to be responsible for the development and review of the scheme. This Director chaired the Environmental Co-ordination Group comprising officers from all departments of the local authority. In addition, the internal audit team took on the role of auditing the proposals put forward by project teams in operational units. To support project teams, an EMAS Co-ordinator was appointed, a key knowledge and status holder. This co-ordinator acted as an environmental assessor to give advice, to check that procedures had been followed and that documents had been appropriately completed. Where necessary, other specialists, such as ecologists, could be called upon to act as environmental assessors. The broad approach was that the project teams within the operational or service units

Figure 5.1   Sustainability Management in the London Borough of Sutton

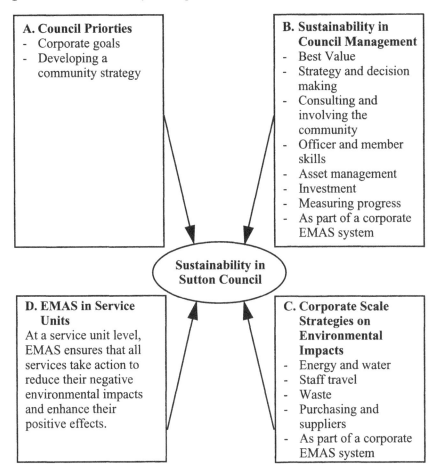

should "own" this environmental management system, advised by specialists. This approach also suggested that the "unit by unit" mode of implementing EMAS would be more appropriate than a "whole authority" way of working.[32]

---

[32] Several arguments were deployed in support of the "unit by unit" approach: (i) a steady development of EMAS produces early results, which can aid motivation and ownership, and reduce the level of scepticism or distrust; (ii) an extended programme provides opportunities to learn from mistakes and to build on successful practices; (iii) the world of local government is not a stable one, so a gradual approach is more responsive to changes in organisational arrangements or service responsibilities; (iv) by focussing on units, staff could be involved in writing EMAS plans, leading to workable action and staff ownership.

In the early years key staff from various environmental backgrounds were trained by consultants on environmental auditing. The EMAS Co-ordinator would be a key member of a project team which would include staff from within the unit and initially staff from elsewhere to provide a broader perspective. However, this latter idea proved less than successful, not least in terms of meeting deadlines, so after the first tranche of submissions, it was decided that project teams should consist only of unit staff acting as "unit EMAS representatives" and the EMAS Co-ordinator.[33] All unit staff received training[34] to improve environmental awareness, whilst the unit EMAS representatives were trained to identify environmental effects and write their EMAS action plans with the help of the EMAS Co-ordinator.

In developing EMAS in Sutton, the Council's existing environmental policy statement had to be amended to cover the requirements of the scheme's regulations. Initially those responsible for implementing the initiative were guided by the 1993 government publication, *A Guide to the Eco-Management and Audit Scheme for UK Local Government*, prepared for the Department of the Environment by CAG Consultants. Whilst this proved to be a useful base from which to start, the Sutton officers responsible found that a lot more work needed to be done to make the scheme operational and robust enough to result in successful verification.

In the project teams the focus was on significant effects, but this proved to be a problem in respect of direct effects, such as energy use, as they could not always be controlled by the operational units themselves. A corporate approach was needed that cut across the unit by unit approach, so this became the responsibility of the Environmental Co-ordination Group, reflecting the overall corporate commitment to environmental improvement. Identification of significant service effects was also problematic, and a flow diagram was created, based on an environmental effects evaluation tool that had been developed by the Local Government Management Board. This flow chart was replaced in 1998 by a tabular format due to pressure from the EMAS verifiers to better document how environmental effects had been identified.

The role of the internal audit team was to undertake an audit six months after EMAS registration. The members of the Group were drawn from the existing financial audit team, who were trained to undertake environmental audits. It was argued that this added value to existing systems. However, following a management review of the EMAS system in 1998 due to EMAS verifiers' reports, units seeking registration were made subject to internal audit prior to verification. A three year EMAS auditing programme has since been established and approved by the verifier. In addition, each unit is expected to report on their EMAS progress to the Environmental Co-ordination Group on a six monthly basis.

From the outset, the London Borough of Sutton was very concerned that the external verifiers of their EMAS submissions should be sensitive to the particular circumstances of local government as well as having the appropriate technical

---

[33] A brief description of the response of Sutton's trading standards service to EMAS registration can be found in Nelson (1996). Greg Nelson is a Senior Trading Standards Officer in the London Borough of Sutton.

[34] At first these in-house events took a whole day. Then they were shortened to half a day and later to two hours.

background to come to sound judgements. The Council looked very closely at the credentials of the verification staff. In 1995, the accredited verifier selected was BSI Quality Assurance, and each of the six environmental statements was signed by Julian Ringer, a key external status holder, on behalf of BSI. With a wider choice of companies accredited to verify EMAS, the Council has reviewed its verifier and chosen DNV, which was already being used for some quality assurance auditing in the Council. Sutton's first environmental statement was a multi-coloured prestige publication of 32 pages, rather longer than many such statements produced by private sector companies. It was argued that the local authority should be providing accessible information to the local community, and its own operations should be put in the context of the developing the Local Agenda 21 agenda. Subsequent annual statements ran to between 20 and 28 pages, with the exception of the September 2000 version, which was 45 pages long. The more recent statements have been available electronically and "a minimum number of copies have been printed". Thus, views changed about the relevance of the environmental statement to the public at large, as opposed to actively interested individuals or organisations. However, the verifier has pressed for a more attractive presentation in the most recent statement. Doubt remains whether the environmental statement makes sense as a document for the general public in its current format. Each year since 1996 the Centre for Environmental Initiatives has produced, on behalf of the Local Agenda 21 forum, a full colour report communicating positive environmental action in the Borough to the public. Following the Local Agenda 21 review this method of communication is being reviewed. The future contents of EMAS environmental statements will need to be reviewed to ensure they complement other publications and meet the verifiers' expectations

A review of the six environmental statements indicates more emphasis on broad intentions to inculcate environmental awareness across the Borough than evidence of continuous improvement in matters of housekeeping. Non-compliances were noted with the comment that many of them were minor documentation issues rather than programme or system non-compliance. The second statement covering 1996/97 did have a section on the progress of units registered in 1996, but this practice was not followed through consistently in subsequent statements. However, in the 1999 statement, the first unit to be re-registered was the Chief Executive's 13 person unit, the first local authority unit to be registered in the UK. It reported an aim to cut paper usage by 10 per cent over three years from 1996 and announced a 50 per cent reduction by the end of 1998. Some of the other targets were less specific, such as ensuring reference to environmental issues in "relevant" press releases, aiming to give high publicity to Green Transport Week, supporting the Local Agenda 21 forum's communications sub-group, or improving the circulation of environmental information amongst officers and numbers. Many of the improvements that were undoubtedly made in the second half of the 1990s related to changes in organisational or management practices to incorporate environmental issues and to enhance environmental awareness. These are less subject to precise environmental indicators than housekeeping issues such as reduction in paper usage.

As noted above, some housekeeping matters were not in the control of the staff in specific operational or service units, and the environmental statements included a separate section on direct effects for all units, such as use of energy and water, travel on Council business and waste management in multi-unit buildings. It did not prove easy to bring together reliable information about the Council's direct environmental effects, but measures were developed which demonstrated a modest increase in energy use in Council buildings between 1995/96 and 1998/99, a small reduction in water usage by metered premises between 1996 and 1998 and an impressive reduction in mileage by Council officers on Council business between 1995/96 and 1998/99. Data on waste were still not reliable enough for comparison over time. A corporate energy strategy was agreed in July 1997, but the need for more effective implementation was recognised in 2000. A corporate waste minimisation strategy was agreed in November 2000 and a pilot of the new scheme started in September 2001. A water use policy and action plan was agreed in December 2000. A green travel plan to focus on Council transport use was developed in 2000. The third environmental statement (1997/98) noted that "it is not always easy to achieve the desired results especially when pressures on resource use combine with adverse environmental developments. The increase of the use of energy in Council buildings as a result of increased computerisation has cancelled out a number of energy efficiency measures in the past."

In evidence to the House of Commons Environmental Audit Committee in April 1998 the Chief Executive of the Borough stated that "the cost of EMAS for us in direct terms is the cost of a co-ordinator at £38,000, to get accreditation in the year is £6,000, and to have our own internal auditors to do the audit is about £8,500 [...]. In addition to that [...] there is the cost of....the directors, the champions and the people in the various groups who are going through the EMAS accreditation and the time they give [...] that we cannot quantify, but we do think it is commensurate with the benefits that are going to come out of it" (Hughes 1998, 173).

5.6.4 Conclusions

The unit by unit approach to implementing EMAS meant that as soon as one unit was registered, the focus turned to the next one. The system may have looked good, but in the early years monitoring was patchy, as units were meant to monitor themselves. Practical implementation was varied. This lack of a monitoring system was addressed in 2000 by requiring the heads of units to report every half year to the Environmental Co-ordination Group. Whether strategic decisions by *status holders* at corporate level on issues such as budget cuts, capital spend or the setting of objectives have been truly underpinned by environmental considerations, let alone ideas about sustainability, remained an open question, despite the formal adoption of *Achieving Environmental Sustainability* as a corporate goal. The "bottom up" unit by unit approach could be undermined if it was believed by the *work holders* that only lipservice was being paid at top management level to sustainability. To address this, a set of actions to integrate sustainability into corporate scale management processes was identified as part of the Local Agenda 21 strategy in December 2000. This has begun to work well within the Best Value process and contract tendering. The challenge is to ensure that project, strategy and budget decision making are influenced by sustainability considerations. The requirement to develop a community strategy needs to be informed by a renewed vision of sustainability, including, but going beyond, specifically environmental matters. In addition, re-

registration over the next three years under EMAS may result in greater expectations on the part of verifiers for a clear top level strategy rather than just evidence of good housekeeping by units.

These pressures from *external status holders* such as government and verifiers are being accompanied by changes in political and executive leadership (*internal status holders*) in the Borough such that the environmental investments of the last fifteen years need to be re-emphasised. The restructuring of the architecture of Local Agenda 21 may help in this task. "If there is a vision for Community Planning and it supersedes Local Agenda 21 goals, it doesn't matter" (EMAS supporter). Whatever the future for sustainability and EMAS in Sutton, the local authority can take great pleasure in the results of a national air quality survey published by the (former) Department for the Environment, Transport and the Regions which reported that the cleanest air was measured at a roadside in Sutton (Lean 2001).

What does the story of Sutton demonstrate for participatory governance? In his evidence to the House of Commons Environmental Audit Committee, the then Leader of the Council Lord Tope, stated that he "would not claim for one moment that we have 170,000 eco-freaks living in the Borough... but I do believe that there is generally a higher level of environmental awareness in the Borough" (1998, 174). This can be seen in part in the restructuring of the Local Agenda 21 arrangements which did identify a mini-electorate of active people associated with the various working groups and projects linked to the Centre for Environmental Initiatives. Public processes of consulting did determine priorities for action.

However, the development of EMAS within the Council has not been a major influence on the population of the Borough as a whole. There is a potential for engagement with pupils in schools, but the main impact of EMAS has been designed to be on the employees of the Council, many of whom have found themselves, willingly or unwillingly, invited to consider both their housekeeping practices and the broader environmental implications of their service responsibilities. This core value of the Council articulated through the corporate objective of *Achieving Environmental Sustainability* is reflected in the induction training for new staff and in the EMAS training for unit staff for the implementation of EMAS as an environmental management system. The key issue is whether this corporate objective does influence the difficult financial issues faced by chief officers and leading councillors. Are the *internal status holders* and the *work holders* singing from the same hymn sheet?

5.7 Summary of the Case Studies

Now that the six case studies have been analysed individually, their results need to be put together to assess the extent of similarities and differences between them. The issues of innovation, sustainability and participation need to be investigated in the context of EMAS. It is assumed that there is a link between participation on the one hand and sustainability and innovation on the other, such that participation leads to a higher degree of sustainable and innovative outcomes. It is also assumed that EMAS is a tool to increase participation. But do the UK case studies support these assumptions?

As has been the case in other countries recently, the number of EMAS registrations in the UK is increasing only very slowly. It is difficult to explain this trend but there are at least two factors that contribute to it. First of all, the concept of EMAS has been poorly marketed by the relevant government bodies. The adoption of EMAS needed to be more positively encouraged and the benefits of the scheme more widely communicated. By late 2001, the UKAS had not issued any guidance on the revised regulation. The IEMA are, however, keen to increase the profile of EMAS in the UK and have been formulating plans to do this. Second, it seems that ISO 14001 has become the norm globally for environmental management systems. A lot of companies -as opposed to local authorities- see the "hassle" of producing a statement as not adding extra value. This attitude is something that could be changed as part of a promotional campaign. One EMAS verifier commented that reporting is a must for any organisation which claims to be serious about its green credentials. Another verifier is of the opinion that organisations do not make the best publicity of their EMAS statements. However some authors, such as Ball, Owen and Gray (2000), have criticised the value of externally verified environmental statements. They have questioned the extent to which these third-party statements promote organisational transparency and the empowerment of external parties. They have raised fundamental questions concerning the independence of verification and the degree of auditee control over the process.

The most important audiences for the environmental statement were found to be *work holders* (staff), *holders of a spatial location* (local residents) and *stake holders* (suppliers and contractors). Hillary (1995) identifies several segments of "the public" for whom the environmental statement is designed: the local community, pressure groups, suppliers and customers, shareholders, competitors, banks and insurers, regulators, the general public and the media. The case studies have shown that other important holder groups not mentioned by Hillary are employees (*work holders*) and universities and students (*interest holders* and *stake holders*) mainly interested in environmental reports for research purposes.

As there are not many EMAS accredited companies or organisations in the UK, it could be argued that those that have decided to go for EMAS are all to some extent "champions of change". A wide range of reasons was employed in the decision to participate in EMAS. Some were mainly focused on identifying the environmental effects of their operations and on achieving environmental improvements. Others saw EMAS as an opportunity to enhance their corporate credentials and to improve external communication and openness. The benefits as a result of going through the EMAS process were also varied. Improved environmental awareness of staff was one of the most often cited benefits. Another benefit was more open communication (usually with *work holders* and *holders of a spatial location*) which had improved relationships between the various holder groups. The case studies also confirmed the conclusion reached by Strachan (1999) that EMAS can act as an effective strategic marketing tool for companies attempting to raise their environmental performance standards and in demonstrating this to important organisational stakeholders. Cost savings as a result of EMAS were generally not very high nor was there a lightening of inspections as a result of the registration.

The UK case studies were made up of two local authorities and four private sector companies. Of these latter four, one was a small company with one site (60 employees). The other three were part of larger companies with a different number of employees at the site (10, 81 and 750 employees). In the case of these latter three companies, the decision to participate in EMAS was taken by senior management (*stake holders* and *status holders*) at the companies' head office. The initiative was however fully taken on by each of the sites, albeit implemented with support from head office. In the other three cases of the small sized company and the two local authorities the initiative to participate in EMAS also came from senior management. Again, this did not seem to negatively influence the implementation of and enthusiasm for the scheme. Despite this, as one of the companies noted, "EMAS has not been very powerful on the people front", meaning that it has not engaged as many holder groups as was anticipated. In only two of the case studies were the *work holders* actively involved in the EMAS process. In the rest of the local studies, groups like *work holders* and *holders of a spatial location* were being kept informed of developments but were not given an opportunity to participate. So EMAS, as implemented in the six UK case studies, cannot be defined as an arrangement for participatory governance. Schemes like Investors in People or the Japanese management system Kaizen have been more important when it comes to encouraging participation, ownership and openness.

All the six case studies first started to become interested and involved in environmental improvement in the early 1990s or in the case of the two local authorities a few years earlier in the late 1980s. All four private sector companies had also registered with ISO 14001, with three of them having registered earlier with BS 7750. There was a noticeable trend towards integration of environmental, quality and health and safety management systems. In the public sector, EMAS and Local Agenda 21 were becoming integrated with the objectives of the Best Value scheme. It is increasingly acknowledged that environmental policies are an integral part of overall management strategies that work in conjunction with economic and social policies.

The local studies have shown that innovation and sustainability, historically regarded as antithetic (Schmitter 2000), are potentially compatible. Innovation and creative thinking can contribute to achieving sustainable development. In several of the case studies the application of new technologies had changed production processes. This resulted in environmental improvements (such as waste minimisation and more efficient use of resources) and financial savings. In a lot of cases these innovations or suggestions for improvement came from the *work holders*. Evidence for innovation in terms of policy outcomes was more difficult to find in the UK case studies. EMAS does not seem to have much direct influence on the transformation of an organisation's policies or procedures. Instead, EMAS status reflects the direction that the organisation is taking rather than taking it in a new direction. As such, the case studies show that EMAS can best be seen as part of larger changes within a company or organisation. EMAS is usually a reflection of a shift to a more environmentally aware organisational culture rather than a change in its own right. In addition, several interviewees described EMAS as an internal instrument and as such it led businesses to engage in self-reflection and to learn about the environmental effects of their own activities. Apart from the environmental statement, EMAS was not seen to be an external instrument.

In addition, EMAS was not seen to be anything other than an environmental instrument. It was used as a vehicle for the measurement of environmental effects and as an environmental reporting system. The environmental report can be an important two-way communications tool that a company or organisation can use as an effective and efficient means to determine not only how well it is communicating its environmental performance but also how it is performing environmentally (Herremans/Welsh/Bott 1999). If EMAS is to contribute to the establishment of more sustainable policies it will need to include the other two elements of the triple bottom line: social and economic concerns. This is where Grafe-Buckens and Beloe (1998) introduce the concept of sustainability auditing. Environmental audits have their limits and hardly address other elements of business performance such as social, human or economic issues. Grafe-Buckens and Beloe argue that auditing techniques and the norms or indicators against which audits are expected to measure sustainability performance should incorporate the interlinkages between these various spheres of operations to mirror the systematic nature of sustainability more closely.

Pressure is mounting for companies (and governmental organisations alike) to widen their scope for corporate public accountability. Many are responding by including social data or examples of social measures in their environmental reports. The next step beyond environmental and social reporting is sustainability reporting which involves integrating environmental, social and economic performance data and measures to produce one report. All six case studies were to a greater or lesser extent involved in the development of sustainability policies. They found it challenging to understand exactly what sustainable development means for the way everyday business is carried out. They were familiar with the Brundtland definition of the concept and were also aware that it included the three interlinking elements of the social, economic and the environmental. Nevertheless, most sustainability strategies focused on environmental improvements. The social and community activities that most companies have been undertaking for some time were generally not included under the umbrella of sustainability policies.

Just as in the private sector, environmental and sustainability strategies are important in the public sector. These strategies need to influence first the local authority's own operations ("getting one's house in order") in a corporate integrated way and second the wider community of residents, businesses, commercial organisations and other public, private and voluntary interests in the locality (Wild/ Marshall 1999). In the public sector, sustainability is usually part of Local Agenda 21 strategies. Local Agenda 21 has enormous potential to green the activities of local authorities, to transform the relationship between the local authority and the community and to create communities in which people feel involved and committed. Wild and Marshall conclude that in practice, however, the impact of Local Agenda 21 has been quite limited. Just as was the case in the private sector, EMAS in local government is not a goal or result in its own right but part of a bigger shift towards environmental improvement. EMAS in local government increases transparency and the "good practice" role of public agencies. It is therefore of interest that EMAS has so far not been adopted by UK central government. Honkasalo (1999) writes that environmental management systems at the national level would open up opportunities for more transparent and effective international environmental policy. It would also make it possible for stakeholders to be more involved. He states that national governments should do no less than they expect from industrial companies.

As was the case in the private sector, EMAS in local government is not so much a vehicle for change in itself, but part of wider environmental strategies. EMAS draws many environmental concerns together under one banner. This can be used to make it a cross-cutting issue within local government. EMAS then does not remain the sole ownership of the environment department but instead becomes more widely owned throughout the organisation (*work holders*). All this can increase the importance and potential policy outcomes of the scheme. But the impact of EMAS on local residents (*holders of a spatial location, interest holders* and *stake holders*) is less pronounced. It is questionable whether EMAS is a tool for local democracy, because – as one of the interviewees in the public sector notes – "it is concerned with continuous improvement in environmental performance and it may be that in order to do this we have to reduce the choices available to the community."

All in all, what seems to be a vital ingredient for the successful implementation of EMAS in each of the case studies is the involvement and enthusiasm of key individuals. Equally, or perhaps more important, is continued commitment from these key individuals and their ability to get others on board in the process. Participation of all relevant holders is an important prerequisite for more sustainable business practices, as is staff interaction and ownership of the improvement process. It could be argued that this is perhaps easier to achieve in small and medium sized enterprises or on small sites with smaller numbers of staff. It is, however, more to do with management cultures and open decision-making processes than with company size. Although this does not distract from the fact that EMAS can lead to improved environmental sustainability, the UK case studies do not seem to support the hypothesis that EMAS is a strong vehicle for the promotion of participation.

Chapter 6

EMAS in Greece

Panagiotis Getimis, Georgia Giannakourou and Zafeiroula Dimadama

6.1 Environmental Policy in Greece: A Brief Overview

Environmental policy is an area of growing interest in Greece as shown both by the passing of numerous regulatory acts and increasing societal concern since the early 1990s.

Traditionally embedded in land use and urban planning policy and legislation, until the mid 1980s Greek environmental policy owes its legal and institutional emancipation from planning procedures to the development of an European Community environmental policy. The need to implement and enforce Community environmental legislation played an important role both in formulating the Greek government's environmental policy and procedures and in accelerating domestic institutional and legal changes (Spanou 1995; Giannakourou 1996).

Greek environmental policy has been so far a strict regulatory one. Indeed it can be argued that the overriding characteristic of environmental policy in Greece is its normative, legalistic and compulsory profile. Nevertheless, despite formal rigidity, common administrative and societal practices suggest a greater flexibility, if not a widespread informal and piecemeal discretion in the policymaking system (Spanou 1996a).

This dichotomy between formal rules and informal values and practices is being challenged by two kinds of complementary factors. First, there has been a gradual withdrawal from the dominant "command and control approach" at the EU level and an adoption of more procedural types of regulation (European Commission 1992b and 1996b).[1] Second, there have been parallel changes in the general administrative, economic and societal modes of operation at the domestic level, as a consequence both of Greece's participation in the EU integration process (Spanou 1996b) and of internal changes in the Greek government's priorities and aims (Getimis 1998) concerning decentralisation, deregulation, privatisation of the public sector and enforcement of civil rights.

Under these pressures, Greek environmental policy is in a stage of transition. The direction of change is the product of complex interactions between the pre-existing

---

[1] According to the European Community itself, environmental policy before the Fifth Environmental Action Programme "was overwhelmingly a matter of attacking a problem with legislation" (European Commission 1996b, 4), based on rigorous regulation and a "command and control approach" (ibid., see also European Commission 1992b).

regulatory approach on the one hand and the new circumstances and challenges implied by both EU membership and domestic modernisation on the other.

6.1.1 The Evolution of Greek Environmental Policy

Although the word "environment" appears in a range of government texts that had no binding force during the period prior to and during the dictatorship (1967–1974), it is recognition of environmental protection as a state responsibility in the Greek constitution of 1975 (Article 24) that forms the starting point for the enactment of a specific environmental policy in Greece.

However, although the creation of a constitutional basis formally accorded environmental policy a high priority among the state's objectives, no specific legal and institutional means for the exercise of this policy were created until the mid 1980s.[2] In 1986 the first broad legal framework for the environment was adopted (Law 1650/1986). Furthermore, at about the same time the new Ministry for Planning, Housing and the Environment (YHOP), created in 1980 (Law 1032/1980), had delegated to it many of its powers and organisational structures (Spanou 1995, 137-153; Getimis, 1993, 114).

The impetus given to environmental policy in Greece in the second half of the 1980s coincided with the adoption of the 1986 Single European Act at European Community level. Under this new legal and policy framework, the EC's role in environmental issues was formally recognised. Moreover, it provided the necessary institutional and legal instruments to launch numerous directives and regulations aimed at the harmonisation of national environmental standards and the convergence of national administrative, business and individual strategies and conduct.

The post Single Act period, known as the command and control period, required the member states to put in place centralised chains of command and control in order to incorporate European environmental law into domestic legislation. In Greece, this requirement reinforced the consolidation of domestic administrative arrangements set up in the first half of the 1980s. The Greek Ministry for the Environment saw its substantive and procedural competences both broadening and deepening either at the central or at the regional and prefectural levels (Giannakourou 1996). A parallel process can be seen in the establishment in 1991 of a specific unit for the scrutiny and enforcement of national and Community environmental law within the Greek Supreme Administrative Court (the State Council). Finally, reliance on Community environmental law (especially the Environment Impact Assessment directive) aided the formation or strengthening of national environmental pressure groups, as well leading to a large increase in individual demands on and appeals to the administration or the courts.

---

[2] Until the mid 1980s, environmental policy was embedded in the system of regional planning that was the responsibility of the Ministry of Coordination (in effect the Ministry of Economic Affairs). The first postdictatorship planning law adopted by the Greek Parliament (Law 360/1976) covered both spatial and environmental planning.

6.1.2 Policy Institutions

Environmental protection and management is seen in Greece as a public sector concern at all levels of the state hierarchy. At the central level, the main institution responsible for the formulation and implementation of environmental policy is the Ministry for the Environment, Spatial Planning and Public Works (YPEHODE). Created in 1985 out of an amalgamation into a single ministry of the former Ministry for Public Works and the Ministry for Planning, Housing and the Environment, this ministry has the main responsibility for co-ordination of national environmental policy and implementation of European environmental policy. However, since it does not have exclusive competence over all environmental issues,[3] it shares the policymaking process with other ministries responsible for particular environmental matters. Among them are the Ministry of Agriculture, responsible for the protection of forests and the agricultural landscape, the Ministry of Culture, responsible for the protection of cultural heritage (especially monuments and archaeological and historic sites), and the Ministry of Marine Commerce, responsible for the management of sea ports and the protection of the sea from pollution.

Apart from these ministries, a whole range of other ministries and public organisations (e.g. ministries responsible for industry, energy, tourism and regional policy) are involved in the formulation and implementation of environmental policy. It is this fragmentation of environmental responsibilities at the central level, as well as the lack of effective co-ordination mechanisms, that has led some European analysts to characterise the Greek environment ministry as a "weak" one, compared to other national ministries such as Britain, where there is a high concentration of environmental responsibilities (Demmke 1997, 55-59).

The situation is aggravated by a lack of decentralisation measures, as well as by the complete absence of independent or "mixed" agencies capable of ensuring efficient monitoring and inspection.

Regional and local authorities, despite being given some specific responsibilities for the implementation of national legislation on environmental protection, are still limited either to a purely consultative or a minor executive role.[4] In any case, they

---

[3] Under Law 1032/1980, by which YPEHODE was established, this ministry has only general competence over environmental matters, as well as the responsibility for coordinating and controlling the implementation of environmental programmes of other national institutions.

[4] Local authorities in Greece are recognised in the constitution as being in charge of local affairs. However, the definition of this concept is explicitly stated neither in the constitution nor in law (with the exception of some second order responsibilities cited in the Municipal and Communal Code). The result of this vagueness, therefore, often leads to disputes with a sustained tendency in government legislation and in court decisions to decide in favour of the central state in allocating responsibilities. Local authorities are divided between a first tier (municipalities) and a second tier (prefectural self-government), while the 13 regions, into which the country is divided, are only administrative units with no proper legal personality, in other words regional branches of central government.

lack the necessary financial, technical and human resources to exercise effective inspection and monitoring.

Because of increasing shortfalls in the implementation of environmental law and policy by existing administrations, the question arises as to whether privatising or transferring environmental responsibilities to independent agencies and semi-private non-profit making authorities would be a better way forward. On this issue, whilst the fundamental principle of public control over environmental policy remains, the government has taken steps to increase flexibility and efficiency in monitoring and inspection procedures. This includes the delegation of some inspection and accreditation functions to independent bodies or voluntary organisations.

The creation of an independent Ecolabelling body[5] or the debate on the establishment of a public association of environmental inspectors[6] are significant developments. In addition, an independent institute for the environment and sustainable development is being planned to act as a national think tank for environmental policy and as the national focal point for the European Environment Agency network. Finally, in an attempt to remedy shortfalls and delays in the implementation of Community environmental policy, the Greek administration has found it necessary to have recourse to non-governmental organisations and independent experts for the internal management and enforcement of some European Community directives.[7]

6.1.3. Policy Instruments

In Greece, the dominant regulatory style for environmental policy, that is the pattern of interaction between administrative and societal actors (Knill/Lenschow 1998a), resembles the "interventionist" ideal type. More specifically, environmental policy in Greece is characterised by the command and control type of regulatory rules which provide clear objectives that leave administrative actors only limited discretion and flexibility.

The basic legal instruments are binding land use plans, installation and building permits and emission, discharge or operating licences. The environmental protection law gives monitoring and enforcement powers to YPEHODE and the Ministry of Development at national and prefectural level. The inspections are mostly carried out by members of the recently established environmental inspectors association, who are responsible for granting permits, inspections, warnings and sanctions. They react to complaints from the public and rarely take an initiative. Non-compliance

---

5   This is the High Council for the Delivery of Ecolabelling, established under the joint Ministerial Decision 86644/2482/1993 (OJ 763B/1993) and known under the official title of "ASAOS".

6   A recent project has been completed under the auspices of the Ministry for the Environment establishing an Environmental Inspectorate and an Independent Centre for the Environment and Sustainable Development.

7   The Goulandris Natural History Museum, a private institution, along with the Greek Biotope Wetland Center (EKBY), an independent research and monitoring institute, were charged with the responsibility for identifying and recording Greek habitats and the flora and fauna species that are included in Annexes I and II of Directive 92/43/EC (Habitat Directive).

with the environmental requirements of permits can lead to sanctions, such as cutting off power supplies, water or telephone, imposition of fines or, rarely, closure of installations (OECD 2000).

A number of fiscal measures are used in Greece, including user charges, environmental taxes and fines. Water supply and sewerage charges are made in industrial areas and municipalities, as well as to households. Raw quarry products are subject to an ad valorem charge, which is paid by the quarry owners to municipalities for the protection of the environment. Municipalities collect fees and other charges for the use of natural resources such as soil, minerals, sea water and grazing land. Transport fuel taxes are collected by the Ministry of Economic Affairs and are administered by YPEHODE. Half is allocated for general environmental protection and the other half for air pollution measures in Athens. Greece has introduced pollution fines for car exhaust emissions. The environmental protection law established a system of strict liability for damage to the environment through pollution or other contamination. A number of ministerial decisions make polluters fully liable to pay compensation for marine pollution or damage caused by used oil, toxic waste or PCBs (OECD 2000).

However, there is no special environmental insurance, nor any special funds for cleaning up polluted sites. Greece has not use deposit refund systems for environmental policy instruments (OECD 2000). Strategic environmental management using fiscal incentives, prizes and awards, market-based instruments (for example tradeable permits) or civil liability is still not practised in Greece (Skourtos 1995).

By the end of 1994, they were about 130 associations, pressure groups, unions and other non-governmental organisations that were unofficially registered (Katsakiori 1994). However, no formal type of intermediation or co-ordination with the state administration has been put in place that would allow for membership of boards or committees, for participation in preparing documents or drafting proposals, for consulting experts or for influencing technical and decision-making procedures.[8]

A rare example of voluntary environmental action on the part of industry is provided by the creation in 1992 of the Greek Recovery and Recycling Association (HERRA) on the initiative of the aluminium industry before any legislation was passed. The Greek government has not entered into voluntary agreements with any industrial or commercial groups to achieve specific environmental objectives (OECD 2000).

In this context, patterns of interest intermediation tend to be formal, legalistic, adversarial and closed, with informal and ad hoc bargaining between regulatory authorities and private interests taking place in the shadow of the law (Spanou 1996a; Getimis 1993, 115, 116). In this system of informal, diffuse and unstable negotiation and bargaining, the main official opportunity for intervention in the

---

[8] The only official form of citizens' participation in the environmental decision-making process is the public hearing procedure during the Environmental Impact Assessment process. Established by Directive 85/337 of the Council of Ministers, this procedure has been transposed into domestic legislation through the joint Ministerial Decision 75306/ 5512/1990 (OJ 691B/1990).

environmental decision-making process becomes the appeal to the judge, in other words, judicial participation.

It is in this sense that the civil, mainly the administrative, judge becomes, not only the basic actor for the enforcement of environmental law but more broadly the basic catalyst for private and societal empowerment in the environmental policymaking process (Giannakourou 1992 and 1994; Siouti 1994).

This kind of participation reflects both the absence of neo-corporatist and pluralist traditions in Greek interest intermediation system (Alexandropoulos 1990; Kioukias 1993), and the low cultural and institutional importance of proactive societal participation in the formulation and implementation of public policies in Greece.

6.2   Implementing EMAS in Greece

Environmental audit, as a means of securing continuous improvement in environmental performance by industry, was practically unknown in Greece before the adoption of the EU's EMAS regulation. This had both a positive and a negative impact on the implementation of EMAS in Greece. The existing institutional and organisational arrangements were not in a position to embrace the environmental management systems required by EMAS. But, on the positive side, the state and industry could directly adopt the new European standard and avoid difficult and time consuming changes and interim provisions.

6.2.1 The Evolution of EMAS in Greece

It is important to note that the lack of effective monitoring systems for the implementation of environmental legislation, as well as the difficulties faced by the economy in Greece, do not encourage companies to take environmental issues on board other than complying with the law. At the same time, companies realise that there have been some commercial pressures encouraging them to adopt EMAS. Under the EMAS regulation, a series of pilot projects for environmental auditing were conducted in several industrial sectors between 1995 and 1998 (see, for example, Christophoridis et al. 1995; Metaxas/Zacharis 1995; Skordilis/Filippeou 1995). Their primary purpose was to inform companies and local authorities about the scheme and to gather data on responses to it, through conducting reviews which provided indicative results for the first period of EMAS in Greece.

The results of these pilot projects were presented to a conference held in 1998 at the Technical Chamber of Greece in Athens. It was noted that the majority of the participating companies and especially the SMEs would not be opposed to incorporating EMAS in their production system. However, there were some crucial factors which were affecting the application of the scheme. They were mainly focused on the organisational and technical capacity of the participating companies. Lack of technical and entrepreneurial competence, lack of finance necessary for the environmental modernisation of the firm and lack of training of the company's staff were some of the major problems the pilot projects brought to light (Alexopoulou/ Nastouli 1998; Giannakis 1998). It took a lot of time before these companies could decide to go for EMAS registration, and at the same time great efforts had to be made by these companies to change their structure and culture.

Furthermore, the degree of EMAS acceptance seems to be strongly linked to the size of the company and its export activities. Large or medium sized companies, as well as companies with a major export profile, seem more positive towards EMAS, mainly due to the pressure from their customers abroad (Dimadama 1998; Alexopoulou/Nastouli 1998). On the contrary, companies operating exclusively in domestic markets, especially local ones, are much more reserved or even hostile towards EMAS. It is likely that the relative low level of consumer awareness in Greece plays a crucial role. In addition, the environmental review tended to reveal that SMEs had to invest in order to fulfil the requirements of the legislation. In most cases, companies were not able to allocate resources to environmental issues in the short or medium term. Furthermore, they were not in a position to cover the system's verification and maintenance costs.

Those responsible in the public administration for the implementation of EMAS were aware of these structural characteristics of the Greek industrial sector, especially the SMEs, which minimise responsiveness to the scheme. This is the main reason which lead the Ministry for the Environment to promote (through the use of the Article 14 of the Community regulation) the application of the scheme, not only to the manufacturing sector, but also to the public and private service sector.[9]

Community-based interest groups and consumers are seen within the EMAS strategy as potentially important partners in transforming the firm into a sustainable enterprise. However, in Greece the absence of a tradition of "green culture" and corporate ethics minimises the learning capacity of industry for environmental management and audit. Moreover, in Greece there has been a delay at national level in implementing information campaigns and encouraging participation in EMAS by offering various kinds of incentives (economic incentives, simplification of procedures, reduction of controls, etc.). As a result, companies, especially small ones, were not motivated to design and implement an environmental management system (EMS). This applied to EMAS as well as to ISO 14001.

The pilot projects demonstrated that most of the participating companies had "weak", if not non-existent or even hostile, relations with the public and the local authorities (Alexopoulou/Nastouli 1998). They were therefore cautious about adopting a system that would give access to the "internal workings" of the firm. On the other hand, both the weak Greek consumer movement and the active non-governmental organisations had not been accustomed to engage in information and participation processes of the kind introduced by the EMAS regulation. Being traditionally involved in the decision-making process in a reactive and ex-post way, especially through judicial intervention (Giannakourou 1992), they are hardly in a position to embrace strategies of a consensual and proactive kind enshrined in EMAS.

There were similar problems with the internal legitimacy of EMAS, that is, its acceptability to the company's workforce. In effect, as the pilot projects have

---

[9] A ministerial plan is under consideration to promote the application of EMAS to every legal person in the private or public sector whose activity has a major impact on the environment.

already shown (Alexopoulou/Nastouli 1998; Christoforidis/Zacharis 1998), staff in the participating companies, especially the smaller ones, were in most cases hesitant, if not negative, towards EMAS. This attitude stemmed mainly from the fear of the loss of jobs as a result of the implementation of EMAS. It also reflected the general lack of corporate ethics in manufacturing industry in Greece. The lack of key information and opportunity for discussion also probably played an important role.

Until 1998 Greece had faced a range of difficulties in promoting and implementing environmental management systems. The dominant regulatory approach was blocking the formation of innovative, voluntary legislative measures which required new institutional arrangements. The absence of other management systems, like the BS system in the UK, was an additional barrier to the adoption of an EMS. However, the main difficulties were not focused on the institutional arrangements but rather on the low degree of public awareness about environmental problems. Greece had not had to face the kinds of problems which other nations, after a long period of industrialisation and de-industrialisation, had had to address through a process of transition and restructuring. On the contrary, the absence of major industrialisation had resulted in relatively modest negative environmental impacts. Greece had begun, in the 1970s and 1980s, to formulate an adequate environmental policy, based on EC regulations. But a reluctance to comply with the demands of EMS procedures by key interests became a major drawback in the 1995/ 98 period. Moreover, neither citizens (and consumers) nor companies had shown much by way of environmental awareness. In addition, there were no specialist staff capable of developing the new systems. Nevertheless, the formation of new structures starting from zero, was an easy task as legislation and existing institutional arrangements were weak and inefficient. However, since 1998 there has been significant progress in implementing EMAS in Greece. The main officials responsible at national level have developed a strategic plan to inform industry and a broader spectrum of interests about the utility of implementing EMAS.

Under the EMAS regulation, the primary roles for the national authorities were publicise the scheme and to establish a flexible and appropriate institutional and organisational framework. These requirements have been met, at least in part, by the Greek state in three ways.

The first step was the foundation of the Greek Accreditation Council (ESYD) in 1994 (Law 2231/94), as part of the Ministry of Development. Its task was to set up a national accreditation system. This Council is the state's official technical adviser on accreditation issues and is also the decision-making authority for such matters. The Council members are representatives from ministries, scientific bodies and societal organisations (Greek Technical Chamber/TEE, Greek Employers Association/SEV) as well as from industry, providing in this way some guarantee of the independence and integrity of its operation.

This Council is supported in its tasks by two General Technical Committees, as well as from Specific Task Technical Committees, the members of which are experts from particular sectors. Before granting accreditation, a team of assessors carries out a site assessment on the premises of the applicant. The members of this team would have a thorough knowledge of the required technical issues, as well as being very experienced in quality management systems audits. ESYD assessors are selected and trained according to strictly defined criteria and procedures, and they have to comply with specific regulations about their independence, integrity and professionalism. They include laboratory based experts, people from certification and inspection bodies and EMAS verifiers.

At the beginning of 1999, the first announcement inviting applications to become accredited environmental verifiers was made, but the interest in participation was not very high. In 2000, there was more interest, and about 20 verifiers have undertaking training to become environmental verifiers.

The second requirement was for Greece to develop a system of registration of participating companies. In 1998 the Ministry of Environment established the EMAS Committee. This Committee comprises four members from the Department of Air Pollution and Noise Control (YPEHODE), the Department of Environmental Planning (YPEHODE), the Department of International Relations and the European Community (YPEHODE) and the Department of Industrial Land Planning and Environment (Ministry of Development). The EMAS Committee has a range of important responsibilities:

- It checks that the environmental report on the site complies with EMAS requirements and consults the Minister of the Environment about registration, deletion or suspension of a registration according to the regulation.
- After verification and a positive decision to register the site, it informs the company concerned and puts the site on the official register.
- It puts forward ideas for projects, such as the promotion of EMAS as an instrument for industry or for local government.

The third – and probably the most important – national role was to raise environmental awareness of both the general public and companies. Since EMAS is a voluntary policy instrument, it is necessary that both companies and the general public are informed and convinced of the potential benefits from the implementation of an environmental management system. YPEHODE entrusted to a private company the planning of a promotional strategy for publicising EMAS and increasing awareness in general. As a result, there were three one day meetings in Athens, Thessaloniki and Volos, which followed public announcements and production of application forms for interested parties. The Prefecture of the Piraeus region and the local development agency (ANDYP) undertook the promotional campaign in the environmentally decaying area of the Piraeus. Furthermore, a database was created, comprising information on 1,500 people or organisations, who were interested in implementing EMAS in Greece. The interested parties were directly informed by post from YPEHODE about new developments. The final objective of this policy was the publication of the formal minutes of the one day meetings, in order to inform other authorities and to help publicise the system.

Participation by small and medium sized companies (SMEs) was an important element of the EU's policy. Greece considered the promotion of EMAS among small and medium sized companies as a vital element for success, because of their large numbers. Therefore, YPEHODE introduced pilot programmes, funded from national and European sources. Through these programmes, the companies could secure the essential expertise to implement environmental programmes and management systems, and to conduct inspections necessary for the preparation and verification of statements (Ioannidou 1999).

At the end of 1998, the Ministry of Development invited the companies interested in ISO 14001 and EMAS to take matters forward. Financial support of amounting to 10 million drachmas (about 29,000 euro) for each company was seen as necessary to persuade them to participate in the programme. Indeed, the level of participation was high, as 109 companies expressed interest and were financed to implement environmental management schemes. 93 companies expressed their intention to register with ISO 14001 and only 16 with EMAS.

At the time of writing (2001) four companies had been registered under EMAS and about 57 under ISO 14001. Moreover further companies were aiming for EMAS and ISO registration (see Table 6.1).

6.2.2 EMAS in Industry

6.2.2.1 A Company Analysis An analysis of companies aiming for registration with either ISO 14001 or EMAS in 1998 reveals differences between industrial sectors, those with and without export markets and the size of the companies measured by the number of employees.

Three industrial sectors were particularly active. The food products and beverages sector has taken a close interest in implementing EMSs. In 1998, three companies were aiming to register with EMAS and 11 with ISO 14001. The chemicals sector listed 12 companies, one of them aiming for EMAS and the others for ISO 14001. Eleven companies in the metal products sector were all aiming for ISO 14001.

Interest in EMAS can be found in the rubber and plastic products sector, as two out of four companies had expressed their intention to participate in the EU scheme. Companies in the non-metallic minerals sector were mainly interested in ISO 14001, as only one out of six had expressed interest in EMAS. Five textile companies intended to implement ISO 14001, while four publishing and printing companies were interested in an EMS (three for ISO 14001 and one for EMAS). The paper sector listed three companies (two for ISO and one for EMAS) and the wood and cork sector also had three companies, all interested in ISO 14001. Finally, two clothing companies, two coke and petroleum companies and two companies from the furniture sector were interested in an EMS, each having one EMAS and one ISO 14001 registration. Two sectors, leather products and electrical machinery, were represented by one company, both aiming for ISO 14001.

There are some sectors that are not yet active in addressing the EMS issue. They include tobacco, basic metal industries, office machinery and computers, telecommunications equipment, medical and optical instruments, vehicles, other transport equipment manufacturers, recycling companies and energy suppliers.

The size of the companies which were implementing environmental management systems indicated another key factor. Measured by number of employees it is apparent for both EMAS and ISO that there has been a much greater likelihood of interest in an EMS by companies which employ between 1 and 499 employees (Table 6.2). These smaller companies represented over 90 per cent of the total. For

EMAS, the companies with 1–99 employees were 54 per cent of the total, whilst for ISO 14001 the figure was 51 per cent.

Table 6.1 The Implementation of EMAS and ISO 14001 in Greece in 1998 by Sector

	Companies Aiming for Registration	
Sectors (by NACE Code)	*EMAS*	*ISO 14001*
11 Oil and gas		2
14 Mining and quarrying		1
15 Food products and beverages	3	11
17 Textiles		5
18 Clothing	1	1
19 Leather and leather products		1
20 Wood and cork		3
21 Paper	1	2
22 Publishing and printing	1	3
23 Coke and petroleum products	1	1
24 Chemicals	1	11
25 Rubber and plastics products	2	5
26 Non-metallic mineral products	1	5
28 Metal products		11
29 Machinery and equipment	1	7
31 Electrical machinery		1
36 Furniture, musical instruments and sport equipment	1	1
TOTAL	*13*	*71*

Source: P 1998

It is clear that the companies with less than 100 employees were particularly interested in an EMS. However, it is the case that, in general, Greek companies are small and employ less than 100 employees. The analysis also demonstrated that the two large companies (more than 1000 employees) were both subsidiaries of powerful multinational companies (Olympic Catering and Motor Oil SA).

Table 6.2   Number of Employees in Organisations Aiming for Registration under ISO 14001 or EMAS in Greece in 1998

Number of Employees	ISO	%	EMAS	%
1–99	36	50.7	7	53.8
100–499	29	40.8	5	38.5
500–999	4	5.6	1	7.7
1,000–9,999	2	2.8	0	0
10,000 and more	0	0.0	0	0
Total	*71*	*100*	*13*	*100*

Despite the difficulties encountered by the companies, EMAS offers a lot of opportunities for improving a company's performance. EMAS is seen as one of the most important tools to ensure compliance with relevant environmental legislation. The majority of participating companies acknowledged the contribution of EMAS implementation to the improvement of environmental awareness amongst employees at all levels. In addition, companies can improve their organisational and managerial systems by adopting EMAS, whilst at the same time integrating environmental issues into their policies, objectives and programmes. One of the most important benefits of the environmental review is that companies can identify areas where better management of natural resources and raw materials consumption can lead to cost savings. However, cost savings tend to be realised only some years after EMAS implementation. Although the comparative competitive advantage which EMAS can offer is considered an important element of the system, only a small number of companies count on it. This is reinforced by the lack of information and knowledge about the scheme and its opportunities. It is only in cases where the companies are export oriented or are owned by an international enterprise, that EMAS can constitute a competitive advantage. On the other hand, for small companies operating only in the domestic market, EMAS does not constitute an advantage at present since they cannot expect market recognition for this initiative.

6.2.2.2 The growing tension between EMAS and ISO 14001   The absence of national standards for environmental auditing in Greece and the lack of a broad awareness of environmental issues was a crucial reason for the limited level of EMAS implementation.

Experience from other European countries, such as Germany, has shown that lack of an earlier national system of standards could turn into an advantage for the implementation of EMAS. However, the experience in Greece shows that this cannot be a generalisation. In fact, most Greek companies actually implementing environmental management systems had shown a preference for the international standard system, ISO 14001. By the end of 1997, ten companies had been certified through ISO, while only seven companies, out of nearly 40 participating in the initial review of EMAS pilot projects, had entered into the next stage (Souflis 1998).

This does not stem solely from the fact that administrative delays and organisational rigidities have made EMAS non-operational in Greece. Two further factors have limited the acceptance of EMAS by Greek firms. First, the ISO system is more flexible and less demanding than EMAS in relation to the provision of information on environmental performance to the public or for the prior internal environmental review of the firm (Souflis 1998). In this sense, it is more adapted to the managerial capacity of the manufacturing sector in Greece. Second, ISO 14001 has links with other management quality systems, such as ISO 9000, already practised in many Greek companies and is, thus, supported by an informal market of professional auditors active in this area. The absence of a domestic system for the accreditation and supervision of environmental verifiers either under EMAS or other environmental management systems, such as ISO, has also played a crucial role.

6.2.3 EMAS in Local Government

The European Commission, which finances many pilot programmes, decided to support member states wishing to implement EMAS in local government.

Within the LIFE initiative, a programme called "Pilot Implementation of Environmental Management Systems" was implemented in Greece in three medium sized cities (Volos, Patras, Larissa) (Karaiskou, 1998). KEDKE (the Central Union of Municipalities of Greece) took the initiative to organise a related programme entitled "Implementing EMAS in Local Government and Companies".

Since 1999, the Municipality of Athens has participated in a three year European programme called "EURO-EMAS. Pan European Local Authority Eco-Management and Audit Scheme". The main objective of the programme has been to improve the environmental performance of the municipalities. Athens has participated through a focus on reducing the environmental impact of the daily operation of its health centres. The increase in environmental awareness on the part of the public servants and doctors constitutes an additional objective. An information campaign was mounted and problems of energy and water use, as well as waste disposal, were addressed. However, the programme had only limited success. The limited funding from the EU and the refusal of the Municipality of Athens to support the programme financially, decreased the scale and the importance of the initiative.

6.3 Networking

Another major effort to support and promote EMSs in Greece has been the establishment of "EMSnet". This network was founded in 1999 by those responsible for the launch, planning, promotion and implementation of EMSs in Greece. The network is supported by the General Secretary of Research and Technology (GGET) of the Ministry of Development in the framework of a national research and development programme, EPET II (1994-99). Those involved in the network are universities, research institutes, relevant ministries (Development, YPEHODE), the Greek Organisation for Standardisation (ELOT), NGOs, consultancies, local authorities (KEDKE) and private companies, such as Siemens and Grecotel. The essential aim of "EMSnet" is to promote and disseminate information about the implementation of EMAS in Greece to help partners to exchange knowledge and to open up dialogue and participation to new partners.

6.4   Conclusions

After five years of implementing EMAS in Greece significant changes can be discerned. The national institutions responsible for EMAS have made major efforts to promote it. However, a range of problems have cropped up. The first was that the unitary Greek state had to overcome the difficulty of adopting a new voluntary system, like EMAS. The long tradition of a command and control system constituted an important barrier to the implementation of new forms of environmental policies, such as voluntary instruments. Moreover, the low degree of environmental awareness of the public and the low participation in EMAS schemes on the part of industry both show a low degree of "green culture" in Greek society.

However, after 1998, the public sector (YPEHODE and the Ministry of Development) initiated a vigorous campaign to overcome these problems and adjust to the basic European requirements for EMAS. The Greek Accreditation Council (ESYD) founded in 1994 finally become operational at the beginning of the 1999, when the first announcement for accrediting Greek environmental verifiers was published and the Council started to prepare special training courses for verifiers (both Greek and foreigner verifiers working in Greece). Furthermore, the recently established "EMAS Committee" and its specific terms of reference have created a new framework for implementing EMAS in Greece. Another significant development is the funding from national sources of pilot programmes with a remit to provide information on EMSs, not only to industry but also to other sectors and target groups such as local authority and consumers.

Chapter 7

The Greek Case Studies

Panagiotis Getimis and Zafeiroula Dimadama

Six companies and the Municipality of Athens were selected to study the implementation of environmental management systems at site level. Three of them were registered EMAS sites (Domylco, Sato and ICR Ioannou),[1] while two others (the Municipality of Athens and Interchem Hellas) had implemented EMAS but had not formally registered. Two ISO 14001 certified companies were also studied because of the low participation in EMAS. The two ISO 14001 companies (Siemens and Elais) had been certified for some years and have shown significant improvement in environmental performance.

7.1 Domylco Ltd

7.1.1 Introduction

Domylco Ltd was founded in 1976, is located in Athens and employs 10 people. The company produces various kinds of concrete and mortar admixtures (chemical admixtures), mould release agents curing compounds, ready-to-use mortars, epoxy resins and a range of other associated products. The company does not export to European or other countries. All products are sold on the domestic market.

7.1.2 Thematic Background

In 1996 the company decided to enhance its competitive position through the implementation of European and international standards. Domylco aimed to persuade its clients that its products were reliable and to promote the belief that Domylco products were superior to those of its competitors. By these means, the company's management team believed that conditions could be created for the improvement of Domylco's market position. They therefore set the following targets which complied with their general policy on quality and environmental matters:

- Implementation and continuous improvement of the quality assurance system under ISO 9001.

---

[1] As pointed out in Chapter 6 there are only four registered EMAS sites in Greece. The fourth (beside Domylco, Sato and ICR Ioannou) is High Fashion. For further details, see Getimis, Giannakourou and Dimadama (2001), 341–343.

- Strict observation of written procedures, as well as relevant national and international standards, by all employees at all stages of planning, production and supply to clients.
- Development of a high technology laboratory in the company for inspection of raw materials and products.
- Implementation of an environmental management system in order to integrate the small number of fragmentary environmental activities of the company.

Before 1999 the company's main emphasis was on quality systems, but since then, Domylco has also paid a great deal of attention to environmental issues.

7.1.3 EMAS Procedure

The owner decided to implement an EMS to improve the environmental performance and the production processes of the company. Other motives were financial savings (water, raw materials, energy) and the efforts of the company to integrate different management systems, for example, health and safety and ISO 9001). These were the main reasons for implementing EMAS, but during the procedure some interesting issues emerged. At the time of writing, the company was awaiting formal EMAS registration.

The role of the environmental manager was very crucial. The environmental manager was responsible for all the management systems and standards in the company. Its small size and the level of trust between the owner and the environmental manager was the main reason why he was managing the whole process.

In trying to identify *innovations* linked to EMAS implementation, it is important to stress that Domylco is a small company that decided to implement an environmental management system in order to reform its internal structure and, more specifically, to achieve "good housekeeping". It is important to emphasise that the company has done some simple things that have had significant effects.

The company produces about 55 different products using 5-6 production lines. Through the EMAS procedure, the company managed to "upgrade" two of these lines to make production more efficient. EMAS implementation expanded the existing knowledge base and promoted new techniques in the production process. The contribution by the external consultant was very important.

Implementation of EMAS helped the company to organise employees' responsibilities in a more rational way. Their participation in health and safety matters and in ISO 9001 enabled them to be more proactive in the whole EMAS process. The company's internal management arrangements after the implementation of EMAS were also made more flexible and rational.

In the course of implementation the company early on arranged for biological cleaning of liquid wastes. This investment helped the company to reduce the level of disposal of "sand" and to reuse some water in the production process. This not only contributed to *sustainability objectives* but also resulted in savings.

Through the biological cleaning process the company managed to save water for washing equipment as well as for the production process. Furthermore, the company reduced the consumption of electricity by replacing the old electricity system. On savings in general, the company is developing a long term plan in order to reduce the amount of raw materials required in the future.

The most important feature of this case study is the environmental policy that the company implemented in transport. The company developed a system for limiting van and lorry emissions. The company collects vehicle batteries and oil in order to reduce emissions to the air and to stop uncontrolled disposal. Batteries and oil are collected and then sold on to third parties.

Additionally, the company has stopped using oil for its small on site trailer, replacing it by gas. In the future Domylco plans to replace some of the fuels used in the production process with gas as well.

Another important initiative was plastic recycling. The company now recycles plastic containers by selling them to other companies as raw materials and by taking back containers from customers, who are mainly other companies.

As already mentioned, the company has been trying to organise its internal structure using environmental criteria. EMAS implementation was crucial for achieving this, but was not so effective in relation to external links. For example, the company has faced difficulties in using environmental criteria for the choice of suppliers and customers. Domylco hopes to introduce some environmental criteria in the future but it is believed that it is unlikely to work for customers. The latter's environmental awareness is limited and most of the time the choice of a product is defined only by its price.

As mentioned earlier, the idea to implement EMAS came from the environmental manager. The owner and the environmental manager then discussed the proposal with the external consultant and decided to go ahead. They chose EMAS because they were interested in presenting the environmental statement to the public through press releases, exhibitions and on the web.

The company had in the past faced some problems with their local authority. These were resolved through the EMS. The most important issue was that the municipality's refuse collection vehicle refused to collect solid wastes on a regular basis. The company hired a private collection company, which now collects the solid wastes. Additionally, recycling and reuse has led to a reduction in waste.

Another problem with local people was that the company's vehicles were parked on a public road causing traffic jams. The lack of parking space on the company's land was resolved by renting land nearby.

The employees have been positive about the implementation of EMAS. The environmental manager informed them about the proposal after the decision had been taken. At first, the employees were grudging but due to the daily support from the environmental manager, they later accepted EMAS as a important tool to improve the performance of the company.

7.1.4 Conclusions

EMAS implementation in Domylco is a "special" case. Although the company is small and has only liquid wastes (which are not particularly important), it decided to implement an environmental management system.

The company was interested in savings as well as in the implementation of a scheme promoting long term thinking. The innovative effects were very limited but sustainability and participation issues were significant. The fact that the company emphasised transportation, recycling and reuse is very important. Through its relations with customers their environmental awareness was enhanced. Lastly, EMAS helped to improve relationships between the company and the local authority.

7.2   ICR Ioannou

7.2.1 Introduction

ICR Ioannou is located in the industrial zone of Inofyta and employs 48 staff. The company produces specialised printers (mainly for paper) and machine parts. Since 2000, a second production line has been in operation, producing rubber plates for flexography. 20 per cent of total production is exported, mainly to the Balkans, Middle East and African countries. A new strategy is planned, aiming to increase sales in EU member states. However, the company estimates that introducing its products into the EU market would be particularly difficult because of the intense competition.

7.2.2 Thematic Background

From 1995 the company had begun to understand the need to develop quality management systems on site. Building on this, the company decided in 1999 to implement ISO 9002. At the same time the company decided to go for EMAS rather than ISO 14001. The external consultant had suggested EMAS and the owner agreed this was a good choice. The owner was also influenced by the availability of a subsidy from the Ministry of Development for implementing EMAS. Support from the consultant was crucial and co-operation between them has continued during and following implementation.

7.2.3 EMAS Procedure

ICR Ioannou decided to implement EMAS – instead of ISO14001 – because EMAS is a European scheme and the company intended to use it for "opening" up the European market for its products. Furthermore, EMAS was perceived, especially by the environmental manager, as a more systematic and integrated EMS than ISO 14001.

During implementation of EMAS the company developed a new product to do with flexography after making some changes in the existing production lines. This was done in co-operation with a foreign company which had introduced environmental criteria for choosing their suppliers. Besides introducing new sophisticated computer software for controlling each phase of the production process, new raw materials were also brought in.

It is important to stress that ICR Ioannou is a company which is following a strategy initiated by the owner to use high technology in the production process. Therefore, most of the company's equipment is very new and high- tech. This technologically advanced equipment and the continuous improvement of technological knowledge and infrastructure has been enhanced by EMAS. During the implementation of EMAS the equipment was tested in order to reach maximum productivity.

The implementation of EMAS has helped the company to develop more flexible structures and a more rational distribution of responsibilities. The first positive outcome after EMAS implementation was "good housekeeping" on the site. The environmental management system has helped employees and the environmental manager not only to address environmental issues systematically but also to reflect on the internal division of labour and responsibilities. This was done in combination with the plans of the company for integration of its policies through total quality management. Currently, the company is organised through separate management systems like EMAS, ISO 9001 and through fragmented health and safety policies. The company has, however, systematically developed a strategy for employees' health protection.

The role of the external consultant was decisive. Although the company hired a new environmental manager who had some experience of EMAS implementation, the company continued to co-operate with the external consultant.

In the future the company plans to organise a separate on site department for environmental and quality matters. Through this, they expect to be able to manage these issues more rationally and systematically.

As already stated, ICR Ioannou produces machine parts. This means that, in terms of sustainability, the company has to face the problem of liquid waste pollution resulting from the production process. The liquid waste contains considerable amounts of copper and chromium. After EMAS implementation there was more systematic monitoring of the use of these materials in order to limit leakage during production and to follow more carefully the existing regulation guidelines. Furthermore, an important move to reduce the problem of liquid waste was, like Domylco, the installation of a modern biological cleaning system. The result was reduction in waste water and the reuse of treated water in the production process and for cleaning purposes.

Before the introduction of EMAS untreated water had been disposed of in the draining tank and some leaked into the soil. Despite the company's location being in an industrial area, there was no water supply nor a sewerage system. Consequently, the company had to buy water in tanks and then store it in its own reservoirs.

Since implementing EMAS, ICR Ioannou has demonstrated a considerable increase in legal compliance, especially in waste management, but also for the company as a whole. The cleaning procedures are no longer harmful to the environment, as there are specific devices that prevent leakages. In addition, the liquids with high copper concentration are reduced and solid copper waste is subsequently sold to a private company. Furthermore, a significant contribution through EMAS for the reduction in solid waste is the reuse of steel rollers. They are now collected and used again in the production process through a new technique, so there are raw materials savings.

The health and safety of the employees have also been addressed through the implementation of EMAS. Responsibilities were redistributed among the staff and inspection was systematised through routine exchange of data. ICR Ioannou have never had an accident nor a complaint about its operations.

As in other case studies, the employees were informed about the implementation of the new system from top management. It is important to stress that the employees, at the beginning, were not very enthusiastic. They had been charged with new responsibilities and they had to participate in seminars – which meant more work. But after the initial negative reaction they gradually accepted environmental management as a necessity. Two one day seminars were held in order to tell employees about the objectives and implementation of EMAS and to encourage employees to take an active role for their own safety and to increase the environmental performance of the company.

The company attempts to inform customers about its environmental policy, although suppliers constitute the key target group for the company's own improvements. In this respect, collaboration with leading firms which have adopted environmental policies has continued, for example, with Dupont and Kodak. However, the majority of customers, which are themselves companies, have not shown an increased interest in environmental issues.

Finally, the local community has not provided any encouragement to the company's environmental activities, although it does publish and promote the results of its environmental policy. On the contrary, the local community was totally uninterested in the environmental problems of the area.

7.2.4 Conclusions

ICR Ioannou is a Greek example of a modern company that is particularly interested in new management systems. The company developed a systematic computer software package for monitoring its production line, introduced EMAS for auditing its environmental impacts, embraced quality system standards (ISO 9002) and took some initiatives in health and safety. It is an SME that is attempting to integrate its different policies into a total quality management system in order to succeed in the future. The company emphasises three features of EMAS: organisational reform or "good housekeeping" in general, reduction in pollution or "green housekeeping" in particular, and increase in exports to the European market.

7.3 Interchem Hellas

7.3.1 Introduction

Interchem Hellas SA was founded in 1968 and is located in Vathi Avlidas, 75 kilometres north east of Athens. With continuous expansion and introduction of new products, the company has steadily grown to become one of the leaders in the Greek chemical industry. Interchem mainly focuses on the synthesis of resins and emulsion polymers for the paint and adhesive industries and polyester resins for powder coatings and construction applications. At the same time, the company has trading activities involving a wide range of chemicals.

The company employs 90 staff and in the year 2000 its total turnover was 21 million euro. Interchem exports about 30 per cent of its products mainly to the countries of the European Community and the Middle East.

7.3.2 Thematic Background

Interchem Hellas operates manufacturing, product handling and distribution facilities. Its activities are significantly affected by a wide range of laws and regulations relating to the protection of the environment, and it is the company's policy to ensure compliance with the law.

From the mid 1990s, the company realised that the nature of its activities was resulting in risks to the environment. Therefore, the company developed a number of one-off responses to reduce air emissions, to curtail the generation of harmful waste and to improve the efficiency of energy use. More specifically, the most harmful output was its air emissions, volatile organic compounds (VOCs). The company attempted to control them by using scrubbers and floating roof systems. In addition, like Domylco and ICR Ioannou, the company had introduced a biological cleaner for liquid waste. Since 1996 the company has publishing an annual report on its general activities, including environment matters, newly adopted technologies and innovation.

Although the company had adopted these environmental policies, an important starting point for the implementation of an EMS was a conflict with the local authority. In 1999 the local authority, Vathi, objected to the expansion of the company and prosecuted it for causing pollution.[2] As a result, the company's expansion was halted and a period of major conflict between the company and the local authority followed.

In this difficult period, the Board of the company, supported by the environmental manager, decided to implement an environmental management system. They chose EMAS in order to inform the public at large about the "real" environmental impacts of the site through the production of an environmental statement.

On the top of this, the company was also accused of sea pollution. The company believed that it was not responsible for this pollution because it had no significant liquid waste. Furthermore, other companies in the same area could have been responsible for the water pollution. Against this background, the company commissioned research from the National Technical University of Athens (NTUA) in order to measure the exact outputs from its site.

The company did, however, decide to implement an EMS under pressure from the local authority. The results of its implementation were encouraging and supported the company's efforts towards total quality management.

In addition, the company took the opportunity, provided by the Ministry of Development, to secure financial support for the implementation of an EMS. One final motive for developing an EMS was its strategy to increase exports to European countries.

---

[2] The trial is timetabled for July 2002.

7.3.3 EMAS Procedure

After the decision was taken, the environmental manager undertook the co-ordination of the procedure. The help of an external consultant was also significant.[3]

In 1994, the company had been certified (by TÜV Austria) under ISO 9001, for the design, production and quality assurance of its many chemical products. This quality management system covered all the activities of the company in order to ensure high quality of the final products, and complete satisfaction on the part of the customers.

Interchem Hellas has understood that a shift in demand towards technologically advanced and environmentally friendly materials would underpin future profitability and growth would be based on technological innovation. For these reasons, the company has invested in research and development (R&D). The R&D Department participates in many European and national research programmes and collaborates with research institutes, universities and both European and Greek companies. Today the R&D Department employs highly trained scientists (chemical engineers and chemists, most of them with postgraduate degrees) making up 10 per cent of the total workforce. Furthermore, the Department has been equipped with state-of-the-art equipment to support its work.

During the implementation of EMAS the environmental manager focused particularly on the production line in order to improve the whole process and to lessen negative environmental impacts. The most significant element in this activity was an increase in the knowledge of the workforce and good use of existing technology. In particular, the company had already introduced equipment for environmental protection, like scrubbers and floating roof systems, but further improvements to the equipment were being planned in order to address the company's objectives more effectively. The following improvements have been made:

- Floating roof systems were installed in the remaining storage tanks not already equipped with this facility. This limits losses of organic solvents into the atmosphere and ensures a safer operation.
- The existing sewerage system for the collection and treatment of waste water was extended.
- A peripheral road was constructed to improve traffic circulation, as well as the loading and unloading of trucks.
- Fire proofing was modernised.
- A fully equipped first aid centre under the supervision of a medical doctor was established.
- A seminar room for training employees was built.

In more detail, the company had 45 tanks, and during the implementation of EMAS floating roof systems for containing air pollutants had been installed in seven tanks.

---

[3] At the time of writing, the company had not been verified, but it was in the final stages. They were awaiting the conclusions of the NTUA project.

By using floating roof systems the VOCs stay in the tank and liquefy again so they are not released to the environment. At the same time the company installed scrubbers in stock tanks and in reactors. In the 30 stock tanks, which contain liquid solvents, although it was not legally required, the company installed 10 scrubbers in order to limit emissions to the air. In the 13 reactors, the company has installed two new technology scrubbers, while the other 11 have old technology scrubbers. The company plans to install new scrubbers in all the reactors in the future. Moreover, the company managed to replace 30 per cent of the oil used in the burners by gas. The result of this replacement has been the reduction of CO_2 emissions by about 5 per cent per year. Finally, the company has introduced an innovative high technology biological cleaner for dealing with its liquid waste. This technique provides an opportunity for the company to reuse water again in the production process and also to use this water for cleaning equipment. There is also a system for collecting all the running water from the site, such as rain water, and directing it through the biological cleaner. The treated water is then pumped into the sea.

EMAS implementation has also helped the company to develop a more flexible organisational structure and a clearer division of responsibilities. The company believes that "good housekeeping" is necessary and this has been a significant impact of EMAS on the site. It fits in with the future plans of the company to integrate its policies into a total quality management system. At the moment, the company organises itself through separate systems, like EMAS, ISO 9001 and through the fragmented policies for health and safety.

As mentioned earlier, Interchem is one of the biggest chemical companies in Greece. The managing director has decided on a new plan for the company's development. One of its main principles is sustainability and environmental protection. The company tries to include new principles in its strategy in order to be competitive and to be accepted by customers, the local residents and the local authority. Against this background, and mindful of the conflict with the local authority, the company decided to give a high profile to environmental issues and to try to improve relations between the company and the local citizens. In other words, EMAS was perceived as both a good marketing tool and – for the local area – a good public relations tool.

The first step in implementing the EMS was the evaluation of the environmental performance of the company. At the same time as the initial analysis Interchem collaborated with a university (NTUA) in order to obtain more systematic and detailed information. However, the results of this research have not been placed in the public domain.

Important technical innovations and their effects in terms of sustainability have already been covered. For instance, the new biological cleaner for liquid waste successfully operated in two ways. First, the company managed to deal with the liquid waste, and second, it reduced consumption by the reuse of water. This resulted in financial savings.

In relation to solid waste, the company is recycling plastics and paper. Over a three year period, the target is a reduction of 5-10 per cent for paper, and 5-12 per cent for plastics. In parallel to this recycling, the environmental department is trying to enhance reuse of metal barrels. Interchem tries to collect metal barrels and plastic bottles from its customers and reuse them, after cleaning them with biologically treated water or by using a specialist cleaning company. In this way, the company's aim to reduce waste management costs through EMAS has been achieved.

The decision to implement EMAS was taken – as in the other case studies – at the top by the owner, but the environmental manager and the external consultant played a crucial role in the decision-making process.

The case of Interchem is special in so far as the local authority had indirectly forced the company to implement an EMS. Indeed, the company is implementing EMAS as a practical tool to convince the "local society" in particular of its sound environmental performance.

An interesting feature of this case study is the company's co-operation with knowledge holders like universities in order to develop better relations with the local authority. The company established an environmental committee, comprising the heads of the company's departments, the environmental manager and a representative from the local authority, where the environmental performance of the company could be addressed with the help of the university based knowledge holders. Therefore one can say that EMAS has opened up the company to the broader society by creating a discussion panel comprising different "holders".

Against this background of the conflict with the local authority, the employees have also been forced to be positive about the implementation of EMAS. Otherwise, they could be in danger of losing their jobs. On the other hand, the environmental manager ensured that employees responded positively to the system because it first secured their health and safety and then addressed the environment.[4] In addition, the employees had experienced the implementation of ISO 9001 so they had become used to the kinds of changes that flow from the application of these systems.

7.3.4 Conclusions

At first, EMAS implementation was a quick reaction by Interchem. The company decided to implement EMAS after the local authority had accused it of sea pollution. However, during the EMAS process, the company stressed that there were significant advantages in its implementation.

EMAS helped Interchem in its "good housekeeping" by specifying problems (pollution) and responses to them. The investments made in the course of EMAS implementation were perceived as effective and efficient. For example, the expansion of biological cleaning has helped to reduce the company's liquid waste and to save water by reusing treated water in the production process.

The company developed new relations with *knowledge holders*, such as universities, and initiated an "open communication" with the local authority.

---

[4]   Most of the employees live near the site. Therefore the protection of the environment is also an issue for them as residents.

7.4   Sato

7.4.1 Introduction

Sato SA used to be a company called Metallon Hellas, which was founded in 1964. In 1974, Metallon Hellas was renamed Sato Office Furniture Industry SA. Sato's core business is to equip offices with furniture that offers both efficiency and functionality.

Sato manufactures a wide range of furniture and office equipment including chairs, for both managers and secretaries, desks, conference tables, filing cabinets, storage units and partitions. The company also produces seating equipment for public areas.

The elements that have contributed so far to Sato's successful development in the last 35 years, and which were strongly espoused by the company's founders, are knowhow, ergonomics, functionality of the products, cost control, which ensures the best prices for the customer, and an emphasis on design and careful choice of materials.

Sato employs more than 335 people in a production area of 33,000 square metres. It owns two factories with twelve units for the production of furniture, showrooms with sales staff and a well established supplies outlet.

Sato has a presence through affiliate companies in the neighbouring countries of Bulgaria and Romania, and Sato exports its products to Cyprus, the Balkans and many other European countries (Germany, France, United Kingdom, the Netherlands, Belgium, Russia, Poland, Hungary, the Czech Republic) as well as Asian markets (Saudi Arabia, the United Arab Emirates, Kuwait, the CIS countries). Exports are managed either directly or through a network of agents, distributors and dealers, who are given direct support and complete service solutions to a continually growing customer list. Consistent with corporate values, Sato products are based on innovation, and customers' satisfaction.

7.4.2 Thematic Background

The production base at Sato comprises two factories, equipped with the latest technology. All this adds up to the philosophy: "Do it right the first time in the most profitable way" (as one of the interviewees put it).

At a very early stage Sato recognised that it needed to establish a system which would ensure quality at all stages of production, quality in sales and in after sales service. Sato had conformed to international standards of quality management by acquiring ISO 9000 certificates (ISO 9002 in 1997, ISO 9001 in 1998).

These certificates strengthened Sato's leading role in the design and manufacture of new products. They assured its customers that the company controls and guarantees the quality of products and services, according to international standards.

7.4.3 EMAS Procedure

Sato is highly sensitive to environmental issues and one of its factories was the first EMAS site registered in Greece. It has initiated the EMAS process for its second site. EMAS is intended to operate in parallel with and be complementary to a quality management system based on ISO 9001 that was implemented in 1997.

In January 1999 preparations were under way for the implementation of a system to handle the environmental management of raw materials, which would bring the company in line with the demands of EU directive 1836/93. The company had already attempted to develop an environmental policy, but it had proved difficult to implement. In 1998 the managers, in co-operation with the owner, considered EMAS a manageable tool to achieve environmental goals. Savings in raw materials was one of the most significant objectives. At the same time Sato was seeking new ways to ensure legal compliance on a continued basis. The company had had some problems with its outputs and it had to find new ways to deal with them. Moreover, the company was interested in continuous improvement of its environmental performance. In order to succeed, they decided to introduce a tool for monitoring and improving the company's performance. Finally the owner decided in favour of EMAS. The main arguments for EMAS (instead of ISO 14001) were that (i) it was a European standard, (ii) it had more demanding requirements and (iii) it placed emphasis on putting information in the public arena through publishing an environmental statement.

Technological innovations have been introduced in all stages of the production process because Sato has aligned itself with a rapidly changing market environment and is ready to seize new opportunities. In the new millennium Sato's aim is to create new markets by anticipating and fulfilling customer needs. The strategy is to bring new ideas to the workplace that will continuously enhance a positive working environment. One thing remains constant: Sato's commitment to innovation and high quality. It is this commitment that has distinguished the company as a market leader in the Greek office furniture market. Input from R&D has always formed the basis for all decisions on future plans.

Since 1999, Sato has implemented, through EMAS, processes for continuous improvement as well as for environmental management and auditing. In particular, EMAS implementation brought about significant changes in production line processes. First of all, there were changes in the production line to address environmental impacts and, second, there were initiatives to modernise the different phases of the production process. For example, the management is considering the production of environmentally friendly or recyclable products.

One important factor was that the EMS helped the company to change its organisational structure. In particular, EMAS defined the role of the environmental department in the total management system. Through the environmental management system, employees have come to understand their specific responsibilities for the company's environmental safety operations. Development of existing technology was another unexpected effect of EMAS implementation.

Furthermore, the company always focused on satisfying customer needs, by improving its products continuously, usually by conducting market research. The customer care department was always available to customers, thereby developing long term relationships of mutual trust. Some of the biggest companies in Greece are among Sato's clients.

The company uses modern computer technology, such as computerised numerical control (CNC) machines for wood and metal processing, for the dyeworks, wallpaper and assembly sections, as well as for large storage spaces.

Turning to *sustainability*, the company has significant quantities of pollutants that can cause considerable problems for the environment. Through EMAS, people were identified who were given responsibilities for air, liquid and solid waste. Every one of them works directly with the environmental manager, the head of the health and safety department, the quality manager and the general manager.

One of the main reasons for EMAS implementation was to bring about a reduction in the level of pollutants. For example, Sato had significant types and amounts of waste, some of which was dangerous and needed special treatment. The most serious problems were the emission of sulphur, carbon and VOCs. Through the implementation of EMAS the company planned to reduce emissions of VOCs by about 50 per cent in five years and of sulphur, nitrogen and carbon emissions by about 10-15 per cent each in 1.5 years. Moreover, in the metal dyeworks chemical emissions were planned to be reduced by about 80 per cent in 3.5 years, and the dust and sawdust (inside and outside the site) were to decrease by about 20 per cent in a year.

All these emission reduction targets were achieved through a range of new techniques which the company installed as part of EMAS. The most significant changes were the establishment of thermostats on the production line, the replacement of combustion chambers in boilers and the installation of high technology equipment for the collection of chemicals. Through EMAS the company was able to collect systematic pollution data and to have detailed information about environmental performance.

As noted earlier, the company introduced an innovative high technology biological cleaner for liquid waste reduction. It also plans to install a new closed system for reusing the treated water. The company also developed an alternative method of waste management, again using a biological cleaning agent for liquid reduction and management. Old water collectors were replaced to reach maximum efficiency.

Some of these changes have resulted in savings. For example, the installation of the biological cleaner reduced water consumption by about 10 per cent and the installation of water meters at every stage of production in order to measure consumption exactly (and to stop leakages) proved particularly useful. Moreover, the consumption of raw materials is expected to decrease through new techniques. Sato has examined the likelihood of replacing chemical colours and material by other more environmentally friendly and recyclable ones. Furthermore, the company has looked at reduction in energy consumption. It has reduced the use of fuel oil for heating by about 80 per cent because it replaced the fuel oil by sawdust.

Sato does produce hazardous wastes. Their separation, identification and safe management were achieved using very high technology procedures in accord with all the necessary safety regulations. The company's strategy for continuous improvements in environmental performance is a long term one (about three years). The management team, which was responsible for EMAS implementation, argues that in the next 3-5 years the company will have achieved very significant positive results.

The decision to implement EMAS was taken by the owner supported by the environmental manager and the external consultant. The decision came from the top. The first reason was to comply with the law. The company had not had any significant problems with the local authority or the state, such as an environmental accident or complaints. However the company did realise that it had to ensure compliance with the law because of its air, liquid and solid pollutants. Following advice from the external consultant the company realised that EMAS could be a useful tool for reducing the use of raw materials and energy. Furthermore, it was recognised that implementation of EMAS could enhance exports to EU member states.[5]

Employees were informed about EMAS implementation through seminars. The environmental manager and the head of every relevant department had – in collaboration with the external consultant – the responsibility for training and the seminars. At first, the employees were hostile, pointing out that they had to do more work and go through an increased number of bureaucratic paper procedures. Although they still perceive EMAS as a bureaucratic system, they quickly recognised that the implementation of EMAS was necessary both for their safety[6] and to enhance environmental protection.

One reason for choosing EMAS was the requirement to publish an environmental statement. The company was trying to "open Sato's door" to society. It was very important for them to inform the local authorities, government agencies, suppliers and customers about their environmental performance.

7.4.4 Conclusions

EMAS had a significant impact on Sato. The fact that EMAS led the company to reduce the level of pollutants was crucial. The company found EMAS a useful tool for achieving environmental performance going beyond reaching performance indicator targets. Moreover, the company introduced some very high technology investments and was able to "escape from easy and limited solutions". For example, the company installed a biological cleaner for liquid waste reduction and management.

In general, the company managed to develop "good housekeeping" and to secure competitive advantage for its exports to the European market. The company had already planned activities for continuous environmental improvement, showing that it had realised that environmental issues require sustained and long term thinking.

7.5   Municipality of Athens

7.5.1 Introduction

Athens is the biggest municipality in Greece with a population of 313,000 people and about 10,000 employees. The municipality is a part of the Athens

---

[5]   According to Sato's environmental statement, the company exports 50 per cent of its products to the European market and 40 per cent to other countries.

[6]   At the beginning the employees confused EMAS with a health and safety system.

agglomeration, which comprises a large number of independent local authorities, totalling about 3.5 million inhabitants.

7.5.2 Thematic Background

Participation by Athens in a pilot project (funded by the LIFE programme of the EU)[7] was crucial for EMAS implementation. This pilot project, called Euro-EMAS (see section 6.2.3), provided a framework for managing and improving the local authority's environmental performance and for integrating sustainable development aims into its policies and measures.[8] The project started in October 1998 and ended in October 2001. The project partners were Newcastle upon Tyne (project co-ordinator), EUROCITIES and EURONET. The partner cities were Newcastle upon Tyne, Athens, Malmö, Birmingham, Göteborg, Helsinki, Leipzig, Palermo and Stockholm.

7.5.3 EMAS Procedure

The Development Agency of the Municipality of Athens decided to implement EMAS in three of the seven medical centres run by the municipality. Following the objectives of the Local Agenda 21 summit and within the framework of sustainable development the decision was taken

- to minimise the direct environmental impacts of all the activities of the medical centres;
- to educate and mobilise the staff;
- to increase environmental awareness;
- to achieve continuous improvement.

The decision to focus on three of the seven medical was mainly due to budget constraints. The three selected were the Kato Patissia centre (12 administrative staff and 15 medical staff), the Kypseli centre (5 administrative staff and 10 medical staff), and the Drougouti centre (6 administrative staff and 16 medical staff). They see about 2,000 patients per month, offering only diagnostic services free of charge.

Those responsible for the administration of the medical centres were aware of the environmental problems, not least in recently developed urban areas, and responded positively to the call for implementing an environmental management system. However, the *innovation element* in the implementation of EMAS in this case focused on the changes to the structure and management of the organisation.

---

7   LIFE is a financial instrument of the EU which funds three major areas of action: environment and nature protection, as well as technical assistance in these areas for Eastern European countries, based on regulation (EEC) No. 1973/92 of 21 May 1992 (OJ No. L 206 of 22 July 1992) as modified by regulation (EC) No. 1404/96 of 15 July 1996 (OJ No. L 181 of 20 July 1996).

8   For further information on Euro-EMAS, see www.uwe.ac.uk/emas/brief (accessed 15 January 2002).

Through EMAS a number of initiatives were taken to bring about environmentally more acceptable working and living conditions. The main aims were:

- to reorganise the interrelationships between the project manager and the managers of the centres, between the centre managers and the heads of sections, as well as between the managers, the head of sections and the staff;
- to clarify how tasks should be defined and shared out between the staff working at these different levels.

The Development Agency was responsible for the flow of all relevant information, updates, staff seminars, and public announcements about this project in co-operation with the managers of the medical centres. The general manager of each medical centre was responsible for the centre's management, the implementation of the environmental programme in close co-operation with the project manager, and the allocation of tasks. The head of sections had to ensure that every action that was taken was in accordance with the environmental policy. Every head of section also had to ensure that all employees within their section participated in the programme and that seminars did take place. The employees (administrative and medical staff) had to accept the procedure and actively participate in the realisation of its objectives by integrating environmental principles in their everyday work. This included implementing all actions laid down in the environmental programme and helping to spread environmental awareness and good practice more broadly in society.

The cleaners together with the managers of the centres had the responsibility for managing the recycling programme. They were required to collect the waste from special recycling bins, to weigh them on a weekly basis, register their quantity and dispose of them according to the municipality's waste management policy. In co-operation with each medical centre manager the work of the cleaners was recorded in order to lay down an environmental programme that would be realistic and suited to the nature of the cleaning service.

Seven categories of activities were designated: photocopying, lighting/hot water, transport of employees, transport of visitors, use of packaging, water use and production of dangerous waste. These activities were weighted on a unit basis according to their environmental importance, the consumption of water and electricity, waste recycling and hazardous waste as well as air pollution. The construction of tables showing the types of direct and indirect environmental effects of each activity and their environmental significance was crucial for deciding on the types of interventions and actions that should be undertaken to improve the centres' environmental performance. By this procedure, they identified the main objectives of an environmental policy as well as the initiatives needed to make up a complete environmental management programme.

The initiatives addressed the management of electricity and water consumption, the implementation of a recycling programme, the organisation of transport and the development of an information and education programme. The actions planned were:

- the introduction of energy saving light bulbs;
- checks for water leakage;
- renewal of taps/flushes and frequent surveillance of their proper function;
- the installation of automatic switches for night safety lighting;
- the installation of a solar powered boiler in the medical centre at Kato Patissia while the management was informed about better techniques in the use of existing electric water heaters;
- the installation of special recycling bins for paper, glass, aluminium and hazardous waste;
- the installation of dispensers in the toilets;
- the development of an information campaign addressed to staff which was intended to inform them about the project's objectives, to raise their awareness and to mobilise them in order to integrate environmental principles into their everyday work.

The design and implementation of the environmental management programme required effective co-operation between and active participation by people at all levels in the company. For this purpose, two types of meetings were organised:

- a monthly meeting between the project manager and staff members to monitor progress and provide updates if necessary;
- four meetings a year for staff which took the form of educational seminars aiming to provide information, raise awareness and mobilise them on environmental issues.

The medical centres produce several types of waste, mainly domestic, as their medical functions are limited to diagnostic services only. The research showed that the greatest amounts of waste were paper, glass, aluminium and plastic packaging. In the first three months, a recycling programme was put in place by installing 40 recycling bins in the three centres and by giving guidelines to staff on how to use them. Batteries were collected separately in special receptacles for hazardous wastes. One important issue for the conservation of natural resources and the reduction of emission was the management of water and electricity consumption. Although this had been addressed by a major awareness raising campaign, it was, nevertheless, quite difficult to change well established habits and practices. The employees did not easily connect the consumption of electricity or water with environmental impacts and had practically no understanding of ways of economising on the use of energy.

In the three centres, efforts focused mainly on implementing a number of one-off actions that would have immediate positive effects on energy consumption, such as the change to low energy bulbs, and educating staff about simple practices and energy saving techniques.[9]

All staff members were told about the environmental management project, its principles and goals, as well as about its progress and possible problems and shortcomings. In April/May 2000, a short information campaign for all employees was organised and they were also invited to join the seminars which were organised four times a year.

The patients of the medical centres were also kept informed about the project and its progress.

On *participation* in the decision to implement EMAS, it should be noted that involvement in an EU pilot project was the starting point, and that the Development Agency of Athens was the main body taking this forward. Thus implementation of EMAS came from the top but had been initiated by a department that was not particularly powerful. In other words, neither the managers in the medical centres nor the key politicians of the municipality (the mayor and other leaders of the administration) participated in the decision to implement EMAS. Furthermore, the choice of the three medical centres proved to be problematic because the centres comprised three different units: public health, the medical office and the secretariat. This made the whole procedure quite complicated because different kinds of employees like health staff (doctors and nurses) and administrative staff (secretaries) with differing qualifications and outlooks were working in the three centres.

7.5.4 Conclusions

It was difficult to implement EMAS in the Municipality of Athens because it was initiated through external funding and was not integrated into the general policy of the municipality. This was reflected in the lack of support and leadership by the central administration (mayor). Overall, the implementation of EMSs at local government level in Greece is limited. Only a few local authorities have started to implement EMAS but none of them have got through the whole procedure and been officially registered as an EMAS site.

Finally, the different kinds of employees in the three medical centres selected for the implementation of this pilot project caused considerable problems because of their varied qualifications and outlooks.

---

[9] Furthermore, in Athens most travel to and from work is made by private car. The use of public transport has increased over the last two years (due to the construction of the metro), but the use of private cars still remains the main means of transport. For the medical centres, the objective of reducing the use of the private car and promoting alternative means of transport remained a broad principle. The main objective was to reorganise the work schedules of the employees. However, this was a long term aim. Consequently, it was more realistic to try to continue to educate and inform staff and patients, while at the same time options for changing the working hours of the centres should be studied.

7.6 Elais

7.6.1 Introduction

Elais has been operating in Greece for the last 80 years. It was founded in 1920 in Neo Faliro (an area of the Athens agglomeration) and has been part of the Unilever Group, one of the largest companies in the world since 1976. In 1998 the company employed 384 people, had a turnover of 158,500,000 euro and made a profit of 25,000,000 euro. It is the biggest oil product company and has carried a leading position in the Greek food and beverage market for years.

Since 1990 Elais has developed many projects in order to integrate its policy within the framework of total quality management. The main plan of the company is to integrate the systems of quality and environmental management. Indeed, Elais has already received several awards for its quality and environmental management policies.

For example, in 1999 it secured one of the first European Awards for Maintenance awarded by the European Academy.[10] In accordance with its corporate strategy, Elais has developed an effective maintenance programme, ensuring efficient fund management and maximising productivity.

Elais also won second prize at the 1999 European Quality Award Competition[11] in the large business category. The award ceremony was held in Brussels in October 1999 at the EFQM (European Foundation for Quality Management) forum. This was the first time that a Greek company had secured such an important award for business excellence. The year before, the company had been a finalist in the 1998 European Quality Awards Competition and winner of the 1998 European Better Environment Award for Industry. These successes reflect the fact that Elais was accredited under ISO 9001 in 1994 and ISO 14001 in 1996.

In 1997, Elais presented for the first time its Contribution to Society Report, which accompanied – as a separate brochure – its annual report for that year. Reflecting the company's commitment to contribute to education, Elais organised a conference on "Education in the Plant" in June 1998 in co-operation with the Greek Children's Museum and under the auspices of the Municipality of Piraeus. Perhaps the most significant contribution by Elais to society was the installation of "tele heating" in a nearby school complex, which provided free heating for 2,000 students. As a result, the social profile of the company was enhanced.

On environmental issues, in 1998 Elais was honoured by being included among the top 15 European companies – out of a total of 85 – in the final selection for the

---

[10] The European Academy is a pan-European organisation aimed at promoting the importance of proper equipment maintenance, to establish the maintenance of equipment as an academic discipline and to formulate a European strategy on this issue. This is based on the idea that proper maintenance in the right place, at the right time and at minimum cost is essential for achieving quality in production and other operations.

[11] The first and second prizes at the European Quality Awards are awarded to companies and organisations with the highest level of commitment to the implementation of the Total Quality Business Management Model and the EFQM Excellence Model.

European Better Environment Awards for Industry (EBEAFI) organised by the European Commission's DG XI and the United Nations Environmental Programme.

7.6.2 Thematic Background

The company initiated its environmental programme in the early 1990s. Because the company is located in the densely populated urban area of Neo Faliro, one of most important aims was co-operation with the local authority and the neighbourhood. Therefore, in 1991 Elais set up an environmental committee, comprising the company's president, the heads of all the departments and the environmental and technical manager. The committee meets every month to set out the main aims of the company's environmental policy and to contribute to the documentation of the company's environmental improvements.

Elais has made a significant number of moves to improve its environmental performance and to reach the goal of sustainable development. They include:

- the implementation of practices that reduce the impact of industrial processes on the environment;
- the continuous improvement of recycling, reduction in packaging, waste etc.;
- systematic and clear information on environmental issues for both employees and the public through workshops, open days and seminars at schools.
- To achieve these aims the company carries out activities such as:
- the introduction of measures to comply with national and European laws on environmental protection, health and safety issues etc.;
- the development of new production techniques in order to achieve the same quality with less environmental impact;
- the publication of continuous online information about environmental issues;
- internal auditing to control the whole process on a daily, weekly or monthly basis by its own or external auditors;
- the creation of a team for preventing environmental accidents.

The implementation of ISO 14001 in 1996 was seen as a way of integrating all the above activities. Elais was trying to reach "eco-effectiveness" (improving the environmental outcomes of manufacturing operations and incorporating environmental factors into the design and re-design of products) and "eco-innovation" by implementing an EMS. Furthermore, Elais interacted with a broad spectrum of interests from outside the company with the aim of working in partnership with key "holders". The Unilever group did not force the company to adopt an EMS but did require it to have a specific environmental performance strategy.

7.6.3 EMAS Procedure

The head of development and quality assurance was the key person responsible for implementation of an EMS. With the company president's agreement he organised the ISO 14001 implementation. First of all, he selected the members of a core team from those departments of the company (production, distribution and marketing) that would participate in the process, and shared out the key tasks between them. This team had control over the whole process, gave advice to the parts of the company affected by the initiative and informed employees about the value and necessity of an EMS. Advice and guidance from an external consultant was crucial.

As noted earlier, the environmental management system was a tool for the company to optimise the available techniques both for environmental matters and for the total management system. This meant that the adoption of ISO 14001 was intended to help the company not only to succeed in "good and systematic housekeeping" but also to offer an opportunity to achieve continuous improvements through organisational *innovation* and technology as well.

Indeed, during and after the implementation of ISO 14001, Elais moved forward by making some changes in production in order to improve the whole process and to achieve better environmental outcomes. Although the company did not introduce any new products, after the implementation of the EMS, every product was thoroughly examined for its environmental implications, such as recycling packaging or minimising bottle weight. All products were analysed for their environmental implications and to identify how improvements could be made in one or more stages. Elais introduced two major changes, first, the continuous improvement of bottle pack design and the minimisation of bottle weight and, second, the avoidance of the use of PVC in bottles. The result was a reduction in bottle weight by 371 tons per year. One of the main environmental principles of the company was that all packaging should be recycled. There has thus been significant success in minimising waste, using best available techniques for reducing the amount of plastic packaging. In addition, there was a reduction in the range of printed cartons from 148 to 17 (blank cartons and labels). This has reduced both costs and waste. Elais has required their suppliers to introduce similar systems for packaging, and it co-operates with the Greek Recycling Company.

One of the most important investments by the company was the installation of a new oil refining technique. The company changed the way of refining oil from a chemical to a physical basis. Through the first procedure oil is combined with some chemicals and water and soap result. Through the physical refining method there is no chemical or biochemical reaction. Instead, the oil evaporates after boiling between 250 and 280 degrees Centigrade. Use of the physical refining system reduced the amount of water and soap produced by about 70 per cent and the use of energy by about 30 per cent. As a result the impact on the environment was reduced.

Although innovation in production processes are important, the direct link between the EMS and an increase in knowledge is hard to pin down. Elais had already studied the capability and limits of its existing technology before applying ISO 14001. Thus, improvements in production processes and environmental consequences have been a result of the general policy of the company for which the EMS has been only *one* instrument. Indeed, the company considered the EMS as a management tool, which helped them to control more systematically the different parts of the production process.

The organisational reforms and the re-structuring of the management were another innovative step. The structure of the organisation was one of the biggest issues.[12] One of the consequences resulting from the EMS was that strict guidelines and detailed documentation was provided, not least reports on environmental matters. The company seized the opportunity to introduce changes. The documentation required by an EMS was seen by the management as leading to continuous learning inside the company. The management recognised that the documentation as well as the opportunities for learning depended crucially on the degree of participation inside the company, not least by the employees.

As mentioned earlier, the Elais management was trying to combine the different tools into an integrated management system. The department responsible for integrating the management systems was the internal audit department. This department co-ordinates ISO 9001, ISO 14001, total production maintenance (TPM), product safety, health and safety plans (HASP), safety for employees, occupational health and the European Foundation for Quality Management's critical process. This department also took part in an IPPC scheme, as a pilot company for the Federation of the Greek Food and Drink Industry and as an implementer of risk management systems. The co-ordination of these systems, together with the associated systematic documentation, is highly demanding, especially for employees, and can be in danger of becoming bureaucratic. Elais, in contrast to Unilever, is trying to limit the bureaucracy by simplifying and making more flexible the reporting process for employees.

Elais is also engaged significantly in public debate on environmental issues. For example, it supported a WWF campaign on the conservation of endangered species in Greece, and the company co-operates and financially supports environmental NGOs.

Decrease in the amount of pollutants produced was one of the most significant elements of the company's environmental policy and contributed to the objective of *sustainability*. Elais had made a number of investments before and during the implementation of the EMS to achieve this objective.

The replacement of heavy fuel oil by city gas (which cost 130 million drachmas) resulted in a major qualitative improvement in emissions to the air. The company succeeded in reducing CO_2 and smoke emissions by about 90 per cent. The innovative way of removing oil smells by physical means reduced the odour by 90 per cent.

Elais has emphasised the importance of decreasing water and solid pollutants. On water, it lowered the level of discharged sulphates by using a closed alkali recirculation cooling system instead using of brackish water from the sea. The use of controlled cleaning techniques kept discharges within legal limits. Elais requires a continuous monitoring and investigation of every leak or accidental discharge of oil to the effluent treatment plant. On solid waste, the company has, for instance, used bleaching powder for cement manufacturing. In the factory, the identification and separation of all materials has been undertaken to support the recycling of steel, paper, aluminium, glass, bricks, stone and plastics. At the same time Elais installed

---

[12] The incorporation of the company into the Unilever group improved its management system through the application of international principles and guidelines. However, the adoption of an environmental management system was also of benefit especially in the organisation of the production process.

systems to control and produce trend measurements for waste and has combined this with TPM techniques which focus on environmental costs.

Elais has had to face a major problem of noise. It therefore installed noise measurement instruments on both working days and at weekends to assess the external noise impact of factory operations. In order to improve working conditions, Elais also implemented (with the help of specialised external consultants) noise control measures inside the factory.

Still in relation to sustainability, Elais also managed to achieve considerable savings in the use of energy, water and raw materials through implementation of the EMS. The energy savings focused particularly on

- the use of an alkali recirculation cooling system instead of brackish water to reduce effluent volume from 200 to 2 cubic metres per hour;
- the use of detailed measuring systems to optimise actual consumption against standards;
- the extensive use of energy recovery techniques particularly in the refinery.

The reduction in water consumption was achieved by a reduction in effluent discharges, due in turn to a reduction of effluent volume and through the extensive use of modern technology, such as recirculation. A very simple and effective way to monitor water consumption was the use of detailed measuring systems in each department. The company had drills to tap underground water but in 1999 they stopped using them because they adopted a new technique to freeze the city's water. This meant that Elais stopped the continuous reduction of the level of ground water. Moreover, using different techniques for liquefying the vapour in oil processing resulted in considerably less use of water from the public water supply system (about 15,000 cubic metres per year). In both situations, the company's managers analysed benchmarking data from similar factories to improve Elais' performance.

As one of the first companies in Greece to have adopted environmental criteria, Elais used them to select suppliers, for instance, of raw materials. An audit of suppliers (their buildings, production processes and products) using quite basic environmental management standards was developed and Elais requested that packaging materials used by suppliers should be made of recyclable materials. This was a fixed principle for the company rather than monitoring the quality of the components of their own products. The implementation of the EMS helped Elais to be more systematic and detailed in looking at their supply chains. The company was also conducting a large number of laboratory tests either in its own facilities or in co-operation with laboratories in other countries (TNO [Netherlands], Leatherhead Food [UK], Doctor Verrey [Switzerland]) in order to identify the ingredients of specific inputs. On recycling Elais is both on the board of and is funding the Greek Recovery and Recycling Association (HERRA).

The environmental commitment of the company sustained the good relationships with its neighbours. For example they have initiated co-operation with both local authorities and NGOs, have hosted visits and contributed to the training schemes of the environmental auditors of the government agency (ESYD) and participated in activities organised by the Ministry of the Environment.

Participation has been a very important element of the environmental policy of the company. Although the company has a hierarchical structure –especially after becoming part of the Unilever Group – it did promote initiatives from employees. It has developed an "open door management" and has put on seminars and training courses for employees. The latter received specific training for implementation of the EMS in general and for their new responsibilities and roles in the system. Employees were also able to put forward suggestions for environmental improvements through the "Opportunity System for Improvement". In the last five years employees have made about 500 suggestions and more than 90 per cent have been taken up.

The company has paid a great deal of attention to its surroundings. It is located in an urban region. Because of this, the company was keen to avoid getting a bad reputation in the neighbourhood at an early stage. It had its own parking places, the loading of lorries took place within the company premises, and local government projects and buildings have been financed by Elais. The most significant offer by the company to the locality was the delivery of "no charge heating" to a nearby school.[13]

7.6.4 Conclusions

Elais is a company which began to address environmental issues very early. It was one of the first Greek companies to implement an EMS and to adopt innovation techniques. Its early environmental awareness might be due to the fact that the company is part of the Unilever group and located in a densely populated urban area where problems of pollution have been very important for the local residents as well as for the local authority. However, Elais managed to prevent possible adverse local reactions through its timely proactive environmental measures.

Moreover, the company developed relationships between different "holders", such as universities, research institutions and government organisations, in order to inform them about the company's environmental policy and to co-operate with them to achieve continuous improvement and for "knowledge exchange".

The implementation of an EMS in the company led to some significant changes in the production process, a decrease in pollutants and an increase in savings, but unfortunately it did not communicate these improvements to its customers.

7.7 Siemens

7.7.1 Introduction

Siemens is a multinational company which has five subsidiaries in Greece. Siemens produces modern high technology telecommunications material for the Greek

---

[13] The company channels, through underground pipes, the warm water that comes out of its production processes, to a nearby school, heating it for free. The air pollutants in the school have reached zero level and the school saves about 15 million drachmas annually. At the same time the company does not use energy to cool the water used for its production processes. It is recycled to the company cold enough to be reused.

market, but also exports its products to the whole world (especially to Western Europe, the Balkans, Cyprus and the Middle East). It employs 2,100 staff in Greece and ISO 14001 has been implemented in all five Greek subsidiaries. The multinational company has a global environmental policy but the way it is implemented varies a great deal in different countries.

7.7.2 Thematic Background

A key principle for Siemens is to follow the ISO 14001 route and not EMAS. Therefore, the Greek subsidiaries have followed suit. Siemens believe that the public in Greece is not ready to engage with the contents of environmental statements. However, the company has implemented ISO 14001 as an internal tool and has selectively publicised some of its environmental activities.

The company has adopted a number of principles on environmental protection in order to promote sustainable development on its sites. Siemens's objectives are as follows:

- The economy, the environment and social responsibility are three key factors that carry equal weight in a world market.
- Through the transfer of knowledge in the fields of environmental management and technology the company supports the dissemination of information about sustainability.
- Continuous activities should be undertaken – above and beyond what the law requires – to reduce the burden on the environment, to minimise associated risks and to lessen the use of energy and other resources.
- Appropriate precautions should be taken to avoid environmental hazards and to prevent damage to the environment.
- An appropriate management system should be developed to ensure that environmental policy is implemented effectively. The technical and organisational procedures required to do this should be monitored regularly and further developed on an ongoing basis.

7.7.3 EMAS Procedure

After the parent company's decision to go for ISO 14001 (taken in 1997) the general manager for environmental protection and technical safety assumed complete responsibility for ISO 14001 implementation in Greece. The manager first chose one company (Siemens Televiomihaniki) to initiate the new system and then moved on to all the others.

One of the most significant advantages that Siemens had was the experience of its managers in environmental management systems. Siemens promoted the exchange of managers between the countries (and sites), so the company managed to spread knowledge through the mobility of its managers.

Siemens puts new ideas to work, creating innovative products and services for the benefit of customers. The company encourages experiments and imagination. Creativity and a willingness to take risks enabled the manager to forge an environment in which promising ideas could be quickly turned into successful solutions. The company has developed an innovation department to serve the Greek companies in co-operation with the international Siemens R&D department.

In Greece, the companies have introduced significant environmental changes to both products and production lines through implementation of an EMS. The company has replaced parts of its products with more environmentally friendly materials. For example, the company has replaced plastic by metal parts. These changes affected existing products, not new ones.[14] These changes in the products led to changes also in the production process. The technologies for plastic were replaced by metal technologies. Some of these changes, which should be seen as investments, were also linked to other sustainability issues, such as a reduction in emissions.

A computer software package was introduced for monitoring every step of the production process. The system was installed before the EMS, but in the course of its implementation, the whole system became more modern and extended to every stage of the production process.

The environmental management system led to reorganisation of employee responsibilities in a more rational way. The systematic documentation, the daily report and the careful audits promoted health and safety, not only environmental protection. Traditionally, the Siemens factories in Greece had a high degree of division of labour, so when ISO 14001 was implemented, Siemens decided to introduce a "horizontal relations management system". This demonstrates the emphasis Siemens has put on development on every site. Within Siemens' global strategy, every subsidiary company has to promote quality systems (ISO 9000), health and safety management systems, as well as risk management and performance indicators. However, all these different activities have to be integrated into a "holistic approach" to be more effective.

The company considered the role of ISO 14001 as an environmental market tool, but it decided in the end to include environmental issues in its general market strategy, reflecting the fact that ISO 14001 was chosen, instead of EMAS, because it seemed to fit better into the company's attempts to develop an integrated management system. The publication of the environmental statement – required by EMAS – was for example seen as an "extra" which did not easily fit into this strategy.

The environmental policy of the company was based mainly on its responsibility to people and to the environment. The purpose of environmental protection and technical safety was to avoid hazards and to minimise risks. The term "environmental protection" covers:

- industrial environmental protection (air pollution control), soil and water protection, waste management and environmental management as well as
- product related environmental protection (environmentally compatible product design, product recycling, avoidance of hazardous substances in products).

---

[14] The global innovation centres of the company are studying the possibility of creating new products, which will be 100 per cent recyclable.

- The term "technical safety" covers:
- radiation protection (avoidance of ionising and non-ionising radiation, transportation of radioactive materials);
- fire protection (in respect of buildings and technical equipment) and industrial disaster prevention;
- transportation of hazardous goods.

The first priority (and a necessity) for the company's environmental policy was compliance with the law. Therefore, the company developed systematic pollution controls in its subsidiary companies. On the Greek sites, there was an emphasis on savings through implementing the EMS. There were only small changes in the level of reduction of air emissions and in the reduction of levels of liquid and solid waste. In most cases, the companies improved their existing biological cleaning mechanisms through newer technologies.

Through better biological cleaning, Siemens were able to reduce water consumption. On some sites the company had problems with the local authority because private drillings had decreased the regional ground water level. In order to resolve this problem, the company adopted high technology biological cleaning. It reduced the amount of waste water and moreover part of the treated water could be used again in the production process. The companies also had to ensure there was the continuous monitoring and investigation of leakages or accidental discharges of waste into the environment.

On savings, the company focused on reduction in electricity consumption and in the quantity of raw materials used. As part of the ISO14001 implementation, the companies replaced dangerous materials which were being used for production purposes. Siemens has achieved the identification and separation of all materials so as to support the recycling of steel, paper, aluminium, glass, brick, stone and plastics.

Siemens has also introduced environmental criteria in order to select its suppliers. However, it is very difficult to find suppliers in Greece who have implemented environmental management systems or environmental strategies. Nevertheless, the company's strategy was to promote environmental issues to all its suppliers and customers in order to "enhance their environmental awareness".

In addition, Siemens concentrated on continuous improvement and long term thinking. In implementing EMS the companies had to make some expensive investments. They realised that these investments would be profitable only in the long term, but some of them were in fact beneficial in the medium term, for instance, the initiative to save water.

Because Siemens is a multinational company the Greek companies had to follow the parent company's decision to implement ISO14001 without an opportunity to take part in the decision-making process. Managers had to inform employees about the new system, with help from an external environmental consultant, whose role was crucial.

It was not senior management who informed the employees about the new system, but the environmental and technical safety managers. Each department and each individual employee had to link up with a nominated environmental manager. Although staff did not participate in the decision on implementing ISO 14001 and although they had only limited opportunities to play a proactive role in the implementation of the EMS, they did get the chance to participate in everyday decisions about the production process as well as about their own working environment. These opportunities were created through the widespread implementation of new management systems and through ISO 14001 as well as through the "horizontal relations management system" which was implemented in parallel with the EMS.

In one of the Siemens companies ISO 14001 helped to improve relations with the local authority. There was some conflict because the company had by private drilling lowered the regional ground water level. Because the demand for water was reduced as a result of ISO 14001 implementation, this conflict was resolved.

Siemens has developed many links with knowledge holders, such as universities, research institutes and governmental organisations, in order to promote its environmental activities but also to become better informed. It is a basic assumption by Siemens that integrity has to guide its conduct with business partners as well as with other "holders" and the general public, and that cultural differences can be of benefit to their own organisation.

7.7.4 Conclusions

Siemens is a multinational company, which implements global strategies in all its subsidiaries. Thus, the Greek companies had to implement ISO 14001 as part of the parent company's global strategy for environmental issues. Nevertheless, every company had to follow specific guidelines in order to comply with the law.

It is the case that, on the one hand, the transfer of knowledge from the company's headquarters has limited local consequences. On the other hand, through knowledge transfer, Siemens has promoted innovation and sustainability and has achieved positive environmental outcomes.

Finally, it has been of importance to Siemens that ISO 14001 has been implemented alongside other management schemes and that the companies are expected to integrate them. This should be seen as part of a wideranging strategy to develop and respond to new horizontal labour relations, business ethics and more local autonomy combined with increased interdependency.

7.8   Summary of the Case Studies

7.8.1 Innovation

Outcomes that are innovative can be grouped into three categories: innovation in *products*, innovation in *production processes* and innovation in *organisational structure and management.*

On product innovation, a modest number of changes have been identified in the case studies as a result of the implementation of an EMS. However, most of the companies (except the multinationals) have been working within an EMS framework for just a few months. This short period has not given them the chance to invest in product innovations. The emphasis has been on production line changes on the one hand and on organisational reform on the other. Improvements in environmental performance can be linked to a reduction in the level of air emissions and the amount of solid waste. Only companies with a longer experience of environmental management systems (since the middle of the 1990s) have made any changes to their products. For example, one of them had developed a new product that was 100 per cent recycled. Another replaced the plastic parts of products by environmentally friendly substitutes.

In looking at the production process, only one case can be identified in which the implementation of an EMS brought changes to the production line. It also led to a new product. This company was also helped in its EMS implementation through co-operation with a major foreign company for the import of raw materials for the new product.

SMEs secured direct and positive innovative outcomes following implementation of an EMS. This shows that the introduction of an EMS can improve company profits through changes in the production process. Moreover, SMEs created new production lines in the course of EMS implementation.

As a result of implementing EMAS, some companies were able to replace the use of oil by gas or were able to bring in more up-to-date technology such as scrubbers or floating roof systems. Furthermore, in some of the case studies, significant improvements in the technology of waste management were introduced, such as a biological cleaning agent, whilst in other companies new investment and innovation took a different form but with the same outcome of reducing waste. These initiatives resulted in changes in the production line. For example, the biologically treated water and its reuse in the production process had an impact on the whole of the company's production line.

SMEs in particular increased their current knowledge about available technology following the implementation of an EMS. By way of contrast, the international companies were already familiar with and using technological initiatives and innovation techniques.

Another important achievement through implementation of an EMS was an improvement in the technical division of labour on the site and the allocation of responsibilities to different departments and to individual employees. EMS implementation contributed, especially in the SMEs, to changes in the organisational structure of the company. The distribution of responsibilities between the employees became clearer, compliance with the health and safety requirements was enhanced and the use of machinery became more efficient. This was usually accompanied by an obligation on employees to prepare regular reports on their activities for the head of the department or for the environmental manager.

Some companies introduced computer software packages to achieve better control of the production chain. By these means, they were able to detect organisational problems and make the necessary adjustments. At the same time, the company's management was better able to control employee productivity at every stage of the production process. Big multinational firms were keen to include an EMS in other management schemes, such as TQM, risk management or health and safety. In small and medium sized companies efforts along these lines were limited but further integration was planned for the future.

In all the case studies, except the Municipality of Athens, efforts had been made to develop strong links between the implementation of EMAS and health and safety.

In almost all the case studies ISO 9001 or ISO 9002 had been introduced before EMAS, so a major innovation in organisational structure had already been put in place. The implementation of quality management systems under ISO had already prepared companies for more flexible organisational arrangements. This led on to increased awareness by owners and managers of environmental management schemes. Moreover, the employees in these companies were at an advantage, because after the implementation of an ISO system, they had gained some experience in the implementation of modern organisational arrangements and were thus more positive and active in responding to environmental schemes.

One of the key features of the management reforms was the plan to use the implementation of an EMS as a modern eco-marketing tool in the future. The publication of the environmental statement and other activities for promoting the environmental profile of the company were seen as a useful way of increasing sales and enhancing the company profile in the eyes of customers and government, both central and local.

7.8.2 Sustainability

In analysing sustainability, three main motives can be identified for taking EMAS on board in the seven case studies.

The first motive was the wish to ease the problem of major or minor environmental conflicts with interests in the surrounding area through implementing an EMS. The companies that faced conflicts with the local authority or, less likely, NGOs, were looking for new ways to resolve them. An EMS was seen as one means of overcoming these difficulties. There were also a few companies that had not faced environmental conflicts because they had adopted environmental policies early on in the 1980s and had anticipated possible reactions. For example, Elais, a company located in an urban area, had introduced its environmental policy at the end of the 1980s. An international company – Siemens – had, since the early 1990s, followed the international strategy of its parent company for environmental management schemes and was able to respond to local government concerns about the level of water abstraction on one site.

The second motive, linked to the first, is the anticipated decrease in the level of pollution. This motive reflected concern for complying with both national and European law. Given the lack and/or the inefficiency of public control mechanisms in Greece, the companies took the view that EMAS would enhance the level of knowledge about pollutants. Subsequently, this argument was used as a marketing and/or public relations device.

The third motive was the desire to make savings, and through this, to deliver financial benefit to the company. Some companies became interested in promoting the successful implementation of EMAS through an environmental statement or through other activities such as a website or workshops.

Within these broad motives, specific actions to achieve sustainability can be identified. A decrease in the level of waste (liquid and solid) was a first priority for the companies. The high cost of waste storage, disposal and management prompted companies to seek new ways to solve the problem. Moreover, where the waste was particularly dangerous, the companies had to handle it very cautiously, so they invested heavily to solve the problem. Most of them had planned to make these kinds of investment in the future, but the implementation of an EMS brought their plans forward and led them to take initiatives to organise their investments through the whole of the production process. Moreover, in some companies the implementation of an EMS led them to adopt very innovative and pioneering methods to bring about a decrease in the amount of waste.

It has already been noted that the state's mechanisms for monitoring and controlling pollution emissions were weak in Greece. Despite this weakness, some companies were keen, through complying with the law, to ensure their long term viability and to achieve beneficial results, such as promoting their credibility and good reputation in both Greece and the European Union. To achieve these outcomes, companies had developed systematic environmental principles and strategies, which were built into the implementation of EMSs. One of the main aims was the reduction of polluting emissions. As a result of implementing EMSs, companies had achieved an impressive 30-60 per cent decrease in pollutants. Each company identified the most significant pollutants in order to make investments for limiting them. For example, some of the companies focused on air emissions, VOCs, nitrogen oxide, CO_2, carbon monoxide and suspended particulates, and had introduced innovative techniques like floating roof systems. Again, some of the companies produced a significant quantity of water pollutants. Through the introduction of high technology biological cleaning agents – or the improvement of old techniques by newer ones – they succeeded not only in reducing the level of pollution but also managed to save water as the treated water was used again in the production process. On the solid waste front, companies were investing in new techniques and better management and were also promoting reuse and recycling.

Reduction in water consumption was one of the most successful results of EMS implementation, not least because of the scarcity of water in Greece. Better water management was introduced by means of detailed monitoring of consumption in each department and by reusing much of the treated water, for cleaning equipment like barrels and bottles. Recycling was introduced for paper/cardboard, plastics and metal objects. Some of the companies sold their plastics to other enterprises for use in a new product.

Another significant saving following the implementation of an EMS, was reduction in electricity use. Companies sought new ways of reducing consumption by changing old systems for new ones with lower voltages or, in some cases, by using gas instead of electricity. On raw materials savings, plans were laid to reduce the level of consumption in the future. Although this was one of the most important objectives, most of the companies found that more time was required to be effective. The multinational companies had already been reducing the use of raw materials in packaging and had been asking their suppliers to use recyclable materials. This was part of the strategy of using environmental criteria for choosing suppliers.

In Greece, the impact of EMSs on the whole of the market chain was relatively limited. Most of the companies planned to introduce environmental management criteria for choosing suppliers in the future. As already noted, international companies had already done this to some extent by, for example, asking suppliers to put in place an environmental policy, and by auditing suppliers against elementary environmental management standards.[15] However, the use of environmental criteria for choosing customers was very difficult. In only one case and then only partially, one company tried to choose customers, which had implemented environmental management systems (EMAS or ISO 14001). The multinational companies preferred to support NGO activities (for instance those of the WWF) and to enhance customers' awareness indirectly. Nearly all the companies believed that it was practically impossible to use environmental criteria for choosing customers. It was unprofitable and the market in Greece was still immature in environmental terms.

To encourage employee support, especially in the early stages of implementation, the environmental management systems were combined with health and safety systems. Only after implementation and after taking part in seminars did they understand and respond positively to the real benefits of an environmental management system.

To increase the environmental awareness of external interests such as local government, local residents and universities, the multinational companies had already developed serious strategies. They organised workshops and collaborated with universities or governmental organisations in order to promote their environmental activities. This included companies which were in conflict with local interests because of their environmental impacts. SMEs and companies that did not face these kinds of problems were more "low profile". They maintained that their EMS was mainly a tool for internal improvements, so it was not necessary to promote it in the public arena.

The companies' efforts were concentrated on specific environmental issues in order to address the most important sustainability criteria (savings, recycling, pollution reduction). It was unlikely that companies would be able to achieve all their environmental objectives, so they concentrated on just some of them. However, the key issue is that they had plans laid down for future environmental developments. They had put in place long term environmental strategic plans, whilst at the same time they regularly updated their environmental plans to achieve continuous improvement. Both multinationals and SMEs need to recognise the importance of sticking to their EMS procedures and to realise that tools like EMSs need time to make an impact. It is not only the short term small scale cost benefits that count, but also the increase in environmental awareness and self-learning, together with a long term environmental strategy.

Finally, on the social dimension of EMS implementation, EMSs were more likely to be found in multinational companies and/or in companies which were located in or adjacent to urban areas (and subject to disputes with local interests). For other companies, the social dimension was expressed indirectly, through general principles of environmental protection.

---

[15] It should be noted that these companies carry out on a regular basis detailed "spot tests" at suppliers sites. They use their own auditors and, where necessary to achieve high standards, they demand improvements in practice.

7.8.3 Participation: The Role of Different Holders

In some case studies good co-operation between all or some of the "holders" was identified, leading to better environmental and social practices.[16] In other case studies there were conflicts between the different "holders". It was very clear that some categories of "holders" were more proactive and prepared to co-operate than others and, moreover, there were obvious differences in the level of participation by "holders" in every case study.

The *owners or general managers* of the sites – usually in co-operation with an external consultant – were in all cases those who decided to introduce an environmental management system. The owners believed that the opportunity to implement an EMS in collaboration with the Ministry of Development, which subsidised the early applications, was of great importance. This state initiative constituted a major incentive for the companies to develop an environmental management system without paying for the whole procedure.

However, the subsidy was just enough to cover the fees of the external consultant and the verifier, so the company had to pay the cost of the investment from its own funds. This could constitute a serious problem because sometimes the owner decided to implement an environmental management system without exactly knowing in advance, how much it was going to cost the company. The result, in some cases, was a significant delay – or cancellation – of the project by the company. Some of the interviewees pointed out that they had not been adequately informed and had to pay more than they thought they would. However, most of them believed that such an investment was necessary to enhance exports or the sustainability of the company. The majority of them believed that the environmental management system was the key for the company's future development in relation to exports, European or national subsidies, customers and marketing.

In some cases, the investment to implement an EMS had already been planned by the company and formed part of its strategic plans. The implementation of an EMS itself helped the company to organise its investments more systematically.

The owner or the general manager was the main person responsible for the EMS. They were the ones who took the final decision to undertake any production changes or to make new investments. However, the roles of the external consultant and of the company's environmental manager were also important. The owner or the general manager tended to have close and systematic co-operation with the external consultant during all phases of implementation, even after the full introduction of the system. It was usual for the owner/general manager and the consultant together to choose the verifier for the company.

Owners or general managers of companies that had problems with the local authority or other interests as a result of sea or air pollution or difficulties with waste suggested that the implementation of an environmental management system could be a good way of overcoming any difficulties.

Moreover, they underlined the attractiveness of national or European inducements for companies that have been verified/certified for environmental

---

[16] For the "holder" concept and the distinction between "holders", see section 2.4.

management systems, such as economic incentives (tax relief subsidies) or other kinds of encouragement, such as more information from seminars and workshops and more export opportunities.

The role of the *environmental manager* in the implementation of an EMS was very important in the Greek companies. In every case the environmental manager was also the person who was responsible for the implementation of an EMS. Consequently, this key individual was more aware of the environmental activities of the company than anyone else. He/she is likely to have studied a relevant science (chemical engineering, environmental studies) and is also likely to have been the main link between the owner, the general manager and the employees. Most larger companies delegated the responsibility for an EMS to the health and safety department. Furthermore, where the company has achieved a quality standard (ISO 9000), the environmental manager is likely to have greater experience and consequently could more easily understand the EMS procedure. It would be the environmental managers together with the external consultant, who would organise seminars and training courses for employees.

In all the case studies, the environmental managers were convinced that co-operation with the external consultant was necessary and without it, the environmental management system could not have been implemented. It was also the case that co-operation with the external consultant did not stop with the formal completion of the verification process. On the contrary, the consultant usually continued to co-operate with the company on a part time base. Companies usually preferred to co-operate with their external consultant following implementation rather than hire another individual for further development of the environmental management system.

The environmental manager of the company, together with the production manager, therefore typically is responsible for the implementation of an EMS. They usually created and chaired an environmental committee, which took on the role of promoting the environmental policy of the company and improving the company's image to external interests (local government, media, NGOs). This committee operated mainly in situations where there were significant local problems or where a company had a long environmental tradition.[17]

The environmental manager supervised all the stages of implementation and reported to the owner or senior manager on progress. He/she would also work with the heads of the departments of production and of research and development,[18] promoting new technologies for continuing improvement of the company's operations. Moreover, it was the environmental manager who took the lead in the allocation of "new" environmental responsibilities among workers and managers.

---

[17] For example, Elais had been involved in environmental matters since 1987, publishing environmental brochures and organising environmental workshops. Elais set up its environmental committee in 1988.

[18] The large companies in Greece were keen to develop R&D departments in order to have the control of new technologies and innovation techniques, adapted specifically to their production systems.

They also had to undertake regular internal audits, publish the company's environmental handbook, keep a record of legislation and a register of environmental effects, which needed to be updated regularly. Another key responsibility was regular briefings to the general manager. It was the environmental manager who would propose changes in the company's environmental policy or suggest new investments to comply with the law and/or to improve the production system.

Because of this wide range of responsibilities, it was very difficult for the environmental manager to have total control over the environmental management system. Particularly when there were problems of production or overproduction, he/ she might be unable to keep good track of the EMS. Consequently, the support of the external consultant was crucial. Some companies used an extra employee for the EMS (an *environmental management assistant*). Some of the interviewees commented that they could do more for their EMS if they had had more time and an opportunity to set up seminars. They also stated that the implementation of an EMS was not a simple procedure that every company would be able to achieve without assistance, and the support from the state was not adequate.

The role of *employees* in the EMAS implementation process was not very great in the early stages. Implementation of a modern management system in a company was a new, different and unknown procedure, which led to increased responsibilities, a need for training and, most important, more detailed attention to the nature of the jobs that had to be done. For example, it was less trouble for employees to allow liquid waste to spill into the drains than to store it in tanks and then use the biological cleaning agent. This had been the attitude over the years.

There was no reason to expect that employees would have an environmental (green) understanding of indispensability of environmental management to industry, and they were unlikely to have high environmental awareness in their private lives. In addition, there were limited opportunities for employees to understand the whole process and the significance of the system, as they had not participated through trade unions or other kinds of representatives in the decision-making on the implementation of environmental management systems. They were informed by senior management about the decision as it was being implemented. In two of the case studies, fears were reported of a more demanding workload and of closer control over productivity by managers.

Although employees were reluctant at the start, their attitude became positive with the implementation of the system. Both the seminars during the implementation phase and the process as a whole (which included a clearer distribution of responsibilities and more systematic safety arrangements) helped them to understand the need for an EMS for purposes of environmental protection and for their own safety. To secure this understanding, the role of the environmental manager was essential, as he/she was in a position to persuade employees that the process was necessary because of their trust in him/her and to a lesser extent in the external consultant. Moreover, employees were more positive where the company had already implemented a quality standard like ISO 9000, because it was not the first time that they had participated in a new system which had had a significant impact on their job.

An EMS in a company, especially in companies which had a long environmental tradition (like Siemens or Elais), had also increased the environmental awareness of employees in their private lives.

In Greece, the *verifiers* had been accredited abroad, were usually foreign nationals rather than Greeks and were employed by verification agencies, like TÜV, Gerling and Bureau Veritas Quality International. Once the company has completed the whole EMS process, it contacts a verification agency, which appoints an expert in the relevant sector (food and beverage, chemicals etc) to assess the implementation.

The verification agency sends a letter to the Greek Accreditation Council (ESYD) informing it of the proposed visit, the name of the company and the name of the verifier. The verification agency also pays a fee to the ESYD (about 100,000 drachmas). The ESYD confirms receipt of the letter and sometimes appoints one of its own auditors to attend the verification process.

The verifiers' role is to assess the general objectives the company has followed. They check the company's compliance with the law, the new environmental procedures and the environmental statement of the company. The most important part of the verification is the thorough check of the contents of the company's environmental statement and specifically the level of employee training provided.

According to the verifiers interviewed, Greek companies have been more interested in implementing ISO 14001 than EMAS. This tendency may be explained by the perception that the Greek public was not ready to take at face value the contents of the published environmental statement. Owners of companies thought that publication could cause problems for the company from local government or the media.

Verifiers believed that Greek companies had not yet understood the main point of the implementation of an EMS. There was a common view that implementation of an EMS could be financed by a state subsidy. It was not fully realised that companies had to invest and spend more money than was provided in the subsidy. This distorted view at times caused problems in taking an EMS forward by putting it off into the future or by simply halting the procedure. This was a significant issue for those companies that were not used to long term planning for their investments.

The Greek case studies demonstrated the importance of *local authorities* in the decision by companies to implement EMSs. Companies perceived EMSs as an instrument to solve conflicts with the neighbourhood and the local state. But an integrated approach was also helpful for identifying not only major conflicts but also minor disagreements between companies and various local interests.

In one case study, there had been a major conflict between the company and the local authority. The latter has taken the company to court for pollution beyond the site. The company had to stop its planned expansion until after the trial. At the same time, the company decided to adopt an environmental policy and to implement EMAS, hoping that through these measures they could improve relationships. It set up an environmental committee on which a representative from the local authority sat. However, this did not persuade the local authority to withdraw the court action.

In some of the case studies, minor conflicts between the neighbourhood and the company were identified, for example, the traffic jams caused by company vehicles (no parking places), or waste disposal problems (failure by the local authority to collect on a regular basis the company's waste). In most cases the companies resolved these problems directly by paying for private services, such as renting land for parking and contracting with private refuse collection firms).

Elsewhere, companies did not face big problems with local government as they had implemented environmentally friendly schemes for the benefit of the neighbourhood. This was particularly true of multinational firms with a long tradition in environmental policy.

Enterprises that did face major conflicts with local interests sought help by collaborating with *knowledge holders* like universities, laboratories or research institutes. In one case, the owner used on a relatively large scale university research centres for measuring and evaluating the environmental impacts of the site. The environmental manager for this company was planning to include parts of the university report into the environmental statement. Furthermore, through the publication of environmental statements, they tried to promote their environmental policies to the public, using a website, local and national newspapers.

Most companies, especially SMEs, used specialist laboratories and institutes for measuring levels of pollution. The multinational and larger companies had their own laboratories for the standard measurements and used external laboratories for special tests. Furthermore, these companies had developed significant links with universities for promoting their activities. Some of them participated in the Greek network (EMSnet) set up to promote the implementation of EMSs. Members of this network included universities, institutes, the union of municipalities and consultancies.

7.8.4 Reflections on Participation

As emphasised earlier, the owner or the general manager took the decision to implement EMS. In all the case studies, they were helped in their decision by an environmental manager and sometimes by an external consultant.

It is clear that power to take this decision lies at the top. The question is: under what circumstances was the decision taken? In some cases, the owners decided to implement an EMS after conflict with the local authority, such as being taken to court. Thus a local factor contributed – to a certain extent – to the implementation decision. Although it started from conflict, the outcome was co-operation between the local authority and the company in a new kind of participation though the establishment of an environmental committee with representatives on it from the local authority and the company. In other case studies, the company had to follow the parent company's strategy with no room for local discretion.

However, once the decision it taken, it is important to address the role of "holders" during the implementation process and note the limited role of the owner during these later stages. The environmental manager (in some cases in collaboration with a team) takes responsibility for the whole process usually with the help of an external consultant. Their role is significant because they analyse the company's problems, its needs and priorities and organise the overall planning of the investment needed. They have to co-ordinate between the various departments of the company and the training of staff.

The latter play a minimal role in the overall process. They have to accept a decision from the top without being involved in any discussions about it. This lack of participation does not refer only to the initial decision-making process for implementing an EMS but also to the phase of implementation. Participation was limited to seminars and specific training courses about their new responsibilities. They were not introduced to the philosophy of the system and sometimes they confused the EMS with health and safety standards. In only a few cases did they play a part on a committee – separate from the EMS – which could reflect their views on environmental improvement issues. However, in most of the case studies, after the early difficulties, the reactions of employees were either neutral or positive.

Although the participation of employees was very restricted, they did nevertheless get the chance to participate in everyday decisions about the production process and their working environment through the implementation of EMAS or through ISO 14001. In all of these cases the new EMS had been a part of a corporate strategy to change the company's organisational structure to bring about more decentralised and more flexible responsibilities so that more use could be made of the potential of new technologies. But on the other hand, these decentralised responsibilities – which implied more autonomy for employees – had themselves to be controlled, so decentralised operations based on new knowledge and approaches had to be systematically recorded. New management systems – including EMSs – are tools for this kind of control and recording.

In multinational companies or companies exposed to local conflict the employees accepted an EMS more easily. In the former, the employees had had similar experiences with other management systems so it was easier for them to understand what was happening. One negative outcome was increased and sustained bureaucracy. Every management system took a different form and it was difficult to comply with all of them. However, where the companies had had conflicts with local interests, the employees accepted the system easily. They realised that an environmental system would support the viability of the company and increase job security.

Last but not least, during the implementation of an EMS, the companies began to collaborate with *knowledge holders*, such as universities and laboratories or through the EMSnet. The EMS process compelled the companies to be more "open" to society at large and to understand the need for co-operation with different "holders" in creating new networks. For the multinational companies, co-operation with interested "holders" was already an essential strategic principle prior to the arrival of an EMS.

7.8.5 Conclusions

In presenting the seven case studies, we concentrated on innovation, sustainability and participation. Every study demonstrated the link between these three issues, and the influence of implementing an EMS on participation was analysed. An attempt was therefore made to identify the positive and negative consequences for governance of implementing EMAS and ISO 14001. In particular, the analysis is intended to identify those particular elements which confirm or deny the hypothesis that EMAS implementation is a vehicle for promoting participation.

7.8.5.1 Positive Consequences   One of the key conclusions from the Greek case studies is that the implementation of an EMS does lead to an increase in sustainability and innovation accompanied by a modest or low level of participation. This low level is a result of hierarchical decision-making in the company, whereby decisions are handed down on implementing an EMAS without any real participation on the part of employees.

In some case studies, the decision to implement an EMS was based on the international strategy of a multinational parent company. Such companies already had integrated global strategies for environmental policies and compelled subsidiaries in different countries to follow a common line. This meant that the EMS was embedded within a broader environmental policy and it did not lead to important changes in the company in respect of innovation, sustainability or participation.

On the contrary the implementation of an EMS in other kinds of companies, such as SMEs, did have an important impact on innovative and sustainability and to some degree on participation. The lack of an existing systematic environmental policy in these cases circumstances provided an opportunity through EMS procedures to enhance innovation and sustainability. Employee environmental awareness was increased by the implementation of an EMS and in part was brought about through participation, even if at a low level.

The implementation of an EMS has promoted improvements in product lines and has helped SMEs to expand their capability for better evaluating their existing technology. Moreover, EMSs have improved "good housekeeping" by companies and have led to the reorganisations of the company's management, particularly by assigning precise responsibilities for specific operations at every level in the company.

In all the companies, one of the most important contributions was the systematic documentation that an EMS requires for a company's internal environmental reporting. This is a significant tool for measuring environmental impacts (levels of pollution, savings), which are in turn linked to sustainability. Sustainability indicators are used to measure changes, such as the lowering of the amount of liquid waste through the use of a biological cleaning agent. Large or multinational companies monitor their innovations through their R&D departments which are independent from EMS procedures. All these procedures are directly linked with companies' attempts to integrate their management systems. Most of the companies have no integrated system so there are parallel and overlapping procedures, resulting in a new internal bureaucracy. Multinational companies that have succeeded in integrating their management systems are in a position to use environmental and social indicators, but elsewhere it is very difficult to measure social impacts through an EMS. Companies can come to a judgement on the social implications of implementing an EMS mainly through noting the reactions of local interests.

Companies that try to integrate different management systems (health and safety policies, TQM, risk management, environmental management systems, integrated pollution prevention and control (IPPC) etc.) seem to demonstrate stronger links between sustainability, innovation and participation. This is a result of new and flatter management arrangements, more decentralised responsibilities plus greater interdependence, flexible specialisation and new business ethics.

As already noted, the implementation of an EMS can also be the result of severe environmental conflict between local interests and the company. In these circumstances, the company is compelled to adopt an EMS as a means of reducing local hostility. Local authorities can therefore play an important indirect role in implementing EMSs in the private sector.

7.8.5.2 Negative Consequences   As opposed to the positive consequences, one of the key features of the case studies in Greece was the problem of using an EMS as a mechanism to address major environmental conflicts. An EMS could prove to be problematic in some situations, simply because a company could promote it as a way of solving environmental conflicts. An EMS could function as a device for manipulating the public and the government, for instance, by demanding lower levels of scrutiny as a result of achieving EMS status.

It is selfevident that compliance with the law applies to all enterprises. Consequently the simple fact of legal compliance is not a step forward. Compliance with the law through the promotion of EMAS should not be used just as a public relations tool. In these circumstances, the EMS is not an environmental policy instrument and in these kinds of situations, the role of the state is crucial.

The same could also be said also for the local authority and other local interests. They could broaden participatory procedures, not only for the implementation of an EMS, but also as part of a local sustainability strategy, working with the main "holders" in both the private and public sectors.

However, the employees of the companies had not had an opportunity to understand the whole process or to be involved in the implementation process. Even in multinational companies where the employees had had experience of similar systems such as ISO 9001 or risk management, the opportunities to participate were limited, even if available through their trade unions. Nevertheless, in these circumstances where the company had significant environmental problems and difficulties with local interests, staff of the company were more prepared to accept the EMS in order to avoid the possibility of losing their jobs as a result of a company's closure.

At the same time, employees were also concerned that the EMS procedure could control their productivity at the individual level and lead to internal competition within the workforce. The need for regular detailed reports from individual workers was seen to be a new form of control in the company. It also increased bureaucracy and got in the way of solving everyday problems, since it undermined flexible decision-making.

Chapter 8

EMAS in Germany

Hubert Heinelt, Tanja Malek and Annette Elisabeth Töller

8.1 History of Environmental Policy in Germany

In line with developments in the European Community, Germany saw in the early 1970s the beginnings of a systematic environmental policy. The federal social democrat/liberal coalition government (1969–1982) laid the foundations for an environment policy in only a few years. The first environmental programme began in 1971, a committee of environmental experts was established in 1972 and the federal environment agency came into being in 1974 (see Weidner 1997). There was, however, no environment ministry at the federal level until the mid 1980s.[1]

Prompted by international developments and internal calls for reform, these early environmental policy initiatives occurred without any major public or party political disputes. They were developed in the exclusive circles of an "environmental technocratic coalition for reform" (Weidner 1997, 2) and permanently changed the face of environmental policy of the Federal Republic.[2] The dominance of scientific and technical expertise, hand in hand with the euphoric planning climate of the early 1970s, produced a style of regulation in environment policy which can be characterised as "end of pipe" and emission oriented. Environmental policy instruments were in the main based on the "command and control" approach (see Weidner 1991 and 1997).

The key enactments, including allocation of responsibility at the federal level, were laid down right at the beginning of the development of the policy. In 1972, under the 30th constitutional amendment, responsibility for clean air, noise abatement and waste disposal was assigned to the Länder level, unless legislation at the federal level took precedence (Müller-Brandeck-Bocquet 1995; Hartkopf/Bohne 1983).

In addition to this legal authority under Article 74 of the German constitution, the federal state was also responsible for framework legislation (Article 75 of the constitution) in the water management, land cultivation and nature conservation sectors. However, the Länder were not completely excluded, as the federal framework legislation had to be supplemented by Länder laws.

---

[1] Environment policy tasks were carried out by the Ministry of the Interior before the establishment of the federal Environment Ministry in 1985.

[2] Examples of early legislation during the social democrat/liberal coalition government include pollution control law (1974), the Refuse Disposal Act (1972) and the law on leaded petrol (1971).

In spite of this early centralisation of environmental policy,[3] the Länder have a number of ways by which they can shape it. For instance, they impact on the process of federal legislation where they have co-decision rights in the Bundesrat.[4] Under Article 83 of the constitution, they are responsible for the implementation of environmental protection policies. There are differences between the Länder on the extent to which these implementation responsibilities are delegated to local authorities.[5]

As a result of the assignment of key environmental policy legislative competence to the federal level and the dominance of the Länder for purposes of implementation, there are great pressures for co-ordination and co-operation. Thus, in 1973 a standing conference of federal and Länder environment ministers was established, at which issues of both legislation and implementation are covered. This conference meets twice a year.[6] Although this level of co-operation in Germany did sometimes lead to environmental protection initiatives, nevertheless there can be seen in the environment sector the same problematic consequences of policy integration in German federalism as can be seen in other sectors, the "joint decision trap". These problems include a lack of transparency in policymaking within the federal system, the vast amount of time it takes to reach a compromise and, above all, the decreasing role of the parliaments at both the federal and Länder levels (see Scharpf 1988).

In addition to vertical integration between the political levels in the environmental policy sector, horizontal integration between private and state interests also developed, particularly in relation to the provision of guidance on the application of the law. In the process of standardisation, technical regulations are often adopted by non-governmental organisations such as the German Standardisation Institute (DIN) or the Association of German Engineers (VDI). This is because the technical aspects of the development of environment policy legislation require the mobilisation of scientific and technical expertise to develop practical measures (see Müller-Brandeck-Bocquet 1995, 137; Voelzkow 1996).

This phase of institutional and legislative developments in environmental policy was followed by a period of "abstinence", as a result of the widespread economic recession in the mid 1970s. However, environmental policy gradually began to be

---

[3] Centralisation of environmental policy legislation at the federal level was clearly in the interests of the Länder as it prevented underbidding in the environmental policy sector being used as a means of regional economic promotion and at the same time the Länder were able to "pass the buck" on unpopular decisions to the federal level.

[4] After some defensiveness on the part of the Länder in environmental policy in the 1970s, since the early 1980s there have been clear initiatives from the Bundesrat to boost environmental policy, for example, on regulations converting gas and oil fired power plants.

[5] Local authorities also address environmental policy issues of local interest autonomously as part of their constitutional right to self-government (see Hucke 1998).

[6] This co-operation between the federal state and the Länder in the form of the environment ministers' conference has been supplemented by a heads of administration conference established in 1986. It comprises the heads of administration from the Länder environment ministries as well as members of various federal/Länder working groups specialising in environmental policy issues.

consolidated in the early 1980s towards the end of the social-democrat/liberal coalition government. The new dynamic in environmental policy was not halted by the election of a conservative/liberal coalition in 1982. Indeed, during the course of the 1980s a number of key environmental policy breakthroughs were achieved, which contributed to Germany's leading international position on environmental matters (see Weidner 1991, 137 ff).

One important reason for this new dynamic was the growing politicisation of environmental policy issues. Environmental policy was gradually attracting the attention of a broader public beyond specialist circles of experts. At the end of the 1970s a number of anti-nuclear power local pressure groups were created. This developed into a broadly based eco-social movement and in 1980 the Green Party was formed. Now over four million people are members of nature conservation and environmental protection organisations (see Hey/Brendle 1994, 93 ff).

Following the electoral success of Green parties at the local and Länder levels at the end of the 1970s, the Greens were first represented in the German Bundestag in 1982. The established parties responded to this growing environmental awareness by the general public by formulating programmes which for the first time addressed environmental problems and possible political solutions. Against the background of a widespread and highly emotional public debate about the dying of the forests, Germany prioritised air pollution policy and played a pioneering role at European level as well (see Weidner 1991, 138).[7]

As a direct result of the catastrophe at the Chernobyl nuclear power plant, the federal Ministry for the Environment, Nature Conservation and Nuclear Safety (BMU) was set up in April 1986. A number of agencies were attached to it, such as the Environment Agency, the Nature Conservation Agency and the Nuclear Safety Agency, as well as two policy advisory bodies. At the same time a Standing Committee on the Environment (Sachverständigenrat für Umweltfragen/SRU) was set up in the Bundestag. The initial priorities of the BMU focused on radiation protection.

Policy instruments on environmental issues in the 1980s reflected the traditional style of ever more detailed legal intervention. Notions of integrating economic and more flexible instruments into German environment policy thinking did not get beyond the status of symbolic policy (see SRU 1994). Indeed, since the early 1990s there has been a return to these traditional ways of thinking, the essence of which is the belief that environmental protection damages the Germany economy, despite the fact that environmental protection was incorporated into the German constitution as one of the state's major objectives in 1994.[8] Against the background of an intense

---

[7] This included tightening up the clean air regulations (1983/1986), the regulations controlling coal, gas and oil fired power plants (1983) and the revision of the pollution control law (1984/1986). The clean air regulations were passed by the German federal government and, subject to Bundesrat approval, have to be taken into account in granting licences to operate.

[8] See the critique by the committee of environmental experts (SRU 1994). There is disagreement on what effects the incorporation of environmental protection objectives into the constitution will have on environmental policy in practice. Whereas some expect a strong boost for further legal measures, others view it solely as a declaration which will have effects neither on environmental law nor on environmental practice (see Schink 1996, 357 ff).

debate on the future of the German economy, both the administrative procedures under environment law and the extremely detailed legislation on the environment have been subject to increasing criticism.[9] Fuelled by criticism from major economic interests, the current debate is focussed more on the efficiency and effectiveness of traditional instruments and the advantages of mechanisms of self-regulation (see Jänicke/Weidner 1997).

Despite ideas about moving from a predominantly reactive environment policy towards more "perpetrator" oriented, flexible measures, such as those in liability law, nevertheless, a policy approach which favours reactive measures and the use of "problem postponing disposal techniques" (Weidner 1991, 151; Malunat 1994) still appears to dominate. Initiatives from the European Community level, for example, environmental impact assessment legislation or free access to environmental data, leading to a change in the existing style of environmental policy to promote greater participation, transparency and self-regulation, have in general been very poorly received in Germany (see Breuer 1997).

In addition, Germany has for a long time been criticised for being one of the few European member states without a national sustainability policy (SRU 2000, 21, 111). The government, which took over after general elections in September 1998 and which, for the first time, has a Minister of the Environment from the Green Party, has focused on environmental tax reforms and on the termination of nuclear power generation. Furthermore, an embryonic sustainability policy is on the agenda (SRU 2000, 21). The idea of a consolidated environmental act which the 1998 government took over from its predecessor has not been successful (SRU 2000, 24). All in all, in a period of decreasing environmental awareness, the 1998 government has set itself quite ambitious targets, some of which have been achieved. It has been successful through application of a mixture of traditional command and control and more recent co-operative approaches. It is, however, too early to draw conclusions on whether there have been fundamental changes in German environmental policy.

8.2   Implementing EMAS I at National Level

8.2.1 Competing Philosophies

The EMAS I regulation required member states to have the systems for the accreditation and supervision of verifiers and for the registration of verified EMAS sites up and running by April 1995 at the latest. The German system, however, missed this deadline, because the bill[10] implementing the EMAS Regulation

---

[9] In 1994 the Schlichter Commission was set up by the federal government as an independent body of experts to develop proposals for accelerating the approval procedures (see Bundesministerium für Wirtschaft/BMWi 1994; Schlichter 1995 and Lübbe-Wolff 1995).

[10] The regulations for accreditation and registration had to be passed as a law because they affected the Grundgesetz (the German constitution). Accreditation affects the basic right to engage in a profession (Article 12 GG) and registration touches the basic right to engage in a business (Article 14 GG) (see Sellner/Schnutenhaus 1993).

(Umweltauditgesetz) was not passed until December 1995. The reason for this long delay was a major dispute between business interests on the one hand and environmental interests on the other, including government ministries. The relative flexibility of the European regulation about which kinds of bodies could be established for accreditation and registration was seen as a means for member states to find solutions which best fitted with their national patterns of administration and interest intermediation. In Germany, however, it was this flexibility, in addition to the fact that there had not been a German standard on environmental management schemes and the fact that a voluntary and integrated approach was relatively new for regulation of the environment in Germany, that led to a major dispute over implementation.

There were two main competing positions, proposing different models for making EMAS work. They reflected two different perceptions of and interests in the scheme and the core of the dispute was the extent of state involvement in the implementation of the scheme.

One position, proposed by the German Federal Industry Association (Bundesverband Deutsche Industrie/BDI) and supported by the Federal Ministry of Economic Affairs reflected the perception of EMAS as a task of self-regulation of the private sector. By placing accreditation and supervision of the verifiers and registration of sites close to the business community, they argued that this would be a strong incentive for participation by companies. A system involving major state interests would have been perceived as an additional regulatory instrument interfering with companies' activities and this would have hampered participation in the scheme. Therefore, in August 1993 they suggested a model that placed the examination of verifiers with the Accreditation Organisation (Trägergemeinschaft für Akkreditierung/TGA) Ltd. This body had been founded and maintained by business organisations. It had been working on issues of quality management, co-ordinating accreditation bodies, but had not carried out accreditation. A sub-unit of the TGA was suggested which would specialise in environmental accreditation and would examine verifiers, whilst accreditation itself (Gutachterbestellung) should be the responsibility of the Federal Agency for the Testing of Materials (Bundesanstalt für Materialforschung und Prüfung/BAM). Even though it was a governmental body, it had carried out certain accreditation functions. It was also suggested that BAM should be the registration body (see Waskow 1994, 70).

The other position, proposed by the Federal Ministry of the Environment and supported by environmental groups and trade unions, saw the administrative tasks connected with EMAS as part of environmental regulation and thus wanted to locate them in the realm of government. They argued that a system which placed accreditation close to the private sector would endanger credibility, based on the assumption that accreditation and supervision would then be less strict. At an early stage in the debate, the federal Ministry of the Environment suggested that accreditation of verifiers should be placed with the Federal Environment Agency (Umweltbundesamt/UBA, see Schnutenhaus 1995, 12). This was also the view of the environmental groups (see Bültmann/Wätzold 2000, 19).

A first compromise, proposed in October 1993 by the federal Ministry of the Environment, suggested that the TGA should examine verifiers, as proposed by the Industry Association, whereas accreditation itself should be undertaken by the Environment Agency. A committee, including all relevant stakeholders, should be created to support the Agency, especially in relation to selecting examiners and developing guidelines for the examination (Waskow 1994, 70).

In November 1993 the BDI and the German Association for Trade and Commerce (Deutscher Industrie- und Handelstag/DIHT) proposed another model whereby, in addition to the accreditation body (TGA), a joint bureau of the Chamber of Trade and Commerce should carry out both accreditation of verifiers and registration of sites (Waskow 1994, 70).[11]

The compromise finally reached in part had its roots in a coalition arrangement agreed after the general election in November 1994. This stated that the implementation of EMAS would enhance the responsibility of the private sector. This solution meant that the Environment Agency would not be involved. In early 1995, the various parties agreed on a compromise which broadly had more elements from the models proposed by business interests, but nevertheless did provide for some governmental involvement (see below). An official draft was formulated and after considerable debate in the Bundestag (Parliament) and in the Bundesrat (Federal Chamber; see Bundestagsdrucksachen 13/1755 and 13/2004) the bill was passed following a conciliation procedure (see Bohnen 1995, 1757 and Bültmann/ Wätzold 2000, 26) in December 1995 (see BGBl 1995 I, 1591), followed by a number of ministerial decrees outlining details of implementation.

8.2.2 Core Elements of the System

There are two main organisations involved in the accreditation and supervision of verifiers. The task of actual accreditation and supervision of verifiers has been given to the DAU (the German Organisation for the Accreditation and Supervision of Environmental Verifiers) which is maintained by various business organisations (the Federal Industry Organisation/BDI, the German Chamber of Trade and Commerce/ DIHT, the Central Craft Association and the Federal Association of Free Professions, Section 28, EMAS law). At the DAU, a committee carries out the examination of verifiers. Another committee was established under the aegis of the Federal Environment Ministry. It is called the Committee for Environmental Verifiers (Umweltgutachterausschuß/UGA) and comprises representatives from various government bodies and from a range of interested parties (in our terminology, "holders"). The composition is as follows:

- six representatives from the business sector;
- four representatives from the profession of environmental verifiers;
- two representatives from the Federal Environment Ministry;
- one representative from the Federal Economics Ministry;
- four representatives from the Environment Ministries of the Länder;
- two representatives from the Economics Ministries of the Länder;
- three representatives from trade unions;
- three representatives from environmental groups.

---

[11] For more detail on the different models, see Sellner/Schnutenhaus 1993; Schnutenhaus 1995; Bültmann/ Wätzold 2000, 18-25.

The most important task for this committee is to draft guidelines on how the ministerial decrees based on the EMAS law should be implemented by the DAU. These guidelines regulate the details of the examination, accreditation and supervision of environmental verifiers. To approve such guidelines, the Committee of Environmental Verifiers (UGA) requires a two-thirds majority. In practice, however, every guideline has been adopted by a large majority (Bültmann/Wätzold 2000, 39).[12] Moreover, the UGA provides a list of examiners for the examination committee at the DAU and recommends members to sit on the appeals committee, which is linked to the Federal Environment Ministry and which has to decide on appeals against decisions by the accreditation body. Furthermore, the UGA gives advice to the Federal Environment Ministry on all issues relating to EMAS. Finally, the DAU has to present to the UGA a twice-yearly report on its accreditation and supervision work.

The EMAS legislation distinguishes between individuals as environmental verifiers, environmental verifier organisations and persons demonstrating a certain level of technical knowledge.[13] In contrast to accreditation systems in most other member states, it is the individual environmental verifier (not the organisation) who becomes accredited. He or she must have a university degree and at least three years' practical experience. Furthermore, a verifier must be "reliable, independent and skilled" (EMAS law, Sections 10, 11, 12, see also Lübbe-Wolff 1996a, 221). The theoretical and technical knowledge of the candidates is tested in an oral examination carried out by a committee, which is composed of different examiners, depending on the subjects to be examined. About 60 per cent of the applicants fail to pass the examination. Moreover, many applicants are accredited for only part of what they were seeking to achieve (Bültmann/Wätzold 2000, 43). If a candidate is successful, he or she is accredited by the DAU.[14] DAU also keeps a list of accredited verifiers. Supervision of accredited verifiers has to take place regularly, at least every three years. Verifiers have to keep all relevant documents on their verifying activities. If a failure to carry out the duties as defined in the legislation (including reliability and independence) is confirmed, the accreditation body is entitled to suspend the verifier partly or entirely from their verifying activities (Section 16, EMAS law; for a critique, see Glatzner 1998).

In the early years the main part of the work undertaken in the DAU was the accreditation of environmental verifiers. By December 2001 there were 243 accredited verifiers.[15] Over time, the focus of the DAU's activities has slowly shifted to the supervision of accredited verifiers. By the autumn of 1999, 507 inspections of verification (witness audits) had taken place.[16] No suspensions were imposed (SRU 2000, 135).

---

[12] By January 2002, seven guidelines had been adopted (see http://www. umweltgutachterausschuss.de (accessed January 2002).

[13] Such people have passed only some parts of an examination for a particular field. They are not entitled to verify a site on their own, but only in co-operation with an accredited verifier.

[14] For details of the accreditation procedure, see Bültmann/Wätzold 2000, 43.

[15] See http://www.ihk.de/dau/Umweltgua.doc.

[16] Data provided by Bültmann/Wätzold differ from these figure (2000, 43).

The task of registering EMAS sites was delegated to the Chambers of Trade and Commerce and, for craft companies, to the Craft Chambers (Section 32, EMAS law). Because the chamber system is a highly decentralised, there is a central site register kept with the central organisation, the German Association for Trade and Commerce (DIHT). Before registering a site, the competent body informs the local regulatory body for the environment, which has four weeks to inform the Chamber of any breaches in environmental law (Section 33, Paragraph 2).

All in all, the German accreditation system is a fairly atypical arrangement in the light of the traditional separation of "the state" and "society" which still shapes most governance arrangements. Because of this unusual approach, some observers (mainly lawyers) initially feared that the EMAS legislation did not conform with the German constitution (see Lübbe-Wolff 1996a, 220). However, the system has managed to enjoy wide acceptance and credibility with the various interested parties.

Whereas other member states extended the scheme to other than industrial sectors at an early stage, the German federal government did not pass a decree on the extension of EMAS to non-industrial sectors (mainly transport, restaurants and hotels, banks and insurance companies, distribution and local authorities) until the beginning of 1998. At the same time, preparations were made for the accreditation of verifiers for these sectors (see Bültmann/Wätzold 2000, 35).

8.3 Implementing EMAS at Site Level

When the EMAS regulation was adopted, it was widely assumed that German companies would be reluctant to participate (see Héritier et al. 1994, 298). In fact, the opposite was the case. German companies have clearly dominated in terms of the number of registered sites in the member states (see Table 1.1). According to the official list of registrations there were over 2,500 EMAS companies in Germany by October 2001. But the increase in the number of registered EMAS sites – shown in Table 8.1 – came to an end in 2001.

Reasons for this recent development will be considered (in section 8.3.2) after the results of a detailed survey have been presented.

8.3.1 1997/98 Survey results

What kinds of companies have engaged in EMAS and why have they done so? A survey conducted in December 1997/January 1998 covered the most important period of EMAS implementation in Germany (see Heinelt/Malek 1999 for details).

The largest number of sites were in the chemical industry which, including rubber and plastics manufacture, accounted for 22.4 per cent of the sites in the survey (21.5 per cent of the registered sites) and in the food industry, which had an 11.8 per cent share (13.6 per cent of all registered sites). In these three sectors 41 per cent of the companies had production processes which required official approval. A comparison with the total number of companies in the manufacturing sector highlights the fact that the number of chemical companies was higher than average among EMAS sites (see Table 8.2). This was also the case for the automobile, recycling, food, electricity generation and consumer electronics industries. By way of contrast, the glass, ceramics, steel and light metal construction, mechanical engineering, wood working and clothing industries were below average in terms of their overall share.

Sites which employed between 200 and 999 staff accounted for 42.3 per cent of all sites in the survey. There were few small sites (less than 50 employees) (12.6 per cent), whereas large sites (over 1,000 staff) numbered just under a quarter of the total (23.5 per cent). By way of contrast with these EMAS figures, official statistics showed that small companies (up to 49 staff) constituted in the manufacturing sector just under half of all German companies (49.0 per cent) and large companies (over 1,000 employees) accounted for only 1.6 per cent. It is therefore not surprising that the survey showed that almost two-thirds (64.0 per cent) of companies with EMAS sites have an annual turnover of 50 million DM or more. By way of contrast, the annual average turnover of companies in manufacturing industry in Germany in 1996 totalled just under 4.5 million DM (calculated from German Statistical Office 1997, 11). Thus, large companies are more likely to take EMAS on board whereas the vast number of small companies are much less likely to go for EMAS (see Table 8.3).

The survey also showed that German EMAS companies were strongly export oriented. Only a small proportion were serving just exclusively their regional market (13.9 per cent), and less than one-third dealt with just the German market (29.5 per cent). Companies with exclusively regional markets were particularly prominent in the food and energy industries.

Table 8.1   Number of Registered Sites in Germany

Date	Number of Sites
October 1997	674
July 1998	1,410
July 1999	2,090
July 2000	2,542
October 2001	2,523

Source: EMAS registration lists.

Half of the EMAS sites in the survey were single site operations. The majority of those with more than one site were German companies. Only 10.8 per cent of the EMAS companies surveyed stated that they were owned by a foreign company.

In regional terms, EMAS companies were concentrated in North-Rhine Westphalia, Hesse, Bavaria and Baden-Württemberg. From the DIHT registration list in October 1997, 70.5 per cent of EMAS sites (and 74.9 per cent from the survey) were located in the Länder. Compared with the general distribution of manufacturing industry in the Länder, companies in North-Rhine Westphalia and Hesse were more likely to have EMAS registered sites (see Table 8.4). This suggests a major commitment to EMAS by companies, which can be explained neither by specific funding programmes nor by subsidy schemes on the part of the Länder. By way of contrast, companies in Rhineland-Palatinate and Lower Saxony do not feature very strongly, despite rules of substitution (see below). There is also less interest shown by companies in the former East German Länder.

Table 8.2   Distribution of EMAS Sites by Sector in Germany (percentage share), October 1997

Sector (NACE code)	All Sites[a]	Registered Sites[b]	Sites Surveyed
Mining & quarrying (14)	2.9	0.4	0.6
Food (15)	10.6	13.6	11.8
Textiles (17)	2.8	2.4	1.8
Clothing (18)	2.0	0.6	0.9
Wood working (20)	4.4	2.4	3.9
Paper (21)	2.3	2.5	4.2
Publishing, printing, recording (22)	6.2	5.3	3.3
Coke and petroleum products (23)	0.2	1.0	0.9
Chemicals (24)	3.8	14.1	14.8
Rubber and plastics products (25)	6.4	7.4	7.6
Non-metallic mineral products (26)	8.2	2.4	1.8
Metal manufacture (27)	2.3	3.1	2.7
Metal products (28)	14.2	7.6	7.3
Machinery and equipment (29)	14.2	8.5	7.9
Office machinery, computers (30)	0.5	1.2	2.4
Electrical machinery (31)	4.9	6.4	4.5
Radio, TV, telecommunications (32)	1.3	2.5	2.4
Medical and optical instruments (33)	4.4	3.1	3.0
Vehicles, trailers (34)	2.2	5.8	5.8
Other transport equipment (35)	0.9	0.4	0.6
Furniture, sports equipment etc. (36)	4.9	5.6	6.4
Recycling (37)	0.4	2.4	2.4
Energy supply (40)	-	1.2	2.7
Total	100	100	100

Source: Own survey 1997.
a)   Calculated from German Statistical Office 1997, 11–21.
b)   The total number of sites in October 1997 was 674. They were categorised on the basis of their core activities.

Table 8.3 EMAS Companies by Number of Employees in Germany (percentage share), October 1997

Number of Employees	EMAS Companies Surveyed	All Companies in Manufacturing Industry
1 – 49	12.6	49.0
50 – 99	10.0	22.7
100 – 199	11.5	14.1
200 – 499	27.6	9.7
500 – 999	14.7	2.9
Over 1,000	23.5	1.6
Total	100.0	100.0

Source: Own survey 1997 and German Statistical Office 1997, 11.

Table 8.4 Distribution of EMAS Companies by Länder, October 1997

Länder	EMAS Companies Surveyed	Registered EMAS Companies[a]	All Companies[b]
Schleswig-Holstein	2.9	3.7	3.1
Hamburg	1.6	2.3	1.3
Bremen	0.6	0.6	0.7
Lower Saxony	7.0	6.4	8.3
North-Rhine Westphalia	26.8	25.4	21.8
Hesse	9.9	9.2	7.0
Rhineland-Palatinate	1.9	2.1	4.7
Saar	0.3	1.3	1.1
Baden-Württemberg	18.8	18.6	18.2
Bavaria	19.4	17.3	17.4
Saxony	2.9	2.9	5.3
Thuringia	2.9	2.9	3.0
Saxony-Anhalt	0.3	1.6	2.7
Brandenburg	1.3	2.0	2.2
Berlin	3.2	2.9	2.1
Mecklenburg-West Pomerania	0.3	0.9	1.1

The federal states with rules of substitution (see below) are shown in italics.
a)   Data for October 1997.
b)   Calculated from German Statistical Office 1997, 11.

A majority of the companies registered with EMAS did not also register with ISO 14001 (59.7 per cent). The reason for this was that the European system had gained wide public acceptance and the high standards of EMAS were welcomed. A sizeable proportion of companies (43.4 per cent) explicitly stated that they had introduced EMAS management and not ISO 14001, as the ISO 14001 was not ready for implementation at the time. Some of these firms wanted to acquire ISO 14001 certification in the foreseeable future. One third (33.8 per cent) of the companies surveyed introduced ISO 14001 at the same time as EMAS. They decided to opt for EMAS certification as well as ISO 14001 on account of the more stringent requirements, the resultant better use of company resources and cost savings. Other frequently mentioned reasons included the fact that ISO 14001 certification had been adopted across the board by the company, and EMAS registration could not therefore be conducted just for one site. In any case, dual certification did not require any great additional investment. A small number of sites (6.5 per cent) had already acquired ISO 14001 certification prior to EMAS participation. These companies explained that this was due to the global operations of the company.

Of the companies surveyed 77.8 per cent stated that EMAS participation had been worth it. Only 2.5 per cent of participants would not reregister (see also the survey by ASU/UNI 1997, UGA 1997 and UBA 1998).

8.3.2 EMAS after 1997

From the end of 1997 onwards, as the scheme developed, there was a clear reduction in the number of new registrations. There was a sustained increase in the number of registered sites up to 2,542 in July 2000. Scrutiny of the number of newly registered sites every half year between the second half of 1995 and the first half of 2001 (Figure 8.1) shows a peak in the first half of 1998 when 444 new sites were registered. Subsequently, the number of new registrations remained at the relatively high level of over 300 in the second half of 1998 and the first half of 1999. Then the number of new registrations fell to 260 in the second half of 1999 and 192 in the first half of 2000. Thus, although new registration rates were still on a relatively high level, there was a significant slowdown in the number of new registrations from the second half of 1998 onwards.

The extent of participation in different industrial sectors also changed over the same period (Figure 8.2). Whereas in the 1997/98 survey the chemical industry and rubber and plastics together had a share of 21.5 per cent of all registered sites, by 2001 they had only a 15.3 per cent share. The food industry share also decreased slightly from 13.6 per cent to 10.5 per cent, and the machinery and vehicle industries which together had a share of 13.7 per cent at the end of 1997 made up only 11.1 per cent in 2001. Following the decree on the extension of the scheme to non-industrial sectors, transport, restaurants and hotels, banks and insurance companies, distribution and local authorities have begun to register with EMAS. By 2001 non-industrial sectors accounted for almost 13 per cent of all EMAS sites.

Figure 8.1   Total of New EMAS Registrations in Germany 1995–2001

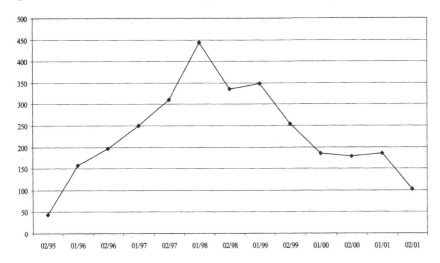

Source: EMAS register.

Figure 8.2   New EMAS Registrations in Germany by Sector, 1995–2001

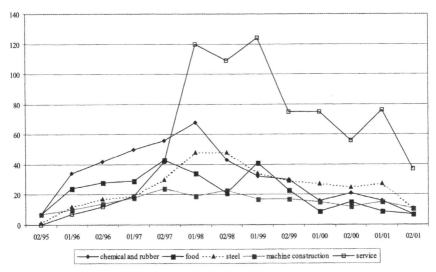

Source: EMAS register.

8.3.3 Reasons for Participating in EMAS

In the 1997/98 survey, German companies stated why they were willing to participate in EMAS.[17] The most frequently mentioned reasons for participation in EMAS were to improve environmental protection strategies and to enhance the image of the company. Other reasons included organisational improvement and legal safeguards. These were followed by cost considerations, staff morale, anticipated competitive advantage and customer or official expectations.

Comments on the reasons for EMAS participation did not directly refer to anticipation of deregulation or the substitution of official controls by company self-regulation. However, the high level of participation by companies in particular industrial sectors, especially those with production processes that required official approval, appeared to support such a conclusion.

However, these reasons did not differ greatly from those given in other member states and therefore could not explain the relatively high level of German registrations. There is, however, a specific development which helps to understand the relative attractiveness of EMAS in Germany.

The scheme is, on the one hand, intended to encourage companies to take individual responsibility for environmental protection, and thereby stimulate and support a self-governing capacity within society. On the other hand, many companies which did become involved expected something in return for their engagement. They were supported in their expectations by their representative organisations. In particular, they sought advantageous arrangements with banks and insurance companies as well as from the state.

From banks and insurance companies they looked for more favourable credit terms and insurance premiums. Their argument was that the introduction of an environmental management system and its verification would improve innovation and the overall future prospects of the company, as well as reducing the likelihood of liability claims. They argued that banks and insurance companies should accept an EMAS statement of participation as a "quality seal of approval", indicating that companies were at lower risk and had a sound economic base. Therefore, insurance premiums and loan interest could be calculated on a different basis than for other companies.

However, developments in practice lagged well behind company expectations. The 1997/98 survey revealed that insurance companies were not offering special reductions for EMAS sites. The banks seemed to be a bit more willing than insurance companies to treat EMAS companies favourably. Furthermore, in the debate on incentives, small and medium sized companies in particular warmly welcomed the idea of favouring EMAS companies in public tenders. However, according to the survey, no such advantage had been secured (see also Deutscher Naturschutzring, 1999).

Thus the fear expressed in Germany prior to the introduction of EMAS that voluntary commitment could turn into "voluntary compulsion" if financial institutions demanded environmental audits (Knill 1998) was not realised.

---

[17] This paragraph is in part based on Töller/Taeger (2000).

Favourable insurance and credit terms and giving preference in public tenders were only part of the story. The expectation of a lower level of inspection and control or even deregulation following EMAS registration, offering "individual responsibility instead of external control", played a far more important part in determining the popularity of the scheme in Germany, because of the very high regulatory burden on German companies in the field of environmental regulation. Representatives of company interests argued that there were four areas where there should be relaxation of controls or deregulation, viz in regulatory requirements, environmental reporting, approval procedures and inspections.

The term "deregulation" refers to a reduction in regulatory requirements. The word "substitution", on the other hand, does not refer to the abolition of environmental controls, but indicates that EMAS could be expected to cover compliance with regulations. Although the issue of deregulation was discussed in the past, it is no longer taken seriously in academic debate as participation in EMAS aims at enhancing compliance with legal requirements and not at abolishing them. A reduction in these requirements for EMAS sites would therefore be somewhat paradoxical (see Lübbe-Wolff 1996b, 174).

Ideas on substitution have been realised to differing degrees and in different ways in Germany (see Taeger 1999 for details, and Bohne/Wagner 1999). At the federal level, the formula that EMAS registration will be "taken into account" has appeared in some regulations. Examples include the federal pollution control regulation and the recycling and waste management laws, as well as the regulations on scrapped cars. Where the internal environmental audit complied with the requirements of recycling and waste management laws and with the regulation on sound recycling principles, the EMAS environmental statement then constitutes the required documentation.

There are more detailed rulings on inspections at Länder level. The first of the Länder to introduce them was Bavaria, through an environmental pact.[18] In autumn 1995 the Bavarian government and business interests agreed that within five years 500 Bavarian sites would be EMAS registered and, in return, self-regulation by companies would be given greater priority in environmental protection (see Böhm-Altmann 1997, 178). In concrete terms, this meant that registered EMAS sites received special treatment in respect of reporting and documentation requirements, inspections and approval procedures. As is clear from Table 8.4 however, this did not lead to overrepresentation of Bavarian companies engaging with EMAS. By mid 1999 EMAS registered sites in Bavaria were still a long way below the target agreed by the business interests in the environmental pact.

Since 1997 other Länder (Bremen, Baden-Württemberg, Hesse, Lower Saxony, Saar, Saxony-Anhalt, Schleswig-Holstein and Thüringia) have introduced rules which require the environmental authorities to treat EMAS sites in a particular way. These rules were based on the idea of "functional equivalence", a situation which is reached when, by implementing EMAS, the same objectives and outcomes can be

---

[18] The pact was drawn up between May and August 1995 by representatives of the Bavarian Chemical Industry Association and the State Ministry for Regional Development and the Environment.

attained as through the conventional application of environmental law (see Lahmeyer ERM International 1998, 26).

Industry representatives have, however, commented critically that (a) substitution rulings have not been found in all Länder, (b) existing rulings have not gone far enough because of a lack of trust on the part of politicians, and (c) rulings on enforcement were likely to be watered down by the opposition of officials in local administrations.

The issue of substitution led to a heated debate in Germany. Lawyers in particular have criticised the "privatisation of state duties" which would come about if extensive substitution shifted the responsibility for securing legal compliance to environmental verifiers (Laskowski 1998). This would have a number of consequences. First of all, the importance of verifiers, their training and their independence would increase. Furthermore, verifications would have to be more complex and therefore more costly to ensure compliance with current environment laws. The costs of enforcement would therefore be likely to be borne increasingly by companies themselves. However, large chemical companies and leading business organisations have made the critical comment, that substitution measures have not gone far enough or tend to fail at the implementation level.

Although those interests stressing the relevance of introducing further substitution measures as an incentive to participate in EMAS are very prominent in the public debate, they constitute a minority of EMAS registered sites. Although deregulation and substitution may seem to be desirable for small and medium sized companies, even if they do not have processes that require approval, this is not a strong enough motive for EMAS participation. As indicated in the 1997/98 survey and the case studies (see Chapter 9), the reason for EMAS participation by small and medium sized companies is that it helps with the development of an environmental management system, with internal arrangements for complying with the law and with motivation of staff. Larger companies are more likely to benefit from substitution rulings, in particular the chemical industry, which has many processes which require approval. [19]

8.4   Conclusions

In the first three years, EMAS turned out to be a success story. Germany was quick to take the lead with regard to the number of EMAS registered sites. This success is perhaps surprising considering that Germany's regulatory tradition does not seem to be compatible at all with EMAS. There are three interrelated reasons for the success of the scheme in Germany. First of all, the fact that the organisational structures for its implementation (i.e. accreditation and registration) were located quite close to business organisations made the scheme acceptable to companies which did not see EMAS as just another set of regulatory interference by the state but as a tool for self-organisation. Second, the relatively high level of environmental awareness in

---

[19] The chemical industry has 14 per cent of the total of EMAS sites, although it represents only 3.8 per cent of the total manufacturing sector (see Table 8.2).

Germany, in addition to the extremely high regulatory burden, made EMAS seem an instrument which could help demonstrate involvement with environmental affairs to employees, customers, neighbours and the general public and at the same time address the ever present gap between regulatory obligations and the reality in companies. Third, it was the relatively low level of involvement with quality management in German companies that made those companies that were willing to go for an environmental management scheme choose EMAS instead of ISO 14001. In addition, there were, for some companies, expectations about reducing the regulatory burden through substitution.

However, more recent developments provide a far less optimistic picture. There are a number of factors that indicate growing disenchantment with EMAS in Germany from the second half of 1999 onwards. In addition to decreasing numbers of newly registered sites, there seems to be an increasing number of EMAS sites that have not gone for re-registration, especially in areas such as North-Rhine Westphalia, because the internal improvements that were achieved in the first audit and verification cycle could not be sustained in the second cycle (SRU 2000, 132, see also Bundesverband der Deutschen Industrie 1998, 1999).[20] Whereas new sectors have increasingly gone for EMAS, the reduced participation rates by those industries traditionally in favour of EMAS (especially the chemical industry) suggest that they are disappointed that substitution rulings have lagged behind expectations. In addition, the public image aspect of EMAS has also been seen as disappointing. Although there is no general crisis in the field of environmental management, these developments lead to the conclusion that EMAS is losing ground in relation to ISO 14001. Two of the leading car manufacturers in Germany now demand an ISO 14001 certificate from their suppliers (SRU 2000, 132), not an EMAS registration, as was hoped by the supporters of EMAS. In autumn 1999 the largest environmental verifier organisation returned its accreditation for verifying EMAS (SRU 2000, 132).

It seems that there is now a clear division between two types of EMAS users in Germany. Many of those mainly larger companies that participated in EMAS, or had intended to do so, for reasons of public image and substitution expectations are now unlikely to register or re-register. However, there are still many companies that see EMAS as a means for internally improving their environmental performance and who have never been strongly interested in substitution issues. This type of company is still interested in EMAS.

One recent development in Germany that is trying to address both the competition between EMAS and ISO 14001 as well as the suspicion between European and non-European countries in this field is a "Memorandum of Understanding" which was agreed in 1999 by the Land Bavaria and the American State of Wisconsin. Against a background of promoting co-operation in the area of environmental law, it is planned to launch common projects supporting engagement in both EMAS and ISO 14001 environmental management schemes (see SRU 2000, 133).

---

[20] At the time of writing, no sound information on this issue was available.

Chapter 9

The German Case Studies

Brigitte Geißel

9.1 Lincoln, Walldorf

9.1.1 Introduction

Lincoln, in Walldorf, Baden, has its origins in the Heidelberg company, Helios Apparatebau O. Weitzel & Co., which was founded in 1910. After merging with Lincoln (St. Louis, USA) and relocating to Walldorf in the 1970s, the firm which had in the meantime been converted into a limited liability company, was taken over by the American Pentair Group in 1987.[1]

The company began with the first patent for a lubricating device for rails and lifts. Ever since, Lincoln has been producing central lubrication systems for the mechanical engineering industry. The aim of central lubrication systems is to reduce friction and wear and to automate manual lubrication processes. With its affiliate in St. Louis, Missouri, Lincoln is the largest manufacturer of lubrication systems in the world. Lincoln employs 280 staff and in 1995 its sales totalled 70 million DM (about 36 million euro). Apart from oil fired plants, there are no processes requiring approval. Production at Lincoln can be divided into three sectors: production, where raw materials, such as steel, are processed; assembly where the individual parts are put together to make finished distributors and central lubricant pumps filled with oil and grease; and finally inspection and testing, where the finished products are tested and calibrated and where new developments are tried out.

9.1.2 Thematic Background

Why did Lincoln decide to participate in EMAS? At the beginning of the 1990s, environmental protection had not come on to the agenda. One significant influence to change this came from the product development sector. In general, central lubrication systems are not without their problems for the environment. As it is impossible to meter the lubricant used to a high level of accuracy, the grease from the bearings of commercial vehicles and construction or agricultural machinery is discharged into the environment and can escape into groundwater from the roads or the soil. However, there are two rival grease systems in use. There are low viscosity

---

[1] In the American Pentair Group, Lincoln can be seen as an environmental forerunner. The American partner is not very interested in Lincoln's participation in EMAS, but it does not hinder Lincoln, as long as the latter makes profits.

grease systems which operate with liquid grease and grease lubrication. The advantage of low viscosity systems is that the low viscosity lubricant can spread throughout the system relatively easily, whereas grease lubrication systems have to operate at very high pressure to move the inert mass to all areas to be lubricated in the system. Lubrication systems which work with low viscosity grease are cheaper, but more environmentally harmful, as the low viscosity grease can be easily released into the environment. As this system was widespread until recently, Lincoln had started to work with this grease, expecting better profits. When the company did not get an expected huge order because of strong competitors, they started to think about a new strategy. Whereas competitors still supply low viscosity grease systems, Lincoln abandoned low viscosity grease technology in 1996 and now produces more high tech, environmentally friendly but also more expensive grease lubrication systems (see 1997 environmental statement, 22). Both systems, with their specific advantages and disadvantages, are rivals in the same markets in Europe and throughout the world. Thus at Lincoln the introduction of EMAS was based more on a strategy of long term competitiveness and product innovation. At a time (1993/95) when business was not thriving, the issue of environmental protection linked to that of product innovation, the profile of orders, the economic future and job security, did take root. It was not viewed as a contradiction to, but was seen as part of, successful business practice (1998 environmental statement, 2). To be a company concerned about the environment meant developing a specific profile and promised an advantage over competitors. Lincoln could argue that their lubricant was more environmentally friendly than that of their competitors. What was born out of a crisis turned out to be an advantage: "Today we can say it was the right decision and our customers appreciate it", as one of the interviewees put it. Lincoln considered EMAS as a way of spurring innovation, as a means of being "ahead of the times", to show foresight.

In addition, a new managing director introduced group work into the production process, a principle which was intended to overcome the Taylorist system of hierarchical work organisation based on the strict division of labour.

In 1995 against this background, an environmental consultant from the Institut für Ökologische Wirtschaftsforschung (IÖW, Institute for Research in Environmental Economics) approached Lincoln and asked if they would be interested in an environmental management system. This external push was crucial for the implementation of EMAS. According to the interviewees, Lincoln would not have taken up until much later the idea of implementing EMAS. Together with the chairman of the works council, they developed an idea for a pilot project called "environment audit with worker participation". They succeeded in persuading the Baden-Württemberg Environment and Transport Ministry to sponsor this pilot project. Participation in EMAS was encouraged by partial funding of the external costs (from 25 to 70 per cent) from state funding programmes. Without such commitment, Lincoln would not have introduced EMAS voluntarily.

The idea of taking part in EMAS was welcomed by the management. Although at first only a few people were interested in EMAS and the participation of employees, EMAS implementation was finally seen as a long term proactive strategy aimed at demonstrating to the market environmental awareness in terms of both products and the production process. EMAS was seen as an aid in introducing a new, environmentally friendly management system together with employee participation.

9.1.3 EMAS Procedure

Following the decision to implement EMAS, groups of employee representatives were established. The main players involved in Lincoln's environmental management system are the environmental promotions manager, the environment team, three "environment circles" and the environmental co-ordinator. The promotions manager is responsible for publicity, for incorporating environmental protection into the corporate strategy, for developing environmental awareness at all levels within the company, for reporting to management and for supporting environmental audits.

The main advisory and supervisory body for environmental protection is the environment team consisting of two members of the works council and two members of management (including the environmental co-ordinator). The environment team acts as the "interface between management and staff" (1997 environmental statement, 13). It has no ultimate authority to take decisions, but still plays a major role in the assessment of environmental effects, the formulation of environmental objectives and the practical implementation of the environmental programme.

The environment team delegates key issues to the three "environment circles". In these specialist bodies "staff bring their knowledge and experience of key individual issues of environmental protection, draw up ideas and work out proposals for implementation". The aim is to "involve staff knowledge on current environmental weak points" (1997 environmental statement, 13). However, most of the work has taken place in the environment team which meets every two months, whereas the "environment circles" meet only once or twice a year.

The environmental co-ordinator has a number of tasks. First, he has to provide information to all those responsible for environmental protection. This includes the development of an environmental information system and preparation and evaluation of environmental data. Second, he co-ordinates and controls the implementation of environmental measures. Third, he conducts internal environmental audits. Fourth, he is also responsible for developing innovative strategies for the future (see 1997 environmental statement, 13).

The issue of staff participation, which is given special weight at Lincoln, was integrated into the environmental management system through a series of special measures on environment related communications with staff. Based on the idea that the actual level of staff activity ultimately depends on their identification with the project, staff were invited to participate in the implementation of EMAS from the outset. EMAS was and is discussed at all works council meetings. Announcements and posters inform and mobilise staff to take an active part in improvements. The system of information provision is therefore of special importance. For example, those who work with e-mail want information via e-mail, those who do not work with e-mail need to be informed by other means. The question was and is to find the right way to get information to people without spending a lot of money. "Knowledge and information must get to the right place." The management and the environmental co-ordinator also tried to make environmental protection and participation in EMAS of interest to staff. According to the interviewees, the high level of information provided had positive results. The introduction and the implementation of EMAS were easy, because most of the employees backed it. Very few refused to take part and the environmental co-ordinator tried to convince them as well. Training of staff is taken seriously. The interviewees agreed that "we depend on the participation of the employees, because we realised how much they can offer to improve the performance." "Creative products need creative employees."

After setting up the basic structure of the environmental management system between October and December 1995 the company's environmental policy was first of all developed by discussions with staff at two workshops and by securing the support of the members of the works council. Managers then approved the environmental policy which had been developed on this basis. The main aim of Lincoln's environmental policy was defined as (i) "a core feature of central lubricating systems [...] from the initial product idea via production and use to disposal" and (ii) reduction or prevention of environmental pollution at source (1997 environmental statement, 11). Working from basic principles, the aim was to save energy and water, to use environmentally compatible materials and additives and to reduce packaging. They also wanted to develop "economic product solutions which are easier to repair and less hazardous for the environment". In future investment decisions and new product and production process developments, environmental criteria were to be employed. In addition to "compliance with all relevant environmental rules, laws and requirements", they were committed to informing staff about company environmental problems, to promote environmentally conscious thinking and action and, in addition, to operate an "active information policy" towards suppliers, customers and the public.

An initial comprehensive survey of the environmental effects of operations at Lincoln was conducted parallel to the development of the environmental policy. As part of an environmental audit, a systematic study of all relevant environmental regulations was conducted. This revealed that there had not hitherto been a comprehensive overview of environment related regulations. The IÖW environmental consultant helped to compile a list and a summary was drawn up of the main requirements based on this overview. The list was managed by the environmental co-ordinator, who was also responsible for updating it, a difficult task in view of the large number of new environmental laws. In Germany there is no single document which contains every important law and which is continuously updated.[2] Although CD ROMs are produced twice a year and there are information services, what is relevant for Lincoln still has to be filtered out. It takes a lot of time to decide what applies to which departments and how to handle new laws.

After the systematic study of all relevant environmental regulations the company, with the help of the external consultant, set up a systematic assessment of the flow of materials and energy and drew up an environmental balance sheet. The IÖW consultant helped to identify the relevant environmental factors and to construct a

---

[2]   Lincoln complained about the amount of legislation, which made it nearly impossible to be aware of all the laws.

systematic input/output analysis. On the input side were the materials and substances used and on the output side both the products and emissions (solid waste, waste water and polluted air).

In evaluating the environmental effects, different priorities were laid down for different parts of the production process. Raw materials and semi-finished products were not included in the detailed analysis as "extremely good recyclable steel is used" and the semi-finished products "are primarily significant to the environment in raw material extraction and pre-production". However, in future the company intends to address the environmental relevance of semi-finished products (see 1997 environmental statement, 20; 1998 environmental statement, 4). Environmental weaknesses were identified in the main in working materials and additives. Paints, adhesives, sealants, grease, oil and coolant lubricants were assessed in terms of environmental criteria. Paints and adhesives used in large quantities were particularly picked out as environmentally harmful. Office material was not environmentally friendly. It was also noted that the amount of general rubbish was still high, despite the introduction of sorting and separation. For emissions, the carbon dioxide content and level of solvent were considered problematic (see 1997 environmental statement, 23).

Following the initial review, the environment team developed preliminary environmental objectives and submitted each of them to one of the three "environment circles". The environmental objectives were discussed and measures for achieving them as well as deadlines for implementation were determined. From these recommendations, a preliminary environmental programme was drawn up which was then submitted to management for their approval. The environmental programme approved by management differed only slightly from the proposals made by the environment team.

Some of the most important environmental objectives in 1996 were the reduction in the use of coolant lubricants by 5 per cent by mid 1999 and the substitution of a noxious adhesive by the end of 1998. In addition, the paint shop, newly installed in 1997, was expected to be able to work with water soluble paints. The company wanted to change to more environmentally compatible office materials following a supplier survey on environmental issues and to raise recycling levels for paper from 6 to 40 per cent. In the energy sector the objective was to reduce electricity consumption by 3 per cent by the end of 1999 by using energy saving lamps. In addition, water consumption in 1998 and 1999 was to be reduced via changes in sanitary arrangements. Finally, waste sorting was to be improved and the volume of waste reduced further by a number of other measures (1997 environmental statement, 26). However, the sorting of rubbish remained a problem although the "rubbish stations" are clearly marked.

Implementation was the responsibility of the staff, whilst the environment team and particularly the environmental co-ordinator had continuously to check whether measures were being implemented (1997 environmental statement, 13). The environmental co-ordinator was also required to undertake "regular variance analysis to check whether the targets set have been achieved. [...] The parties concerned are immediately notified of variances in target performance" (1997 environmental statement, 13) and to intervene to address failings. Annual internal audits were conducted in order to assess the robustness of the procedure and to weigh up experience to date. Adjustments to the environmental management system, to environmental policy and to the environmental programme were made in response to problems identified or changing circumstances.

EMAS implementation was finally assessed by the environmental verifier in March 1997. Although the assessment was "based on partnership and co-operation", the environmental verifier had a number of reservations, and Lincoln was registered only after these reservations had been addressed. At the same time as EMAS registration, Lincoln also acquired ISO 14001 certification. Since then annual internal audits have been conducted and short environmental statements have been published. The latest verification at Lincoln was in 2000 and Lincoln plans to be verified again in the next cycle.

The costs of introducing EMAS were estimated to be about 250,000 DM (about 130,000 euro), of which the Baden-Württemberg Environment Ministry provided 70,000 DM. Investment in environmental protection measures accounted for a further 180,000 DM. This expenditure did not include the initial measurable savings. Savings by 2001 amounted to no more than 20,000 DM. The interviewees stated that it was nearly impossible to achieve a cost benefit balance, particularly because some benefits would not be realised for another 10 to 15 years. According to normal standards of cost benefit analysis, "you can forget EMAS, but we take future developments into account", as one of the interviewees put it. For example Lincoln is taking part in a new project supported financially by the Ministry of Education and Research ("Service Partner Industry"), which would not have been possible without the investment in EMAS. The philosophy is long term, not oriented towards short term cost benefit analysis.

Lincoln's environmental statement described very clearly how the environmental management system was working, the results of the first environmental audit and the assessment of the environmental effects. Weaknesses were openly addressed. In the short 1998 environmental statement, implementation of environmental policy measures in 1997 was clearly recorded. Considerable progress had been made in some areas. For example, between 1995 and 1997 the volume of coolant lubricant employed was reduced by a quarter. Solvent emissions were cut by 50 per cent (1998 environmental statement, 3).

Conversion of the paint shop to water soluble paints had not come about, as it had not proved possible to produce water soluble paints with a sufficiently high particle density. Between 1995 and 1997 energy consumption did fall, but the short environmental statement pointed out that this was not entirely due to environmental initiatives, but was partially caused by a shift of some production processes to another plant. In 1997 a slight reduction in oil usage was recorded and there was an increase in water consumption. A reduction in general waste was recorded.

A number of adjustments were made to the management system in 1998 to improve communications within the company. In the short environmental statement, there was reference to the failure to achieve "a system through which the environment related regulations can be monitored continuously" (1998 environmental statement, 5). A new set of environmental objectives was introduced for 1998, such as drawing up a checklist on supplier firm conduct on Lincoln premises and commissioning a new energy saving compressor (see 1998 environmental statement, 5).

Clear potential for improvement could be seen in terms of co-operation between different departments. Future improvements include packaging, the volume of coolant lubricant employed and waste disposal. Since 1997 Lincoln has voluntarily undertaken to accept the return of its products, in order to gather experience before this becomes a legal requirement.

Internal changes have occurred at Lincoln as a result of the introduction of EMAS. For example, there has been greater visibility of environmental protection issues as well as a clear increase in their credibility. The acceptance of environmental protection concerns, the transparency of the work and the achievement of real improvements have convinced even the sceptics in management and on the shop floor. Increased awareness by staff of environmental issues is shown by the fact that more frequent queries are being addressed to the environmental co-ordinator and the promotions manager. These queries concern, for example, the exact requirements of the regulations in relation to specific manufacturing processes or to the composition of materials used. More proposals for improvements are being submitted by staff. Those interviewed believed that staff had begun to address opportunities for environmental improvement outside the workplace as well. Staff had become committed to EMAS in part through the environmental statement which documented what had been achieved for them. This helped them identify with the project.

Yet EMAS also produced conflicts. As the co-operation of each member of staff was required, implementation of EMAS undermined the traditional company hierarchy. This was not always viewed positively by those who had lost influence and prestige as a result, such as the master craftsmen. According to the report by the IÖW consultant, the participatory style introduced via the environmental management system had to be learnt at Lincoln despite the fact that group work was already well established.

There had been a clear shift in attitude to complying with the law. Although some problems had still not been satisfactorily solved as a result of EMAS, the relevant environmental requirements were being systematically documented for the first time. What then happened was a "spillover" into the production side of the company. In 1998 a product innovation group was employed to develop new business areas, new products and potential product uses. In particular, the reduction in lubricant use in central lubricating systems clearly had its origins in environmental concerns.

There were also severe problems with drawing up the eco-balance sheet. The first eco-balance made was very informative, but rather unstructured. It was not published, because it contained too much information which could have been useful for competitors. Lincoln is required to publish the input and the output under EMAS rules, but because of the problem of competitors it remains unpublished. This is still an unsolved problem.

A number of changes were made to external company relationships, such as a classification of suppliers. This led to suppliers with EMAS or ISO 14001 being given priority where price and quality were the same.

EMAS site registration and ISO certification were well received by customers (1998 environmental statement, 4). In the future Lincoln hopes to achieve concrete advantages in customer relations. This will become more significant when certification is made compulsory for suppliers to the automobile industry. So far, there has been no change in relations with banks and insurance companies in the light of the certification. The same applies in relation to the regulatory authorities, although the company is hoping that controls will be eased.

In contrast to most other companies Lincoln also does public relations work, for example, it invites schools in to demonstrate how companies can be environmentally friendly. Lincoln also provides an award for students.

The disadvantages of EMAS are that it is very formal and is time and money consuming. The interviewees stated, that they think "it is more important to live EMAS, than to follow every paragraph".

The concept of sustainability did not play a role when EMAS was implemented. The discussion about sustainability emerged later.

As well as the external consultant, other external knowledge holders played a part in the early stages of EMAS. A committee was established consisting of different groups, such as the "Bund für Naturschutz" (Alliance for the Protection of the Environment), or the "Industrie- und Handelskammer" (Chamber of Industry and Trade). This advisory committee provided helpful comments.

The interviewees stressed several times that participation by employees led to several innovations. For example, the systems of transport, storekeeping and distribution were improved and the lighting system was changed. Awareness of environmental issues changed dramatically. According to the interviewees, without participation the employees would not have developed the level of awareness about energy saving or rubbish sorting that they have today. Also interest in training has increased. People now want to be qualified, to become informed about environmental issues.

Although the quality of the product and the price still count for more than anything else, in future the automobile industry is expected to make either EMAS or ISO 14001 certification compulsory for its suppliers throughout the sector. Sites like Lincoln could then see themselves as a strong link in the supply chain. There are further examples which show the benefits of EMAS or ISO 14001. For example, the Netherlands have strict environmental rules and Dutch companies expect their partners to be environmentally friendly. Because of this, Lincoln secured several orders. These practical benefits finally convinced those employees who had been sceptical at the beginning.

9.1.4 Conclusions

Innovation, participation and environmental protection are closely linked at Lincoln. According to the interviewees, participation is a condition for (process and product) innovations, and innovations in turn provide the conditions for continuous improvement through environmental protection. In particular, innovation in production processes helped to fulfil the objective of continuous improvement. Whereas product innovations could not be expected from employees, they did have a lot of ideas about decreasing the environmental impact of the production process, for example, how to reduce the use of energy and water.

Sustainability did not explicitly play a role when EMAS was implemented. Although environmental, economic and participatory elements were clearly involved, the company emphasised the environmental and economic dimensions.

Lincoln fulfils most of the "important criteria" for sustainability. They take environmental protection seriously and they address the consequences of complying with the law. They have a long term time perspective, although couched in terms of the future of the company as a whole. The issue of local links and relations with third parties seems to be less crucial in the decision-making process. As far as it could be assessed, Lincoln tries to put increasingly emphasis on the "life cycle of products". The environmental management system seems to work well and does not need constant attention.

Participation has been limited to the workforce, experts and internal status holders. Participation by employees has meant primarily the provision of information, training and mobilisation of innovations through "local knowledge". The external consultant participated by initiating the idea of implementing EMAS and by supporting the project in different ways. Now they are no longer necessary. Local residents and other stake holders participated to only a small extent in the advisory committee in the early stages. Those who did participate, such as the workforce, changed a lot. Environmental knowledge, identification with the site, motivation at work and environmental consciousness increased substantially.

What factors made the implementation and the success of EMAS – especially in respect of participation – possible? To start with, the management as well as the works council members believed in co-operation and employee participation. Both of them, especially the chairman of the works council, were aware that economic and environmental factors represented both competitiveness and safety at work. They believed that EMAS could be used to transmit this knowledge to others in the company.

The external consultant was experienced and thus he was able to support the implementation of EMAS within Lincoln. He also favoured a high level of participation by employees and had enough knowledge to encourage and assist the company in their attempt to involve the workforce.

Finally the government provided financial support for the introduction of EMAS at Lincoln, because it wanted a pilot project that showed the benefits of the scheme.

Implementation of EMAS was made easier by the special way work was organised at Lincoln. There was only a low key hierarchical system and staff were involved through working in groups. The same is true for the philosophy of the company, which is more oriented towards long term success, rather than to making quick profits. In this environment, EMAS and a real opportunity for worker participation could flourish.

Last but not the least, there was the commitment of the environmental co-ordinator and the environmental team. Both were convinced of the importance of an environmental management system and worked hard to involve the whole workforce.

9.2 Mitteldeutsche Druck- und Verlagshaus

9.2.1 Introduction

The Mitteldeutsche Druck- und Verlagshaus GmbH & Co.KG (MDVH) is a print shop, where the newspaper *Mitteldeutsche Zeitung* is produced with a weekly print run of 350,000 copies. Additionally, the company produces advertising and other printed material, so that altogether, 1,400,000 items are printed per week.

MDVH is located in Halle (Saale) in Saxony-Anhalt, in eastern Germany. 750 people are employed at the main plant and with all the branches MDVH employs over 1,000 people. Most of the employees at MDVH worked there before the Berlin Wall came down. The site had printed *die Freiheit* (Freedom), one of the newspapers of the SED, the ruling party of the former German Democratic Republic. At the beginning of the 1990s the MDVH became part of the Verlagsgruppe M.DuMont Schauberg (MDS), a publisher located in Cologne. MDS became the parent company and supported the development of MDVH. Both sites are in close touch with each other. They exchange skills and information about the best available technology.

9.2.2 Thematic Background

Traditionally printing is one of the industries with a relatively high impact on the environment, because the materials needed for printing, such as ink or solvents, are very toxic. The MDVH site was particularly dangerous, because before the wall came down, the machines were old and the available technology was not environmentally friendly. The destructive impact of the print shop was huge. After the demolition of the wall, MDVH changed the name and the content of its newspaper, the *Mitteldeutsche Zeitung*, and also wanted to change its former image as a destroyer of the environment.

MDVH has been active in environmental protection since the beginning of the 1990s. When they demolished the building, in which the printing took place during the time of the German Democratic Republic, they did this in an as environmentally friendly way as possible. When the new plant was built in 1992, they ensured that eco-friendly technology and procedures were adopted, for example, installations to recycle ink and solvents. Since 1992 MDVH has used special aluminium printing equipment, which is environmentally more friendly than the old machinery. Since 1995 it has used digital systems, which are also environmentally more friendly.

9.2.3 EMAS Procedure

The push to start EMAS at MDVH came from outside. First, the parent plant, MDS, had started to discuss the possibility of introducing EMAS, initiated by contacts with the Martin-Luther University of Halle-Wittenberg. Second, an external consultant asked MDVH whether it would be interested in implementing EMAS. According to the interviewees, these initiatives were welcome, because the area around Halle had suffered terribly from environmental destruction in the 1970s and the 1980s, and people were interested in improving the environment in which they lived. Another reason for thinking about EMAS was the idea that it would lead to better management, because it would bring about better insight into costs as well as into who should be responsible for what. It was thought it would make control easier. MDVH also felt that it should develop an environmental image and became a kind of role model, so that, through its newspaper, the *Mitteldeutsche Zeitung*, it could more easily criticise environmental destruction at other sites. With the decision by top management to go for EMAS, it was then implemented in both MDS and MDVH at the same time. Before then, there had been no management system. The question on whether there was a choice between EMAS and ISO 14001 was not discussed, because EMAS was the only environmental management system known to MDVH.

To implement EMAS, the management set up an environmental commission which consisted of representatives from the environmentally crucial units of MDVH and the works council, plus a safety representative and an environmental manager. The commission and the environmental manager were responsible for the initial review and the development of an environmental manual. They were supported by an external consultant, who turned out not to be the best person for the job. According to the interviewees, he had difficulty in communicating environmental problems and wasted a lot of time, so he was replaced by another consultant. The initial review took some time and effort to complete, because some of the information needed had never been documented. Detailed research and inspections were required to collect all the data. The result was in a number of ways surprising. It turned out, for example, that the non-productive units had a bigger environmental impact than expected, for example, the administrative units needed more energy than the commission thought. They also discovered that much of the equipment was already old fashioned and that newer technology was available. Furthermore, problems which were known about but had remained unsolved became even more visible and had to be addressed.[3] Based on the results of the initial review, on external inputs and in the context of anticipated investments, the commission developed environmental objectives and identified ways of measuring whether they had been achieved. Some objectives were achievable, though others were not, because, for example, the technology needed was not yet available or because external bodies were not supportive.[4] The environmental manual was developed with the help of an external consultant, who wrote the framework. This framework was revised by the environmental commission working with colleagues at the Cologne site (MDS). EMAS at MDVH was verified in 1997. Since then there has been an annual internal audit, and every three years MDVH has to be re-examined by an external verifier and then registered.

What role did employees play in the implementation of EMAS? Environmental topics were (and are) discussed at the monthly meetings of the works council and at

---

[3] At a later stage, MDVH introduced a computer system, which made the review easier to conduct.

[4] In one case MDVH needed special expertise and they tried to work with a university to get a dissertation written about a special problem. But this did not work out, so the innovation was not introduced.

all company meetings. Employees get information from a notice board, from papers they get with their pay cheque and from a house journal. Each employee also gets the environmental statement. When the separation of waste was introduced, many employees became aware of EMAS for the first time. They have accepted EMAS, but are mostly interested in measurements only within their own immediate working environment. For the last two years employees have had the opportunity to put forward suggestions to improve environmental performance, and the management even offers financial incentives. Nevertheless, employees usually have small scale ideas which have only a modest environmental impact. Some of their ideas have already been acted upon, but important innovations usually stem from projects, rather than from employees.

MDVH has adopted a number of environmental principles. Environmental protection is a crucial part of the company's policy. Wastepaper is used as much as possible in the printing process. The choice of procedures and materials is oriented towards environmental friendliness. On site activities aim to reduce the consumption of energy, emissions, sewage and waste. Continuous improvement is another objective. Suppliers are to be selected based on their environmental performance. The environmental awareness of all employees should be improved. MDVH tries to avoid any impacts on the environment through accidents. Through documentation, the environmental policy, the principles and the results are easily accessible to the public. The public should be informed about changes and developments. An open dialogue should be conducted with the public.

EMAS also demonstrated qualitative results. MDVH uses a lot of paper, including as much wastepaper as possible. But wastepaper is not suitable for all print work. MDVH chooses procedures and materials (for example ink) which reduce environmental destruction as far as possible. According to one of the interviewees the site is "lucky" in respect of continuous improvement, because printing is an area with continuous technological improvement and innovations. So it is not difficult to demonstrate such improvement. Compliance with environmental laws is more difficult, because both the law and regulations change fast. An external consultant helps to keep track of new laws and regulations. Clear responsibilities were laid down and improved the structure of the organisation. MDVH has also tried to influence its suppliers and partners and expects them to improve their environmental performance. One outcome is that all the company's big suppliers are EMAS certified. According to the interviewees, investments are now more often environmentally friendly. EMAS has helped managers to see the complex relationship between investments and environmental performance. MDVH has also developed a network with other sites, which had already implemented EMAS. Nevertheless, profit has to remain the main aim, which means that the demands of the market have to be taken seriously. For example, customers demand coloured printed material, which uses more energy. Participation in EMAS does not deflect the company from responding to this trend.

Training of employees and enhancing overall environmental awareness is a crucial objective. MDVH informs its staff of all initiatives and puts in place environmental training to achieve this objective. One reason for the training is, according to the interviewees, that a corporation, like a print shop, can only work according to environmental principles, when all the employees take the principles seriously. The management also had hopes of financial benefit from this training, because it could result in closer control and greater savings in resources. Nevertheless, MDVH has not found it easy to achieve its own objectives. It is planning that all employees should get some special "fitness training for environmental protection" with crash courses, workshops and different kinds of events. The idea has been approved in principle. MDVH will be supported by a unit of the German Federation of Trade Unions (DGB) to undertake this training which will consist of three steps. First, the top managers as well as the members of the works council and the environmental commission will be trained. Second, the managers of all the smaller units will receive environmental training and third the employees will get all the information necessary for their particular roles. Parallel to this training information is to be provided on special topics. At present, this plan is on hold, due to lack of staff capacity.

Cost benefit analysis is not possible. The quantitative impact of EMAS on the environment is difficult to measure, because MDVH has increased its production in recent years and so consumption has also increased. Nevertheless, compared with the former times, MDVH has improved across the board. Although the amount of paper has increased, the amount of ink used has been stable due to better techniques and better ink quality. Consumption of water has decreased. Some environmentally necessary investments have been made, which will pay for themselves in the long run. MDVH is convinced that from a long term perspective there will be savings.

Dialogue with the public did not turn out as expected. For example, the readers of the *Mitteldeutsche Zeitung* or the company's customers, were not particularly interested in the implementation of EMAS at MDVH. The latter had expected a more positive response. Nevertheless, visitors, for example, guided site tours for school children, have shown interest. Contact with public authorities remained unchanged. There is no regulatory relief because of its participation in EMAS.

The interviewees stated that EMAS was a helpful tool, because, compared with other sites, which perform only internal audits, MDVH have to be disciplined and complete all the tasks laid down.

9.2.4 Conclusions

Sustainability did not play a role in the decision to go for EMAS in 1996 and is still not an explicit topic in MDVH.

How did the different stakeholders participate? The most important "holders" were the owner, the top managers, the parent company (MDS) and the external consultant. Top management was involved from the beginning. It provided the necessary financial and staff resources and presented EMAS to the public. The external consultant was important when EMAS began, but later his main task was helping to ensure compliance with the law. Neighbours or other kinds of citizen interest holders did not play a role, because MDVH is situated in an industrial area. Consumers, i.e. readers of the *Mitteldeutsche Zeitung* or customers, who put advertisements in the newspaper or have material printed at MDVH, did not participate in the EMAS process. Employees were told about everything that was happening. They have experienced some training and have been asked to suggest improvements in environmental performance. They are also involved in the sense that they have to carry out the requirements laid down in the environmental manual. In this sense, participation is quite limited.

Technical innovations have been very important for improvement in environmental performance, whereas other kinds of innovations have had less environmental impact.

The interviewees did not mention a connection between participation, innovation and sustainability. Innovations, which had led to a reduction in negative environmental impact, were of a technical nature. The innovative ideas of the employees mostly had only a small impact on the environment.

9.3   Dr. Kade

9.3.1 Introduction

Dr. Kade is a pharmaceutical company, founded in 1886. It has two sites, one in Berlin and one in Konstanz. In total, 250 people work for the family owned company, 110 in Berlin and 140 in Konstanz. The company produces drugs for medical use, for example, painkillers or for use in the field of proctology. It also does research for new drugs in several fields. The sites in Konstanz and in Berlin are both located in industrial areas.

9.3.2 Thematic Background

Dr. Kade has a long tradition of environmental awareness. When, for example, a new building was constructed in the 1980s, technical possibilities for saving energy were taken into account.

In 1994 environmental protection became one of the objectives of the company. Since then, Dr. Kade has developed environmental guidelines, stating, for example, compliance with the law, continuous improvement and transparent communications with both stakeholders and the public.

The senior management decided to implement EMAS, because it hoped to get some relief from the "bureaucracy of authorities", as one interviewee stated. The company also hoped to secure some benefit in the market by being registered. Furthermore, it was a relatively small step to implement the environmental management programme, because Dr. Kade already had a long history of environmental protection. The implementation of EMAS did not mean much work. It was also underpined by financial support from the state (*Bundesland*). This financial help was an important incentive.[5] In 1996 Dr. Kade, Berlin, was one of the first sites to be registered in Germany and in 1997 the site in Konstanz had also implemented EMAS and had been registered.

---

5   Without this financial help, they would not have implemented EMAS. The interviewees stated that most small companies would not be able to implement EMAS without such support.

9.3.3 EMAS Procedure

In July 1994 preparations for EMAS began on the site in Berlin, followed several months later by Konstanz. The procedure was similar at both sites. The introduction of EMAS was initiated by top management. The Chamber of Industry and Trade had asked Dr. Kade, together with an external consultant, whether it would be interested in implementing EMAS. The reason was that the Chamber wanted a pharmaceutical company to implement EMAS. The first step was to identify the responsible people and to set up an environmental committee. The manager for quality control became also the environmental manager. He became responsible for the maintenance of the environmental management system. The environmental co-ordinator manages, together with the environmental manager, the implementation of all environmental tasks. The environmental committee consisted of 12 people from different hierarchical levels and included the key managers and co-ordinators, the people responsible for other issues, such as waste, emission or safety, and employees. The "Umweltausschuss" (the environmental committee) developed the guidelines, the objectives and the environmental programme. It developed good information flows. It is an advisory committee that works on the preparation of decisions. The employees were informed about the whole process after the basic structures were established.

The main EMAS players following implementation were the manager of the environmental management programme, the co-ordinator for EMAS, and the environmental committee. In the early stages, the external consultant was extremely helpful and played an important role.

The *environmental management handbook*, which lays out what should be done, was produced, but according to the interviewees, hardly anybody ever looks at it, because it is too complicated and too complex. Nevertheless, the requirement to inform all employees is explicitly described in the handbook.

The environmental management system has been combined with quality management and with safety in the work place. An internal audit has to be conducted every three years.

In the company, several toxic materials are used. The main environmentally relevant results of the company's processes are, besides emissions like CO_2, CO, nitric oxide, sewage and solid waste, some special wastes, such as solvents or grease.

There is a paucity of information about how the initial review took place. In the environmental statement several reasons were given why the initial review was important, but reports of the actual procedure were vague. It seems that the environmental manager and the external consultant collected all the important information.

The company had detailed objectives in its first environmental programme. For example, it took care that a container for toxic material was placed on the site, that the number of waste sorting stations increased and that there were changes to the cooling system. What were the outcomes of these developments? The environmental statement provided very detailed information about different materials and substances, resources and energy used in the company. It explained the increase and decrease in the use of each in detail. The amount of water was reduced in later years, but the amount of energy increased, partly due to the fact that high quality pharmaceutical products need air conditioning, but mostly due to the fact that a new building was brought on stream. The company tried to reduce the use of non-environmentally friendly materials, like PVC or aluminium. Waste was sorted, which also had positive economic effects.

Employees have been informed every six months about changes in the company as a result of EMAS. At the meetings employees can complain or offer suggestions for improvements. These innovations are usually very modest and have little effect, though sometimes employees get very involved and make helpful comments. The outstanding example were the store room employees, who got very involved in the discussion about dangerous, toxic materials and suggested useful innovations.

9.3.4 Conclusions

The decision to take part in EMAS is not easy to explain. It seemed that the company used EMAS partly as an instrument to improve its organisational structure and to strengthen corporate identity in the company. What also seemed to be important was the long term view about likely changes in health management in Germany. The company also wanted to comply with environmental standards before they were set by law, for example, with respect to packaging.

Sustainability did not play a role in the decision to implement EMAS. The company is more engaged with the idea of "Responsible Care", rather than issues of sustainability. Environmental and economic concerns played a role, though the social dimension was not seen as important. Most of the criteria of sustainability (see Chapter 2) were, however, fulfilled. Knowledge about legal compliance was provided by the external consultant who made sure that the company was made aware of the implications of new laws. Continuous improvement was attempted, but was found to be difficult to sustain.

In Berlin, the case study site, participation was understood as using and improving the awareness and understanding of employees. The latter were not, however, enthusiastic about implementing EMAS. When they were told that the company intended to introduce EMAS, they just took it for granted. Nevertheless, they did not refuse to co-operate. According to the interviewees, implementing EMAS did not mean more work, because the company already had a high level of environmental protection. Employees were involved via the environmental team and informal discussions. There were no formal announcements, posters or other forms of communication, as at Lincoln (see section 9.1).

The environmental committee consisted of employees from different levels of the company hierarchy and, according to the interviewees, everybody had a chance to be heard in the meetings. However, in Berlin, the production and packaging plant, the staff were not highly qualified.[6] Thus, the topics discussed in the environmental committee were too difficult for most people to understand, as environmental protection is in the main a technical issue, requiring technical solutions. However,

---

[6] By way of contrast, in Konstanz, the other company site, where most of the research is done, staff are highly qualified scientists. The environmental manager, who is responsible for both sites, stated that the staff were more involved in Konstanz because of their higher level of education.

the works council was trying to be better represented on the environmental committee and was pleased that the EMAS II regulations require higher levels of participation (see section 3.5).

Other stakeholders, such as neighbours or customers, did not participate in the EMAS project.[7] Customers, such as physicians, were not interested in the environmental management of the company.

The external consultant was important, as was the verifier, who acted more like a consultant and made useful suggestions.

Innovations were to be found in the field of new technologies. They were rather expensive, but it was recognised that they would pay for themselves over time. These innovations were put in place because the company expected more restrictive environmental laws to be brought in, so they wanted to ensure they already met these likely future requirements.

According to the interviewees, there were hardly any links between participation and innovation. Some very modest innovations were suggested by employees, but the environmental consequences were insignificant compared with the technical innovations.

The management style in the company was described as like a family. The hierarchy was flat and the owner of the corporation expected the managers to have a paternalistic attitude towards the staff. Turnover was low.

How did this change? The structure of the organisation was improved and the production process became less opaque. None of the interviewees mentioned improvement in environmental awareness on the part of employees. They said that awareness was already high before EMAS was implemented. The company was disappointed about the lack of reaction by regulatory authorities and insurance companies to the fact of EMAS registration.

9.4   Hamburg Airport

9.4.1 Introduction

Hamburg's international airport is located in the northern outskirts of the city. Part of the 564 hectare site is situated in the state of Schleswig-Holstein. It is located about 9 kilometres from the city centre and the adjacent city suburbs are densely populated. In addition, Hamburg airport features a cross-runway system with four flight paths. As a result many people are affected by aircraft noise during take off and landing.

Hamburg is Germany's fourth largest airport. In 1998 there were more than 150,000 aircraft movements and more than 9 million passengers. It offers direct flights to 150 destinations, mostly European cities, particularly in north-east Europe. There are a few long haul flights. Most of the passengers come from the metropolitan areas of Hamburg and the neighbouring German states. In addition to the passenger traffic, some 65,600 tons of freight and 17,000 tons of air mail passed

---

[7]   Some interviewees mentioned that in Konstanz, communication with the city had improved.

through the airport in 1998. On peak days 40,000 passengers and 500 aircraft leave from or arrive at the airport. By 2010, it is expected to cater for nearly 14 million passengers and about 172,000 aircraft movements. The airport is therefore planning to increase the existing 42 stands in a three phase expansion programme over 15 years. For example, new apron positions and passenger departure lounges are due for construction. Planning permission for this project was granted by Hamburg City Council in 1998.

The airport is run by Airport Hamburg GmbH (FHG), whose principal owner is the City of Hamburg. FHG supplies the infrastructure needed for air traffic, i.e. the take off and landing system, runways, taxiways and aprons where the aircraft are prepared for departure, plus the terminals and other buildings. FHG also provides other services such as baggage handling, aircraft supplies and cleaning, and also runs a waste disposal service for the entire airport. It is responsible for operating facilities like power generators and other technical installations. FHG sales revenue for 1998 totalled 360 million DM. Some services are managed by FHG subsidiaries or other partner companies. FHG has several subsidiaries, which belonged to FHG before they were outsourced during the 1980s and the 1990s. The FHG has an influence on its subsidiaries, its business partners and all the companies located at the airport in relation to implementing environmental standards.

In 1998 FHG employed 1,369 staff. Apart from FHG and its subsidiaries, about 220 other firms operate from the airport, generating a total of 11,600 jobs.[8]

9.4.2 Thematic Background

Due to the location of the Hamburg airport, environmental issues have been of central concern for many years. Being a city airport means taking the protection of the environment very seriously, because residents, journalists, different commissions and politicians can all observe what happens at the airport very closely. Thus FHG has to be aware of its environmental responsibilities and needs to pursue environmental objectives. The management at all levels, the works council and most employees, many of whom live near the airport, agree that environmental protection and an "environmental image" are important because of the airport's location. Environmental protection measures are seen as a way of securing jobs at the airport.

Because the airport is so close to densely populated areas, noise is a special problem. Due to improvements in engine technology, modern jets make less noise than older aircraft. However, the number of flights is increasing and engine test runs, heavy goods vehicles and feeder traffic also contribute to noise pollution in the surrounding areas. Other problems are air pollutants and emissions that destroy the climate. Refuelling and de-icing in winter also produce contaminants that get into the water and the ground. Noise protection, the protection of soil and water and waste management are therefore key responsibilities of FHG. Other issues, such as

---

[8] FHG subsidiaries include the Special Transport and Ramp Service GmbH and the Hanseatic Security and Service GmbH. FHG partners at Hamburg Airport include shops and service companies.

efficient land use and measures to protect the natural environment, are also regarded as crucial objectives of the company. Even before the implementation of EMAS, FHG had laid down these objectives. EMAS and ISO 14001 were seen as a way to implement them systematically.

At Hamburg Airport, EMAS is not one specific environmental protection project, but is embedded in a long history of environmental protection. In 1989 an environmental centre was set up to co-ordinate corporate environmental protection activities. The task of the environmental centre is to advise and support all operational parts of the airport in planning, implementing and improving environmental protection. It is also responsible for dealing with queries and complaints on environmental matters. The environmental centre started with two people. By 2001, it comprised seven people, the centre manger and six staff responsible for noise abatement, environmental management, waste, the natural environment, water protection, bird strikes, and also training.

Night flight restrictions at Hamburg Airport are particularly stringent compared with German and European requirements and also compared with other German airports, such as Frankfurt or Düsseldorf. The restrictions were tightened still further by Hamburg City Council in 1998. Practically no flights are allowed between midnight and six in the morning. Only two mail planes, aircraft in emergency situations or other kinds of major problems can use the airport between these hours.

Hamburg airport was the first German airport to be certified under EMAS and ISO 14001.

9.4.3 EMAS Procedure

EMAS was initiated by the environmental centre and by the management. The latter supported EMAS mainly to build and strengthen the airport's environmental image. The environmental centre regarded EMAS as a helpful tool to systematise the activities of FHG and to meet external pressures. There was unanimous agreement to introduce EMAS. The environmental centre and the management planned to be certified with both EMAS and ISO 14001, because they wanted to demonstrate their environmental concerns in both the international and the European context.[9] The environmental centre was responsible for the implementation and secured the finances to undertake it. An external consultant was not required.

The initial review did not take much time, because all the necessary data had already been collected in an environmental impact study.[10] Because of the

---

[9] When FHG was planning EMAS implementation, service industries had not been included in the EMAS regulations in Germany. So FHG certified with ISO 14001 first, and then later with EMAS.

[10] The environmental principles of Hamburg airport are reflected in the fact that environmental protection is an essential component of company strategy, which means, for example, saving energy and resources and investing in best available technical innovations. Environmental protection extends beyond statutory requirements and is understood as a continuous process. Informing employees is taken seriously and staff are asked to suggest ways of improving environmental performance. The company holds discussions with local interests, providing information and asking for comments.

expansion of the airport, an environmental feasibility study was carried out, which – on the basis of numerous expert assessments – provided profiles of the airport's current and future effects on the environment. Thus the environmental centre could quickly produce an environmental manual to provide a basis for verification and certification under EMAS and ISO 14001. One member of staff of the environmental centre with responsibility for EMAS produced the environmental manual by asking the managers of all operational units to write down their work procedures. Most managers agreed without arguing to do this work for a range of reasons. First, top management had decided to implement EMAS and second, most staff were aware that the location of the Hamburg airport required a policy for environmental protection and a strategy to enhance its environmental image. Third, EMAS helped to increase safety, because it clarified responsibilities. The manual laid out the structures of the organisation and the assignment of duties. One interviewee said that employees were worried about the possible concerns of the environment officer from the City of Hamburg before the implementation of EMAS. There was a degree of uncertainty. But after the implementation of EMAS, everybody was informed about work procedures, responsibilities and rules to enhance environmental protection and consequently felt reassured. The manual was put on the intranet and every new employee is given all that they need to know. One person in each unit is responsible for the implementation of the manual.

The environmental centre has also developed and constantly improves the environmental indicator system, which provides data about specific processes, such as energy and water consumption per passenger. This information also provides a basis for regular analysis and helps to document and scrutinise trends and developments. In addition, the indices can be used to compare the environmental performance of different airports.

In 1998 a training course on environmental management was established. This now forms a part of the company's internal training programme.

The works council was not heavily involved in the EMAS process, but was informed about all the stages. The main focus was on security and safety issues. At the time the works council was preoccupied with the process of outsourcing at the airport. It had started in 1994 and the employers were concerned, that working conditions would deteriorate in the independently operating units. In fact, a lot of mistakes were made so that the works council had a lot to do to ensure that employee interests were taken seriously. Nevertheless, the works council fully supported EMAS and ISO 14001, because they regarded it as an instrument to improve job security. However, the environmental management system was not seen as the most important topic. The works council believed that the environmental centre would do "the right thing" and did not pay too much attention to it.

The management of the company and the staff of the environmental centre did not consider participation by employees as very important. No special working groups or meetings were set up. Employees were considered important in so far as they had to carry out the processes detailed in the environmental manual. They were also asked to provide information and to suggest improvements, which some of them did. Depending on the operational unit, the impact was modest or rather more substantial. Some interviewees regarded the suggestions of employees as important, some did not. It seemed that technicians and scientists regarded technological innovations as more effective, whereas the works council members considered employees' suggestions as important.

The members of the environmental centre stressed that participation in different working groups and networks outside the company was important, for example the Working Group for German Commercial Airports and the Hamburg Noise Abatement Commission *(Lärmschutzkommission Hamburg)*. The Hamburg Noise Abatement Commission, which had been set up in 1969, is an initiative to solve the noise problem by considering both economic and social factors. The continuous and detailed dialogue between different interests is aimed at guaranteeing solutions "everybody can live with". Regular meetings, 4-5 per year, are held and representatives from pressure groups, local residents, the general public, politicians including some mayors of districts and villages, state agencies, airlines, the Chamber for Industry and Commerce and FHG, particularly from the environmental centre, – altogether nearly 30 people – take part. The Commission makes recommendations to the local authorities. FHG also tries to keep in touch with its neighbours. It participates in local events, publishes a quarterly newspaper for local residents, financially supports clubs in the nearby areas, produces and distributes videos about the airport and an information folder for school classes. FHG arranges events at the airport like "Hamburg Airport Classics" with aircraft from the past or exhibitions.

Members of FHG also keep contact with government offices, local authorities and politicians and co-operate with universities and research institutes on the environmental aspects of airports. Several dissertations have already been written on environmental protection at the airport. FHG is active in different national and international airport networks, exchanging information about environmental protection and environmental management systems at airports.

Noise is the most important problem so different means have been developed to deal with it. Apart from technical innovations, the prohibition of night flights (with some exceptions) and sustained dialogue with all stakeholders, FHG imposes higher fees on noisier aircraft. Noise is also continuously measured and FHG publicises the results. It is planning to introduce special software to make it easier to identify sources of unnecessary or excessive noise. To reduce noise pollution caused by engine test runs, FHG is also planning to build a new noise abatement hangar, large enough to contain all the aircraft types regularly landing in Hamburg. There is no comparable construction of this kind anywhere in the world and therefore the construction turned out to be difficult. Another technical innovation was a success: the use of mobile warm air supply units. This means that stationary aircraft do not have to keep their engines on to provide air-conditioning. The supply units can do this and they use less energy, as well as making less noise.

FHG also records information from local residents to find out which areas are especially affected by noise. Residents who were heavily affected received special help to improve noise insulation in their homes. FHG gave more financial support than the law required. Since the summer of 1998, it has spent 10 million DM for different kinds of noise insulation. To keep noise down, a fourth noise abatement programme has been put in place. It provides funds to install insulated windows in local homes. Additionally, residents can apply to have insulated ventilators installed free of charge in children's rooms and bedrooms.

FHG handles all waste disposal at the airport. The separation of waste is a major objective. In 1998 1,866 tons of waste and 1,124 of recyclable materials were generated at the airport. This included the waste collected cleaning the aircraft cabins and the terminals as well as from the shop floor. FHG's strategy to separate waste can work only if staff, companies and passengers co-operate. Effort and information is needed to convince these groups and it also requires user-friendly logistical systems. Before the introduction of EMAS, it had already proved to be possible to reduce the amount of non-recyclable waste per passenger by a third.

FHG supplies the airport with electricity and heat from a combined heat and power (CHP) plant. Half of the power has to be supplied from the city grid. Power stations like the CHP plant at Hamburg Airport have a very high level of efficiency. FHG is planning a study that will analyse energy use at the airport to see where savings can be made.

The environmental centre deals with legal compliance by training, reading and going to conferences. Because of the large number of relevant laws and regulations, it is sometimes difficult to keep track.

What was the impact of EMAS? Environmental performance, environmental awareness on the part of employees and in part also that of passengers all improved. Suggestions from employees were taken seriously, and there were financial incentives. Nevertheless, according to one of the interviewees, most employees do not know much about EMAS.

Relationships with the authorities improved, and public opinion was also influenced by the implementation of environmental management systems. Public support is much greater than at airports without a positive environmental image. FHG also won a prize for its implementation of EMAS and ISO 14001, the Flight International Reward. Most people were interested in the problem of noise, though other topics are seldom mentioned. Saving money was not the first target for EMAS. Therefore a cost benefit analysis was not undertaken.

9.4.4 Conclusions

What kind of role did sustainability, participation and innovation play in the EMAS procedure at Hamburg airport? Sustainability hardly featured at all. In fact some interviewees stated that other airports pretended to be sustainable, but did not improve their environmental performance. The concept of "sustainability" was looked upon with mistrust, as it could provide the opportunity to disguise low environmental performance. This mistrust could also be due to the fact that the staff of the environmental centre were without exception scientists or technicians, who were all committed to technical means for enhancing environmental protection. Overall, the environmental and economic aspects of sustainability have been taken seriously. The social dimension has not been specifically taken into account.

Nor did participation play an important part in the EMAS process. The employees were not important sources of data for the initial review and were only partially relevant in the production of the manual. They were involved in some units, depending on the perspective of the unit manager, but no structure was introduced to compel their participation. The staff who were involved from the beginning were the managers of the different operational units. How EMAS ideas were disseminated in the units depended on the leadership style of the managers. Several stakeholders were involved in discussions about noise protection, but this was not a result of EMAS. It had happened years before EMAS was implemented.

Being a high tech company, innovation at Hamburg airport was seen to be encouraged by technical means. Some improvements were based on employees' ideas, but these innovations were relatively modest compared with the technical solutions put forward. The pressure groups focusing on the protection of residents from noise or the Hamburg Noise Abatement Commission demanded innovation, but they were not innovative themselves.

There seemed to be no important connection between innovation, participation and sustainability. All the innovations were technical; other kinds of innovations did not play a significant role. The impact of other stakeholders, particularly local residents, was limited to noise protection. Other aspects of sustainability did not feature. Participation in the form of pressure can be seen as a crucial incentive for technological innovations. Nevertheless, this form of participation existed before EMAS. EMAS was useful for systematising all operations and for strengthening the image of Hamburg airport as one of the leaders in environmental protection.

9.5 The City of Augsburg

9.5.1 Introduction

Augsburg is a wealthy city with a 2000 year history. It is one of the leading Bavarian towns, has more than 250,000 inhabitants and is both an industrial and a trading city. In Augsburg there are several administrations and authorities, universities and other educational units, galleries and opera houses.

9.5.2 Thematic Background

Augsburg has a long history of environmental involvement. In 1988 it called for a comprehensive survey of its environmental situation and the measures needed to protect the environment. In 1992 the city council decided to take the results of the survey forward. Other decisions and initiatives demonstrated concern for environmental protection and sustainability. Local citizens and NGOs started a Local Agenda 21 scheme in 1996. As a result of this, an environmental advisory council was established in 1997, comprising different local groups. This environmental advisory council gave advice to the city council about Local Agenda 21 and sustainability. Furthermore the Environmental Competence Centre of Augsburg-Schwaben is located in Augsburg. In 1999 the city council agreed a set of evironmental guidelines, stating, for example, that

- the city is committed to a programme for improving environmental protection,
- environmental protection is a task for all staff,
- managers have the responsibility to initiate appropriate measures and programmes,

- the city administration encourages all staff to implement environmental protection through the provision of information and training, and
- a Local Agenda 21 scheme is in operation.

The guidelines also mention the implementation of an environmental management system as a part of a reform of the administration. A year previously, in 1998 the mayor had directed that an eco-audit system should be implemented in all the administrative units of the city. The initial impulse to implement such a system came from the Department of Environment and Health. An external consultant – a professor at the University of Augsburg – asked this department whether it would be interested in embarking on the EMAS process. Up till then, no environmental management system had been implemented in the city. The consultant was already familiar with EMAS and was able to demonstrate environmental as well as management competences. The director of the Department of Environment and Health was sympathetic to this idea. In the end, he and the external consultant convinced the mayor that EMAS would be an appropriate vehicle for environmental protection and a good instrument to give the city a positive environmental profile.

9.5.3 EMAS Procedure

The city first developed the idea of introducing EMAS across the whole administration and then pilot projects were selected. Heads from different departments and offices were asked whether they would be interested in taking part in a new project. At the beginning, practically nobody knew what EMAS was. Therefore, the director of the Department of Environment and Health, together with the external consultant, started to brief the departmental heads and managers about EMAS. They were helped by a study undertaken by one of the external consultant's students. The student found out from all the heads of departments what they knew about EMAS and whether they were interested in implementing it. As a result, EMAS became a topic of discussion in the administration.

A letter was sent to all departments, stating that the city was seeking administrative units, which were interested in participating in EMAS. Five responded and were validated and registered in 2000. They were the emergency services department, the environment department, the Education and Social Centre, the sewage plant and one of the administrative buildings.

In the rest of this case study, general findings about the implementation of EMAS are outlined and the introduction of EMAS into two of the five units is profiled.

The key people responsible for the environmental management system in Augsburg were the director of the Department of Environment and Health, a "core team" and the eco-audit team. The responsibilities of these key players overlapped. The director was responsible for co-ordinating public relations, taking care of the implementation of the system, planning the eco-audits and the reviews and publishing the environmental statement. The "core team" consisted of the director and delegates from two departments. Its task was to co-ordinate EMAS, to keep contact with other internal environmental organisations and to develop the concept and its implementation. The eco-audit team comprised the "core team" and the managers of the pilot projects. They were involved in the implementation of EMAS, the collection of data, the exchange of experience and the development of new ideas. For specialist ideas in relation to, for example, water or rubbish, expert teams were to be established.

Environmental management systems and EMAS were established on two levels. On the one hand, there were the comprehensive and management oriented structures with a general overall concern for the whole administration and, on the other hand, there were decentralised units undertaking eco-audits in particular parts of the administration.

Broad targets were binding for all units, but each of them had to develop its own objectives reflecting its particular tasks. The framework for the whole administration offered a set of rules, for example, the rule that staff should be informed or that the Local Agenda 21 process should be taken into account. It also included guidelines for every important sector, energy, water and so on. One of the purposes of the framework was to transfer experiences from one unit to another.

The managers of the five pilot projects, which were to implement EMAS in the first cycle, were briefed several times about EMAS. These meetings were strongly supported by the external consultant and his assistants, for example with workshops or local expertise. After a full briefing, the managers in turn briefed the staff members of their units. The city had provided funds to pay the external consultants[11], but did not finance additional staff.

The next phase of EMAS was the initial review. The different units undertook this task in different ways. According to the interviewees,[12] the quality of the reviews depended on the personality of the managers and their leadership abilities. In some units the managers had a good overview of what data was needed; other units had difficulties in collecting the relevant information. The same was true for the establishment of the management system. In some units the management system was implemented without problems, whereas in other units difficulties were faced.

After internal validation by auditors trained within the administration, all the units were verified by an external verifier. A second verification will be required after three years, whereas internal audits take place every year.

Some interviewees considered EMAS a good tool for systematising measures for protection of the environment. One positive result of the implementation of EMAS was, for example, that awareness about energy saving was raised markedly. A negative factor was that resource savings were not always possible in the same way in all the units. It was difficult, for example, in rented housing, where the landlord was not prepared to improve the insulation. Most interviewees also regarded EMAS as too expensive and too time consuming. Costs were likely to be too high and there would be likely to be too many formal procedures.

---

[11] After the external validation of the first pilot projects in Augsburg, external consultants were no longer required. The managers and other staff involved were trained to be consultants.

[12] Interviews were held with the environmental representative, the director of the Department of Environment and Health (the environmental co-ordinator), the manager of the Education and Social Centre (already verified) and a staff member from the administrative office, which was soon to be verified.

The interviewees agreed that the five units which had first implemented EMAS were the forerunners and that their unique experience with EMAS would not be able to be transferred to other units. The forerunner units were the mostly highly qualified in terms of staff, were well organised and had good leadership. It is planned that all the city's offices and departments will implement EMAS in the future. The interviewees believed that the next stages would be more difficult and that the units at the end of the process would make the most trouble ("These are the units that do not work anyway"). In the second stage, four more units were identified for implementing EMAS.

In the rest of this section, the implementation of EMAS in two units is profiled: the Education and Social Centre, which was registered in 2000, and the administrative building, which, at the time of writing, had reached the stage of internal review.

The *Education and Social Centre* is a meeting place for different groups and organisations. Many different kinds of events take place there. There are 13 staff, several main users of the centre and half a million visitors per year. The centre is based in a building which has several other tenants.

The EMAS procedure started with the letter that the city sent to all departments telling them of the idea of implementing EMAS. The manager of the Education and Social Centre was personally involved in environmental protection and supported the idea. Another reason for taking part was that the centre received some funds from the city which it was responsible for managing. Any money saved through EMAS could be used for other purposes. Thus environmental and financial reasons led to the implementation of EMAS. The choice of EMAS rather than another environmental management system was because the city had already decided to adopt EMAS.

In several meetings at the Department for Environment and Health, the managers of the pilot projects were briefed about EMAS. Following this, the manager of the centre set up a meeting of the 13 staff to discuss the introduction of EMAS. As the numbers were small, it was easy to persuade all of them to take part. In other offices with larger numbers of staff, this process was more difficult.

The initial review caused few problems, because most of the data was already collected. Environmental protection had been a kind of a "hobby" of the manager and he had been gathering environmental data since the 1980s. An additional advantage was that the centre administered its own building, so it did not have to ask others for information about the consumption of water, electricity and so on.

The management system was put in place by dividing the centre into different sections with one person being responsible for each section. The environmental performance of each section is checked every two months.

The relevant laws and regulations are updated continuously. Compliance with the law is a problem, because it is difficult to get a good overview of all the new laws. Use of consultants and a great deal of reading are necessary to keep track.

Different *stakeholders* participated in the EMAS process: the manager, the staff, external consultants and the general public, or, to be more concrete, those people who use the centre. Everybody received environmental information and was asked to participate in enhancing protection of the environment. The other tenants of the building were invited to co-operate, which they did in part.

Participation was limited to the provision of information. Staff were informed at meetings. Every one of them and all the main users of the centre were given a handbook on environmental management, in which all the responsibilities and processes were laid out. The visitors to the centre and the other tenants in the building were informed about EMAS, for example, through an exhibition about EMAS organised by the centre. This exhibition attracted a lot of attention and some local employers asked for more information in order to implement EMAS themselves.

The staff were very co-operative. They all agreed to implement EMAS and considered it in a positive light. This positive attitude was due to the small number of staff who furthermore exhibited a high level of corporate identity and a high degree of mutual trust. They liked the idea of being forerunners and they continued to seek out weaknesses and to improve overall environmental performance. The centre manager also encouraged staff through incentives. For example, he promised the cleaning women a new environmentally sound vacuum cleaner if they could manage to save money. Both the manager and staff believe they are respected for their activities within and outside the centre and this has sustained their enthusiasm.

Sustainability was, according to the interviewees, a crucial reason for implementing EMAS. They argued that environmental protection had to be a long term strategy. Before the implementation of EMAS the centre had been active in environmental issues, although in a less systematic and structured way. The financial benefits were smaller than in those units which just started with environmental concerns. Although continuous improvement in environmental performance had its limits, some money could be saved and used for other purposes. Social issues were not on the agenda.

Innovations initiated by the staff or by visitors were modest, but there were some ideas which enhanced environmental performance, for example, a fountain was turned off, because it was hidden in a place where nobody could see it anyway. The air conditioning system was turned off, because staff preferred to open the windows. The centre was also helped by some innovative ideas from external consultants, from experts (knowledge holders) and from other units ("diffusion of best practice"). There were in the main improvements, but innovations which needed more money could not be put in train.

The introduction of EMAS in one of the administrative buildings took place in the second round of implementation in Augsburg. This unit comprised several sections and had about 100 staff. Implementation was full of difficulties. The process started in the middle of 2000 with a meeting for the staff, supported by external and internal consultants. Then in every section a responsible person was appointed, who had to collect the data for the internal review and to encourage the staff. However, the latter did not co-operate in the initial review. They were afraid that EMAS was just a new method to control them and felt they would be disadvantaged.[13] Most staff felt that the implementation of EMAS was a top down regulation and tried to be as little involved as possible. They were convinced that EMAS just meant "more work" with few results. They argued that, in contrast to

---

[13] One interviewee said that staff were, for example, afraid that their coffee machines would be taken away for environmental reasons.

private industry, public authorities would get no benefits from implementing EMAS and that not even the general public was interested. They also complained that the city appeared to have money to implement EMAS, but refused to find funds for other urgent problems. They argued that the first pilot projects received money to improve their environmental performance, but this would not be the case in the later stages of implementation.

One main problem was that staff did not respect the leadership of the unit. Distrust was high and nobody was doing anything about it.

In contrast to the employees in the Education and Social Centre, the staff of this administrative unit did not feel pride in implementing EMAS. They felt that they would not get any praise or recognition, and regarded the process as useful only in part.

One benefit perceived was the provision, after the initial review, of more data about the consumption of energy and water. Another benefit was that the unit learnt more about its own consumption behaviour. Again, knowledge of environmental legislation became more widespread. Small scale changes happened, too, like the sorting of different kinds of rubbish, but, according to the staff, there were no noticeable innovative changes.

9.5.4 Conclusions

Different "holders" actively participated in the EMAS process in the City of Augsburg. External consultants played a crucial role to start the process and to provide the necessary input. The professor at the University of Augsburg and the Institute for Management and the Environment were the initiators and *knowledge holders* at the beginning of the EMAS process. In the next stage the managers of the selected units were involved *(status holders)*. Staff and in part some members of the general public were engaged in the following phase, although this participation was limited to collecting data and the provision of information.

The innovative capabilities of staff were used in different ways in the different units. Interviewees agreed that EMAS was more successful in those units which involved staff from a variety of hierarchical levels. In one or two cases the manager or other senior member of staff acted without involving employees and without setting up a team. EMAS did not work well in these units. For example, in one case the EMAS process focused only on the pet interest of the manager involved, or, to give another example, there was no continuity after a key individual left. In units where staff were put in the picture from the beginning, where teams were set up and where employees were encouraged to have innovative ideas, EMAS worked better. Although the interviewees agreed that the capabilities of staff would have been a valuable asset, nevertheless, participation was not a priority. In contrast to the Division of the Environment in the City of Hannover (see section 9.6), the importance of a co-operative management style was downplayed.

According to the interviewees, sustainability did play a role in the implementation of EMAS, but the social dimension was not emphasised.

What changed with the implementation of EMAS? Business management concepts were introduced, a bridge to quality management was built, communication about environmental issues improved and the administration understood that protecting the environment could help to save money. Overall awareness about environmental matters changed.

The conditions for successful implementation of EMAS, including sustainable development, innovation and participation, were threefold. First, a tradition of participation and a culture of trust were crucial. Participation was more likely, when there was a high degree of trust between the key players (director, manager, co-ordinator) and the staff. Without this trust, staff members refused to co-operate, because they felt EMAS was part of top down control. Where staff had experienced in the past some disadvantage from the implementation of a new programme, they were less likely to welcome another one. Where trust between the unit manager and staff was non-existent, participation and innovation on the part of staff was very unlikely to occur.

Second, a tradition of environmental protection was also helpful, if not necessary for success. Organisations reflecting such a tradition had fewer difficulties in implementing EMAS.

Third, according to the interviewees, good leadership was necessary for success. It should comprise a sustainable vision together with an ability to convince staff to take sustainability seriously. EMAS worked only in well organised units with good leadership. Those units with internal problems were not able to implement EMAS successfully.

Thus, it can be said that there is high path dependency. If EMAS fits into the path of traditions and cultures, it will be successful. If not, success is less likely and transfer of "best practice" is hardly to be expected.

9.6 City of Hannover, Division of the Environment

9.6.1 Introduction

In August 1988 the City of Hannover decided to establish an environmental services directorate headed by a Director of Environmental Services. Following this initiative, a Division of the Environment was established. The different sections and units of the Division of the Environment had existed in other departments in the past and were combined under one director. At first, the Division of the Environment had two sections, one for environmental protection, including, for example, city planning and green spaces, and the other for environmental monitoring. Later, new sections were added like environmental planning for Expo 2000 and the regional headquarters for energy and climate protection.

Since the beginning of the 1990s the Division has been reorganised several times.[14] 60-80 staff work in the Division of the Environment, about 60 of whom are permanent staff and 20 are freelance, trainees or other kinds of "non staff". The staff are highly qualified specialists. Most of them have a higher education or university

---

[14] The Division of the Environment will be re-organised again in 2002. The environmental services directorate will get a new name, part of its current responsibilities will be handed over to other recently created offices and new responsibilities will be taken on board.

degree. The objective of the Division of the Environment was and is to seek, through environmental planning, to avoid environmental degradation. It has developed plans for the city to improve its environmental performance. It collects and analyses relevant data and is also responsible for environmental advice. One section is responsible for nature conservation, others for environmental monitoring. The regional headquarters for energy and climate protection develop co-ordinated measures to reduce CO_2 emissions. Since 1996 the Division has been located in two buildings that are protected by historical monument regulations.

9.6.2 Thematic Background

Hannover was one of the first German cities to address environmental issues. It is a member of the International Council for Local Environment Initiatives (ICLEI), the Cities for Climate Protection, the Climate Alliance and the international Local Agenda 21 network. In 1995, Hannover signed the Aalborg Charter, a charter of European towns and cities promising sustainability. In 1997, a meeting of all the heads of administrative departments agreed basic guidelines in the spirit of ISO 14001 and EMAS. These basic guidelines included, for example, resource savings and continuous improvement.

The city has been governed throughout the 1990s by a red-green coalition. From the mid 1980s environmental policies have been a high priority in both the political arena and in the administration. Some interviewees[15] stated that there was a network of environmentally concerned people (staff and managers) in the administration. Initiatives for new forms of environmental protection come from the politicians as from the administration. Since the mid 1990s sustainability has become a topic as well.

At the beginning of the 1990s Bad Harzburg, a city near Hannover, implemented environmental monitoring. This led the city council and the Division of the Environment to become interested in environmental monitoring systems and to think about implementing some kind of environmental management system. The city council, the Deputy Chief Executive the director of the environmental services directorate took the view that environmental management systems were a useful tool for environmental protection and for the modernisation of the administration. In 1996, the Environment Committee and the Finance Committee held a hearing of ten experts on environmental management systems. This became a starting point for wider discussion.

The city council considered implementing an environmental management system throughout the whole administration of Hannover, but decided that the different units were too varied to integrate them all into a single uniform management system. Nevertheless, the city council wanted to set an example and passed a resolution that a pilot project should be undertaken. One division should implement an environmental management system. This pilot project would develop basic competences and

---

[15] Interviews were held with the director of the environmental services directorate, the director of the Division of the Environment, the environmental representative from the Division of the Environment, the overall co-ordinator (a member of the environmental services directorate), a member of the staff council.

produce a handbook for implementing an environmental management system that other units could build on. Staff from the Division of the Environment asked that they should be the pilot project. They wanted to be the forerunner because the Division of the Environment was supposed to disseminate information on environmental topics and so it would make sense to build competences there. The Division had been innovative in the past and was interested in improving environmental protection. In 1996 the city council passed a resolution that the Division of the Environment should implement an environmental management system which covered the requirements of both EMAS and ISO 14001. They laid down that the process should be integrated into the administrative reform of the city and should be undertaken in accordance with the rules of Local Agenda 21.

The objectives of the EMAS project were to support the "role model" position of the administration, to ensure compliance with the law, to identify and reduce environmental costs, to improve the safety of the work environment and the capabilities of the staff and finally, to improve the quality of service performance.

9.6.3 The EMAS Procedure

The Hannover Division of the Environment was one of the first administrations in Germany to implement EMAS and 14001. At the time (1996), EMAS was concerned only with manufacturing sites. Administrative units and service industries were not part of the first Eco-Management and Audit Scheme. Nevertheless the Division of the Environment was interested in EMAS, because it was a broadranging idea, not only targeting a few short term effects, but continuous improvement as well.

Because the regulations for EMAS and ISO 14001 were designed for the manufacturing sector, they did not fit with the special procedures and needs of an administrative division. Thus the Division of the Environment designed a new environmental management system based on ISO 14000 and EMAS but including elements of "6E", a programme designed by the Swedish Labour Union (TCO). This programme covered work environment issues such as ergonomics, the psychological dimension and security. A participatory bottom up approach was seen as essential.

Being a pilot project raised special problems, but it also had advantages. On the one hand, the Division of the Environment was "thrown in the deep end", as one of the interviewees expressed it, because it could use an established environmental management system adapted to the administration. In the beginning the director and staff of the Division of the Environment were not quite sure how EMAS would work in an administrative setting. None of them had previous experience of such a system. They had to "invent" an appropriate one for it to be registered under the rules. On the other hand, as a pilot project, the division got special support from the city council. External consultants were funded by the city so that the Division of the Environment did not have to pay for them out of its own budget.[16]

---

[16] One interviewee stated that one person responsible for the EMAS process in a unit of 80 employees was too lavish and too expensive. This kind of support was provided because of the frontrunner role of the division and could not be provided for other units.

The staff of the Division of the Environment reacted to the decision of the city council to implement an environmental management system in different ways. The director and other key actors were pleased that a political decision both forced and supported the Division of the Environment to implement an environmental management system. Without this political backing it would have been difficult to do what they considered to be necessary to improve environmental performance and the associated administration. Large numbers of staff of the Division did not approve. They felt that the city had forced the system on them. Therefore the initial staff council meeting, at which the idea for an environmental management system was introduced, was a difficult one. First, the staff were afraid that such a system would simply mean more work or perhaps reduction in staff numbers. Second, the consultants used terminology that most staff did not understand. So, at first, most staff appeared not to be interested in the process, even though most of them, working in a Division of the Environment, were already involved in environmental topics. Fortunately, two committed members of the staff council were able to influence the debate. They made it clear that there would be no staff redundancies. They promised to enhance not only protection of the environment but also the working conditions for all staff.

The first step was to develop competence within the Division of the Environment. In 1996 several courses were held to teach staff about environmental management. Some staff also trained as auditors. As it turned out, the building of competence was crucial for the success of the whole procedure.

The division employed two experienced consultants in 1997, who played a crucial role in modifying the environmental management system.

The first phase began in October 1997 and was planned to last nine months. Eight project teams were established in the division. The director decided the topics the teams should work on after consulting experts and the staff. Teams consisted of staff, the director and the managers. Nobody was forced to take part. People could join the team they preferred. The teams turned out to have a mixed composition in several ways. First, people often chose teams focusing on topics other than those they worked on. This was seen to be very helpful, as the staff would learn about the work of other parts of the division. Furthermore, they might be able to see possible improvements that the staff working in a particular area could not see from an insider point of view. Second, hierarchies did not play any role. The staff, the managers and the director of the division worked together as equals. For example, the managers did not necessarily speak for the teams. The fact that hierarchies were not reproduced in the teams led to a feeling that every member of staff was important. In contrast to some other divisions and to some manufacturing sites, there were no problems on this score at the Hannover Division of the Environment. As the director pointed out, the philosophy of management in the division was more co-operative and collective rather than hierarchical.

The teams worked relatively autonomously. Every team met several times during working hours. About half of the staff were members of project teams. They had a strict time limit of nine months, because the Division of the Environment wanted the system to be established by its tenth anniversary. This time limit turned out to be helpful. Working against a deadline meant that the meetings were taken seriously and thus were very effective. At regular intervals, the teams met to present the results of their work. They could also ask for help, if they had serious problems, from an external consultant or from a conflict manager.

One of the first tasks of the teams was to undertake the initial review, which meant collecting data about energy, water and electricity use. They also described all the working procedures of the division, discussed the objectives of the environmental management system and developed ideas for improvements. This initial review turned out to be rather more complicated than expected. The teams had to find out what information was relevant for EMAS and how to collect it. One of the consultants was able to help, as he had had previous experience with EMAS and could provide a checklist. Nevertheless, it was not easy ("We checked everything we could find and think of."). They also collected information from schemes elsewhere in Germany and in other countries, for example England. It was "a process of finding the right way" and they had to identify different sources to collect the data they needed. One positive outcome of this participatory procedure was that some improvements occurred as early as this first phase.

One of the teams also developed an anonymous questionnaire to ask staff about possible improvements, about working conditions, about colleagues, the director and so on. 80 per cent of the staff completed the questionnaire. The results were very helpful, for example, in identifying which processes should be worked on further. Many of the ideas suggested in the questionnaires were implemented.

The efforts of the teams were very successful. Nearly all working procedures were described, analysed and improved. Furthermore most staff recognised that the changes were advantageous for them. For example, management never made a decision without consulting the staff council, so the staff no longer feared that people would lose their jobs. The process was anonymous so that it was not possible to find out details about a single person ("We analysed process, not people."). Furthermore, the flow of information and the transparency of the decision-making process was enhanced by such means as publishing the reports of the meetings of the managers on the intranet. By the end of the nine months, nearly everybody was taking the process very seriously. It was apparent to all that EMAS really made a difference, for example, in energy usage, and that money could be saved.

In the beginning, the division had worked within the ISO 14001 and EMAS rules which were oriented towards particular issues. However, with the help of experienced consultants and based on the experiences in the teams, it was seen that this approach did not work and so it was decided to develop a system that was oriented towards process, which was more suitable for application in an administration. Thus, the unit that was documented, analysed and improved was the process.

A later move was the appointment by the director of one of the staff to be the environmental representative. He became responsible for the implementation and maintenance of the environmental management system. His work was supported by a workgroup for environmental management, which met twice a year and as and when necessary. This workgroup took over most of the work the teams had done before, sometimes with the help of internal experts, for example experts on particular issues, or other officials. Some tasks were given to individuals or small groups. The management system ceased to need external support. At the time of writing, the key players were the staff, the director of the Division of the Environment, the environmental representative and the workgroup.

All the rules of the management system were compiled by the environmental representative into a handbook, the first version of which became available in 1998. Other administrative units could use this handbook as a source of ideas which they could adjust to suit their own needs. The handbook is revised every year. Training ensures that all staff know their responsibilities.

In 1998 the system was put in place. The first internal audit was carried out a few months later and the first external audit took place in June/July 1998. The Division of the Environment was certified under ISO 14001 in 1998. Several months later, the service sector was brought into EMAS and this included public administration. Therefore, the Division of the Environment decided to be validated and registered under EMAS. It was the fourth EMAS in the service sector in Germany.

The Division of the Environment introduced yearly internal auditing – in addition to the external auditing – with trained auditors within the administration. It had started to build up a pool of auditors by training interested people.

The interviewees mentioned three problems implementing EMAS. First, the process of involving most of the staff was extremely time-consuming. The amount of regular work was not reduced, although it was somehow expected that those not involved should support those involved. This difficulty was eased when managers also became involved in the EMAS process, because they then gained more understanding of its implications for the staff involved. Second, some staff did not understand why they should have to describe their working practices. They felt they were under surveillance and distrusted the process. Some staff remained sceptical, but after realising that control was not the main issue and that their involvement was taken seriously, most employees changed their outlook. Third, little information was available about how resources were used in previous years.

Another difficulty was the fact that the division was located in a house owned by a private person. The landlord was not willing to implement energy saving changes.

The city council had decided that the environmental management system should be part of the Local Agenda 21 process. Even before that formal decision, there had been a close connection with Local Agenda 21. The Local Agenda 21 people were in the same building and the Division of the Environment had an overall responsibility for their activities. Some of the division staff were engaged with Local Agenda 21 process, so ideas could be easily incorporated into the EMAS process. One Local Agenda 21 group was working on social and environmental issues in the business sector. Some of the outcomes of its discussions had been included in the EMAS process. The link with Local Agenda 21 underpinned a connection to the general public, even though there seemed to be little interest in EMAS as such. For example, only experts, companies and researchers were interested in the environmental statement.

By adjusting an environmental management system to the needs of the administration and the experience of implementing it, staff gained a great deal of knowledge. Some were trained to become auditors, so these kinds of competences are now held by staff of the city administration. A new approach to training was also developed to improve the competences of administrative staff. This was particularly important as the city council had decided that every unit in the administration of the city had to introduce some kind of environmental audit within three years. The staff of the Division of the Environment were in a position to help those responsible for developing these audits.

There was some criticism that the amount of work and the costs of registration were too great. Therefore only some of the divisions are to be registered. Most of the interviewees stressed that it was more important to implement the idea of EMAS than to be registered. It was more important to implement the instrument than to be verified.

Besides the local knowledge gained by staff (work holders), exchanges with other units and divisions working with environmental management systems were important as well. By co-operating with others "best practice" could transfer to the Division of the Environment. For instance, it took on board some ideas developed through the implementation of EMAS elsewhere, such as in England.

The question whether EMAS or ISO 14001 were formally implemented was not central. Both systems were basically a vehicle to do systematically what the Division of the Environment had previously done unsystematically. Interviewees saw disadvantages in EMAS such as the formal procedures, which took up a lot of time and work, and validation was seen as very expensive.

Most of the interviewees agreed that the benefits of EMAS could not be calculated on a cost benefit basis. The time and the number of staff involved were never calculated. The validation and registration processes cost about 50,000 DM (about 25,000 euro).

On innovation, the interviewees agreed that when staff are taken seriously and when they accept the management system, there always are innovative ideas put forward. In the Division of the Environment it was not surprising that process and technical innovations were far more widespread than product innovations. It was, for example, the idea of a member of staff to install an energy saving device on all computers. As a result, there were savings of 15 per cent on electricity.

The advice from interviewees to staff in other divisions starting EMAS was:[17]

- make the management system process oriented;
- involve all staff and the staff council from the beginning;
- use existing structures and flows of information (staff council meetings);
- in large administrations, start on a small scale;
- adjust the system to your needs ("It makes no sense to implement a norm/ standard, when there are no benefits for the company", as one of the interviewees put it);
- develop competences in the staff;
- descriptions of the process should be written by staff not by external consultants;
- implement only when the director and all managers support the idea;
- get in contact with other companies or administrations, which have already had experience of EMAS.

---

[17] Other divisions have begun to adopt the system, but there have been problems, because some directors and managers have not been supportive. Where no financial benefit is identified, the key actors lose interest.

To sum up, the decision to implement EMAS was not a purely bottom up decision. In the mid 1990s, when EMAS was being discussed in Hannover, only a few members of staff were keen to implement it. After the EMAS process got under way, it became a participatory bottom up one.

9.6.4 Conclusions

Participation took place in different phases and took different forms. The city council was a crucial player in initiating EMAS. The key players for the first phase of implementation were the *internal status holders*, the *work holders* and, in the early stages, *external knowledge holders*. The general public and *other stakeholders* were only partially involved through the Local Agenda 21 process. In general, the ordinary public was not aware that the Division of the Environment had implemented EMAS and showed very little interest.

Innovation, sustainability and participation were closely linked. All interviewees agreed that the participatory approach had led to technical and procedural innovations which made improvements in environmental performance possible. They also agreed that there had been a continuous process of innovative ideas for improvements, which was only possible because the staff were involved in a participatory process.

What role did sustainability and the Agenda 21 play in the introduction of EMAS? According to some interviewees and the documents scrutinised, the concept of sustainability was crucial. One of the interviewees stated that the debates about sustainability and about the implementation of an environmental management system ran in parallel. Discussions about sustainability started about the same time as the debate about environmental management systems. Economic, environmental and social issues were taken into account as well as a focus on a participatory approach.

The impact of EMAS on sustainability can be found in the four environmental, economic, social and participatory fields:

i.   Usage of water, electricity and other energy sources could be reduced. Sorting of rubbish had been accomplished. The amount of paper, however, could not be reduced but rather increased. Although the amount of resources saved was not huge, it was a beginning and the Division of the Environment remained eager to achieve constant improvement.

ii.   The economic effect was small. All those interviewed stressed that EMAS was not implemented primarily for economic or financial reasons. EMAS was a pilot project, targeted to make the environmental management system more systematic and to produce a handbook that could be used in other divisions.

iii. The working environment, for example, the flow of information, was improved.

iv.   All interviewees considered it vitally important that staff should be involved from the beginning. They pointed out that it was enormously helpful if staff developed ideas for improvements themselves, supported by consultants. The participatory approach had had a tremendous impact on the *work holders* taking part. The fact that their ideas and wishes were taken seriously changed their attitude towards their workplace in a range of ways. They developed a corporate identity. They gained an overview of the whole division and its objectives. Each individual understood how his/her work fitted into the complex working processes of the Division of the Environment to achieve these objectives. Staff recognised that they were able to contribute and thus were able to take seriously their responsibilities for saving resources. Their knowledge about environmental and management issues was enhanced and they acquired environmental and management competences. Furthermore, the system of training was substantially improved. For all interviewees, these seemed to be to be the most important changes ("We would not have needed the system just to save some water").

The crucial criteria of sustainability (see section 2.2) were fulfilled. Long term thinking was being taken serious and the division wanted to be a forerunner and a role model. As the division had only a limited impact on the environment there was no need to consider effects on a larger scale.

Why was the implementation of EMAS at the Division of the Environment so successful in terms of innovation, sustainability and participation? The answer was found in the way it was handled. Top down and bottom up approaches worked hand in hand. There were people at all levels seriously committed to sustainability. In this context, synergy effects could be identified which made innovation possible. Internal factors included committed staff, a committed staff council, a committed environmental representative, who was accepted by both the managers and staff and a convinced director. External factors included a convinced city councillor, experienced consultants and a "sustainable climate" in the City of Hannover. Instruction from the city council was necessary to initiate EMAS. If just one key player had refused to take the process seriously, if, for example, the city council had not instructed the Division of the Environment to implement EMAS or if the staff had refused to play their part, the process would not have worked (for a counter example, see the case study of Athens in section 7.5).

9.7   Summary of the Case Studies

This final section summarises the findings of the German case studies in terms of the key issues of participation, innovation and sustainability and draws some final conclusions on the circumstances under which participatory governance can lead to sustainable and innovative outcomes.

9.7.1   Participation

Following the analytical framework of the research project, it is possible to distinguish between different actors (internal and external to the sites) and to identify varying forms of direct and indirect participation in different phases of the process of deciding upon and implementing EMAS. How did the different "holders" participate?

One of the distinctive features in Germany is the formalised participation by employees (*work holders*) through the works or staff council.[18] This council has to play a part in most decisions made by the employer to ensure that employees' interests and needs are taken into account. Nevertheless, participation by council members and by the employees involved in introducing and implementing EMAS varied across the case studies in terms of its extent and nature as well as in relation to the stages of the policy process.

Work/staff councils were usually simply informed about EMAS and were not in a position to influence whether or not it should be introduced. The councils were – with a few exceptions such as Hannover – just informed about the decision and passed this information on to the staff. Nevertheless, in all the case studies members of the works/staff council were included in the environmental teams.

Participation by employees usually occurred in the implementation phase which generally meant acting on the decisions of the owner, the manager and the environmental teams and attendance at in-company environmental training sessions. In a few cases, employees played a part in data collection (the initial review), environmental policy formulation, the definition of objectives and the means of measuring progress. In only one local administration (Hannover) was active involvement of employees and the staff council considered as essential right from the beginning of the EMAS cycle.[19] In the Hannover case study, employees were involved in all the initial introductory stages and had a huge influence on implementation, for example, in setting priorities, targets and identifying appropriate measures.

EMAS did have some impact on employees in all the case studies. Environmental awareness was enhanced, environmental knowledge was increased and often motivation increased, as well as identification with the company. These outcomes usually corresponded with the level of participation. A high level of participation, in the sense of being a part of the decision-making process, tended to lead to a major impact, whereas a low level of participation was associated with little by way of change.

Verifiers *(external status holders)* by definition participated in the EMAS process, because their involvement was required under the EU rules.

---

[18] The establishment of works/staff councils is based on a 1952 federal law (Betriebsverfassungsgesetz/BetrVG) (modified in 1972, 1988 and 2001) which has its roots in a 1920 law (Betriebsrätegesetz). The law explicitly aims to limit "managerial prerogatives" to secure the liberty and human dignity of workers. The councils have broad ranging competences in social issues, for example, the regulation of work time and wages agreements at the enterprise level, and also in relation to the social services provided by a company. On these social issues, agreement between the employer and the works council is required. There are lower levels of competences for personnel issues (recruitment, dismissal, job placement, promotion, qualification etc.) and for other matters, such as organizational change. However, formal consultation is required by law. The members of works/staff councils are elected every four years by the employees (for details, see Streeck 1984; Keller 1997: 79-88).

[19] Even in this case the employees could not themselves decide whether they wanted EMAS or not.

Owners and/or top managers *(share holders)* were also crucial in all the studies. They were the people who decided to implement EMAS and they were usually involved both in the formulation of environmental policy and in the definition of objectives and associated measures.

This was also true for the external consultants *(knowledge holders)*. In most cases their contribution was crucial. With the exception of one company (Hamburg Airport) all the case studies were supported by external consultants throughout the EMAS cycle. They kept links with the organisation, for example, to provide specialist knowledge or to keep track of relevant environmental laws. In just one case study, the external consultant was changed during the process, because he turned out not to be the best person for the job.

The environmental managers had a crucial role in setting up, implementing and maintaining the management system. They were usually responsible for drafting, publishing and distributing the environmental statement.

Heads of departments, works/staff council representatives and work safety representatives were members of the environmental teams *(status holders)*. Their involvement and participation differed greatly, depending on their status, the environmental impact of the company's activities and the importance of EMAS within the company for the protection of the environment. Work safety representatives were, for example, less important where there were few toxic or other dangers (local administrations), whereas they were very active on sites where such dangers were present.

Neighbours *(holders of spatial interest)* participated in only one case, Hamburg Airport. The participation was indirect, being about reduction in the level of noise. Even this level of involvement was not due to EMAS, but had been initiated many years earlier.

Other kinds of *interest holders* were involved in three of the case studies. In Hannover and Augsburg the Local Agenda 21 activists influenced EMAS indirectly, and at Hamburg Airport some interests had been involved in an advisory committee, even before EMAS was introduced. Their main concern was to influence environmental targets, for example, reduction in noise levels, but as this was not a result of EMAS, it cannot be said that EMAS led to any indirect participation on the part of *interest holders*.

Stake holders such as customers or suppliers and the general public can in principle participate indirectly, because an environmental statement has to be published. Nevertheless, *stake holders* and the public were in fact not interested in EMAS. There was rarely any public feedback. Interest was expressed only when environmental impacts were perceived as irritating or very disruptive.

9.7.2 Innovation

Innovation is multifaceted. There can be innovation in products, in product lines or in the way organisations are structured and managed. Depending on the issue, different kinds of innovation may be necessary. In the German case studies most innovation was to be seen in the structures and management of the organisation. New products did not appear. There were some innovations in the production process, a new product line or better use of existing technology, but in the main these had been put in train before EMAS was introduced.

The kind of innovation needed for sustainable development has to depend on the kind of environmental problem thrown up on the site. In the local administrations, product innovation could hardly be expected, whereas in some industrial sites, such as Lincoln, product innovation would be one of the main features. The same was true for the noise problem at Hamburg Airport. Such an issue could not be addressed through organisational changes within the company. Technical innovations had to be the answer in these kinds of circumstances.

9.7.3 Sustainability

At the time when EMAS was being introduced, sustainability with its three dimensions (environmental, economic and social) did not play a role in most of the case study organisations. They were mainly interested in the environmental and economic dimensions of sustainability. Only one site, Hannover, incorporated the social dimension in the form of improving working conditions, partly to encourage staff to agree to the introduction of EMAS. In some of the case studies, the notion of sustainability appeared during the implementation of EMAS, parallel to the general emergence of this idea in society at that time. Most companies continued to focus on the environmental and economic aspects and were successful in this regard. Environmental performance improved in all the case studies and most of them also enjoyed some economic benefit, in the shorter or longer term. Nevertheless, if sustainability is defined as the incorporation of environmental, economic and social dimensions, then the majority of the case studies could hardly be described as sustainable.

However, defining sustainability in a different way leads to a different conclusion. If it is defined as proactive, holistic, long term and with a three-dimensional perspective (space, time and third parties) then most of the case studies could be seen as sustainable. They are proactive insofar as they do work on continuous improvement of their environmental performance going beyond just compliance with the law. They also act in a holistic way, as they address the issue of environmental impact in a principled way and take all the relevant features into account: reduction in pollution of the air, water and the ground, energy and other resource savings, avoidance and recycling of waste, training of employees and so on.[20] Last but not least, most case studies were not focused just on short term targets, but took their decisions from a long term perspective, taking space, time and third parties into account.

In all the case studies the decision to implement EMAS was top down, but the actual implementation took place within different governance arrangements or management styles. In some cases the style was a co-operative one. Two companies had relatively strict hierarchical top down management styles. One was strongly influenced by the parent company, so there was a top down management style

---

[20] In addition, the environmental awareness of different "holders", especially the work holders, increased at all sites and in local administrations.

within a top down management style. Another could be described as paternalistic, because the managers behaved more like "parents" taking care of "their children", reflecting the owner's philosophy.

9.7.4 Conclusions

How are innovation, sustainability and participation linked together and what kind of governance arrangements support any linkages there may be? There were two main findings in these German case studies. First, there is no inherent interconnectedness between innovation, sustainability and participation, but the relationship is multifaceted. Second, any interconnectedness depends on the kind of problem which has to be addressed. A solution to a problem through sustainable innovation cannot be considered without an in depth knowledge of the presenting problem. The problem determines the kind of innovation, which can lead to a sustainable solution. And the kind of innovation which is required sets out what kind of knowledge is needed and which stakeholders can provide this knowledge. Thus the nature of the problem determines the kind of participation that is needed and from which "holders". In some situations sustainable innovation did not need the participation of many "holders", but simply required a technical solution. In this context, the management style was not important, because technical solutions can also be identified in a hierarchical environment. If there were innovations which needed the competences of different "holders" in order to be implemented, then participatory governance would be required which, in turn, would encourage participation by the appropriate *knowledge holder.*

PART III
REFLECTIONS

Chapter 10

EMAS: An Instrument for Participatory Governance?

Hubert Heinelt, Britta Meinke and Annette Elisabeth Töller

EMAS was adopted in 1993 as an innovative policy tool and has been shown to be an instrument for improving the environmental performance of companies. It has also helped to achieve sustainability objectives. The scheme was revised between 1998 and 2000 with widespread participation on the part of relevant "holders". The revision was intended to improve the instrument itself, as well as extending its scope to all sectors, clarifying its relationship with the international standard for environmental management, ISO 14001, and increasing the participation of interested parties.

Against the background of the findings presented earlier, we will address the question whether or not and under what circumstances EMAS has contributed to or fostered participatory governance leading to sustainable and innovative outcomes. In the first part of this final chapter we will summarise the findings of the case studies on the selected EMAS sites (enterprises and local authorities). In the second part we will address the "design" of EMAS at the EU level, because the decisions on the structure of this policy tool have undoubtedly had an important influence on how it has worked. Another influence has been the very varied contexts in which EMAS has developed in the member states. These differences and their effects on the implementation of EMAS will be summarised in the third part of the chapter, and in the last part policy recommendations will be presented.

10.1 EMAS, Participation, Innovation and Sustainability

In respect of EMAS, we have to discuss participation in a specific environment, in a company, which is per se not structured in a democratic way. It was clear from the beginning that EMAS involves a specific set of actors and a clear boundary – the enterprise level. So participation is restricted to the site's social environment (employees, staff members, neighbours, consumers, customers, interested actors). Local authorities, however, are different in this respect because the decision to implement EMAS can be influenced by the democratic decision-making process of local government. In addition, the implementation of EMAS can be part of a broader strategy aimed at, for example, implementing Local Agenda 21 (see Voisey 1998; Beuermann 1998).

In thinking about new arrangements for governance within an enterprise or a local authority, one has – in Schmitter's words (2002, 58) – to reflect on three questions: (i) *What is the purpose of delegating power in this way?* (ii) *Who should participate?* (iii) *How should they participate?*

The first question is linked to the reasons why EMAS has been taken seriously. Two broad sets of motivations can be identified.

First, there were firms mainly interested in the "label", the verified environmental statement which could be used for marketing and/or public relations strategies. In some cases clients, suppliers or customers demanded it. Additionally, these companies hoped to use the "label" to stress good will in the event of new environmental regulations laid down by government. A sub-category of this group also perceived EMAS as a way of enhancing links with the surrounding area, for example, Hamburg airport and some of the Greek sites.

Second, there were organisations which were mainly interested in the environmental management dimension of EMAS. By focusing on the management aspects, they hoped to achieve a better environmental performance by cutting back the emissions of pollutants, which would result in financial savings and a more efficient and more environmentally sound production process. In addition, they hoped to improve the overall management performance of the organisation in general. This required a better understanding of what was really happening on the shop floor, as well as a better picture of interrelationships on the site and a clear(er) distribution of tasks and responsibilities.

This distinction between an "outward looking" and often tactical orientation on the one hand and an "inward looking" strategic long term perspective on the other hand, is important for the second and third questions, i.e. who is important and should (therefore) participate and how should important actors be involved. This is particularly the case in respect of broader staff involvement. The latter is crucial for the "inward looking" organisations but is of minor importance for the "outward looking" ones.[1]

On the second question, we can distinguish participants who are internal and external to the site. Using Schmitter's "holder" concept (2002: 62-63), we can focus on share holders (owners), status holders (directors, departmental heads, works council representatives, health and safety representatives), work holders (employees), knowledge holders (external consultants), external status holders (environmental verifiers), holders of a spatial location (neighbours), interest holders (those persons or organisations who have an awareness of the site's activities and an interest in participation), and stake holders (all those that could potentially be affected by what happens at the site, regardless of who and where they are).

On the third question, "holders" can participate *directly* in implementing EMAS or they can participate *indirectly*, for example by making demands on the organisation or by reading the environmental statement and responding to it.

Moreover, participation by "holders" can take different forms depending on the status of the participant. Participation at site level is limited to its social environment and within the site it takes place against a background of hierarchical decision-

---

[1]   Needless to say, some kinds of actors (holders) are important per se where EMAS is concerned. These are the share holders, some status holders and – depending on the individual case – some knowledge holders.

making on the part of owners or senior management. Internal participation will most likely be limited to consultation and some form of negotiation. Consultation means that participants have the right to express their opinions and try to convince others of their argument, but decisions are taken by a group of (pre-)designated "notables" or even by a single "authorised" person. In negotiations, participants are consulted, can exercise bargaining power by using the threat of the exit option or refusal to agree and decisions are reached according to some "qualified procedure". This applies especially to status holders who possess a specific entitlement to participate which is "designed" by political decision and can be enforced by the authorities ultimately responsible for decisions. In other words, status holders possess a bargaining power based on the politically accorded "right to represent a designated social, economic or political category" (Schmitter 2002, 63). The crucial point is that such a status and the bargaining power derived from it can limit the discretion of share holders. However, the concrete definition of a status – which can vary between countries (as in the case of employee representatives) – can determine the way in which bargaining power and particular forms of participation can be secured.

Furthermore, participation can take place at different stages of the EMAS cycle, starting with the decision at the site to participate in EMAS, followed by environmental policy formulation, the initial review, the definition of objectives and measures, the development of the management system, the drafting of the environmental statement up to the point of external validation and verification, the registration process, followed by publication and distribution of the environmental statement. Participation in EMAS should be seen not only in respect of the implementation of the scheme as such, but also in respect of the everyday operations of the site management system in general. It is possible that a management and audit scheme of this kind could open up new options for participation – not least for the work holders – through its everyday use. In other words, EMAS could alter the pattern of daily interaction on a site and it could also create new horizontal sets of interrelationships resulting in a new form of governance on a site. *Governance within a site* is a new form of governance that could be created through EMAS. In particular, EMAS II, the revision of the directive concluded in July 2000, focuses specifically on the participatory aspects of internal decision-making and implementation of environmental systems within a firm or a site.

However, these legal requirements based on EMAS II do not have sufficient influence on governance arrangements at a site to allow work holders or their representatives to achieve a strong new *status.* In other words, through EMAS II work holders or their representatives do not became *status holders.*[2] In most of the case studies innovations were introduced without meaningful participation by employees. With the exception of German works council members who can be labelled as status holders (see section 9.7.1), employees have not participated in the decision to implement an environmental management system. However, the case

---

[2] On the political "design" of status holders and their importance for structuring governance arrangements, see section 2.4. Without going into detail, one can only conclude that in the case of EMAS the "generic design principles" for governance arrangements developed by Schmitter (2002, 60-66), are not adequately met.

studies did show that, even though the decisions were taken by owners or managing directors, sometimes in close co-operation with external consultants, acceptance by employees was considered important for the successful implementation of the scheme. In addition, in a majority of the sites, changes in the organisational structure did occur, leading to a clearer organisation of work and a more detailed assignment of responsibilities. In some sites, the owner or senior management not only created new means for information sharing but also created structures to generate knowledge that led to improvements or ideas from the employees about sustainable innovations. Not only had the management style been modified to allow for more participation by interested employees, but new arrangements, such as environmental committees, were introduced, indicating a formal kind of partnership between higher management and employees.

We can therefore conclude that there are signs that the development of new forms of governance on a site has increased opportunities for enhanced understanding through debate, and has served to bring about higher levels of trust among those who participate, which in turn has led to a longer time-horizon for thinking about possible sustainable innovations. It is likely that all this will continue in the future. Furthermore, we can conclude that a co-operative management style does lead to the best results. Some case studies also clearly demonstrated that staff participation was more likely when there was a high level of trust between the leaders (director, department head, environmental manager or co-ordinator) and the employees. Without this trust, staff refused to co-operate, because they feared top down control. Where staff had experienced problems following the implementation of a new programme, they were less likely to welcome another one. Where trust between management and employees had been destroyed, there was little by way of participation by staff. Linked to this, good leadership can be seen as a condition for success, and good leadership requires a sustainable vision and an ability to convince staff to take the sustainable elements of EMAS seriously. In summary, EMAS works best where sites well organised with good leadership.

Another one of the purposes of EMAS is to foster a new form of *government-market governance*. On the one hand, this purpose can be related to "governmental" objectives to bring about self-regulation on the part of enterprises and to complement the existing regulation of companies through command and control policies. But on the other hand, these "governmental" objectives can aid business interests as well. Traditionally, command and control policies, which make up the core of most countries' environmental regulations, have sought to bring about environmental improvements by setting strict emission limits as well as prescribing the industrial technologies and processes needed to meet these limits. These policies are typically supposed to be applied uniformly across industry regardless of local economic or environmental conditions, and they impose stiff penalties on violators. Governments intend to create stronger capacities for self-regulation by playing a more limited role in the EMAS "beyond compliance policy".

The questions "Who should participate?" and "How should actors participate?" in relation to *government-market governance* are both easy and difficult to answer. On the one hand, it seems reasonable to argue that participation by enterprises has to be voluntary because it would not to make a great deal of sense to prescribe self-regulation. On the other hand, the scheme has to demonstrate public credibility to permit reduction in control by the state. Prescribed procedures are needed as well as the involvement of actors who are engaged in their implementation (knowledge holders) or who are verifying their implementation (verifiers and registration bodies as status holders). Finally, a wide range of other "holders" may be in a position of some influence by being given the opportunity to response to the environmental statement.

How EMAS has contributed to the development of new forms of *government-market governance* in individual member states is addressed below (in section 10.3). At this point, it is important to stress that the picture varies a great deal, because of (i) the different governance structures and (ii) the different arrangements introduced by the member states for the implementation of EMAS.

Finally, EMAS can be used to bring about a new form of *market-civil society governance*. In principle, by putting the general public, or more precisely the consumer, in a "observatory position" in relation to the individual firm/site through the publication of the environmental statement, EMAS can increase the transparency of the registered sites in general. One could argue that it is the aim of EMAS to enforce better environmental performance in the light of public scrutiny. Again, this can be seen to be a "governmental" objective which can also meet the intentions of companies.

It can be argued that the environmental statement required under EMAS can be used as a marketing tool only if it is able to have an effect on competitiveness. If the implementation of EMAS is required by suppliers, customers or clients, it can result in competitive advantage, and if the environmental statement does impact on the decisions of consumers, it can lead to financial benefits. However, to have a positive impact on competitiveness through the publication of an EMAS environmental statement, there would need to be informed and environmentally aware recipients of the document.[3]

How an environmental statement is seen by those recipients also depends on how government-market relationships have been organised in the past. Environmental policy in Germany, for example, can be traditionally characterised by strident, public disagreements between industry and environmental groups with government regulators caught in the middle. While the government has on a number of occasions caved in to pressure from business interests, it has also consistently shown a willingness and an ability to pass stringent environmental laws despite protests from industry. Lack of trust between industry, environmentalists and the informed public could imply that environmental statements resulting from a voluntary instrument like EMAS would be regarded with suspicion by environmental groups as well as by some elements of the public.

For the UK, similar conclusions can be reached but for completely different reasons. The traditional British approach to environmental policy has been to negotiate with those responsible for the pollution in order to agree on the level of pollution that could be countenanced, the "best practical option". This consensual approach was based on the view that improvements would be best achieved by

---

[3] This explains the fact that in countries lacking environmentally aware suppliers, customers, clients or consumers (as in Greece), companies have preferred to implement ISO 14001, the EMS which does not require the publication of an environmental statement.

working with rather than against industry (Weale 1996). Therefore, the environmental statements of individual companies – the result of the implementation of a voluntarily agreed environmental instrument – were likely to be accepted by potential recipients. But how far environmental statements resulted in concrete benefits or led to competitive advantage cannot be measured.

We can also conclude that EMAS is a means for moving towards new forms of *market-civil society governance* by creating and strengthening relations with the neighbourhood. It can contribute to "openness" and to a public debate about any environmental problems caused by the site and the measures required to deal with them.[4] Although some sites which have a major environmental impact on their neighbourhood had already established contacts with spatial and interest holders before deciding to implement EMAS, additional ideas were introduced through the environmental management system, usually in the form of a more specific long term measure. In other words, EMAS is a tool that can lead towards a new form of *market-civil society governance*, but the instrument needs to be embedded in an overall strategy.

To sum up, EMAS can be a means for moving towards new forms of governance in three different ways. But have such new forms of governance – with their opportunities for participation – really contributed to innovation and sustainability?

In general, the distinction already made between being "outward looking" and "inward looking" is important. The more EMAS was used for "tactical" reasons, such as a marketing and/or a public relation instrument, the more sustainable innovations were achieved as (welcomed) by-products. However, the more EMAS was implemented strategically or for long term purposes, innovative and sustainable outcomes were less likely. The long term objectives can be linked explicitly to specific dimensions of innovation and sustainability (the ecological, the economic and/or the social, see section 2.2).

In analysing the innovations achieved, one can, as noted earlier, focus on changes in the product (new product/product improvement), on changes in the production process (production lines, increased use of existing technology) or on the reform of the overall structure and management of the organisation.

In respect of the latter, the outcome was clear. As an instrument EMAS did not lead to changes in policy. Instead, it reflected the direction that an organisation was already following. As such, the case studies showed that EMAS should be seen as a part of broader changes within a company or an organisation. The adoption of EMAS was usually a reflection of a shift to a more innovative (and environmentally aware) organisational culture. The degree to which EMAS contributed to innovative organisational changes depended on organisational culture, i.e. an openness to address challenges, to manage uncertainty, to absorb new knowledge and to encourage new forms of participation. In this respect we have to be aware of the possibility of a high degree of path dependency. Nonetheless, there is evidence from the case studies that EMAS can speed up and alter the direction of innovation processes beyond what was originally intended. In other words, once implemented

---

4   See for this kind of "Umweltkommunikation" (environmental communication) Braun et al. (2001).

as an instrument to achieve certain objectives, EMAS can have an unplanned effect on organisational change. This effect is linked to the specific nature of the innovations themselves.

This effect can also be seen in relation to the other two forms of innovation (innovation in products and production processes). Becoming aware of previously unknown production processes or product improvements may be one of the reasons for taking EMAS on board. Evidence from the case studies shows that this was more likely in smaller companies than in larger transnational ones because the latter were more aware of the available options. Larger companies had usually already developed a systemic capacity to address innovations of this kind before thinking about the implementation of EMAS.

There was a major difference between innovation in products or production processes and innovations in the overall structure and management of an organisation. Whilst the latter depended on a given organisational culture, the former did not require this condition. EMAS can be used as an instrument to achieve such innovations systematically, but in a purely "technical" way, i.e. improvements in the raw materials used, better use or reuse of natural resources, reduction in emissions, the introduction of new technologies or the reorganisation of the division of labour. This "technical" approach towards achieving innovation in products and production processes – instead of a "social" approach in relation to organisational innovation – implied that innovation processes were more controllable. Key "holders" (the owner or senior management) were able without difficulty to decide on the direction of the innovation process or were able to stop it at a certain point. But it should be emphasised that, depending on the nature of the "technical" improvements, the degree of controllability of such innovation processes did vary. The more technologically demanding they were and the more they relied on certain knowledge holders (not least qualified work holders) the more dynamic they became. This in the end resulted in a blurring of boundaries both between innovation of products and consequent necessary changes in production processes, and between innovation in production processes and changes in the overall organisational structure of company or other kind of organisation.

The link between innovation and participation of work holders was another key feature. For sites with major environmental impacts, the innovations which counted most in decreasing or avoiding these impacts were in the main achieved by investment in new and more environmentally friendly technology or by improvements in the production process. The case studies demonstrated that those generally responsible for these kinds of innovation (managing director, research unit, technical department, environmental manager) had actively participated in their development. Participation by employees was limited to seminars and training, focusing on learning to work with new technologies or to work within improved production processes. In the longer term the new skills learnt by the workers and their active involvement in production processes were crucial for the achievement of the outcomes expected by those who initiated the improvements.

This can be linked to another finding. On sites employing staff with high qualifications and with high technical production processes requiring a high level of flexibility, participation was sought through a broad spectrum of opportunities for discussion and knowledge generation and was helped by a change in the organisational structure ("intelligent jobs need intelligent employees"). On sites where these kinds of jobs did not exist, the importance of participation by employees varied. Whilst the provision of information through training and seminars took place in all cases, employees played different roles in different phases of developing and implementing EMAS. They may have been responsible for the collection of information, they may have participated in information meetings about objectives and the identification of measurements or they may have provided ideas for future improvements to environmental performance.

We can therefore conclude that, in reflecting upon *new forms of governance within a site, the type of innovation matters.*

The issue of *sustainability* can be addressed by examining a number of topics (see section 2.2) such as decrease in the level of pollutants, recycling, re-use of materials, use of renewable energy sources, use of environmental criteria in the choice of suppliers, influencing the environmental behaviour of customers and consumers, environmental neighbourhood effects, transport, increase in environmental awareness both internally and externally, legal compliance, continuous improvement and long term thinking.

Sustainable innovations resulting in a decrease in pollution, recycling, re-use of materials or use of renewable energy sources were identified on all sites. Additionally, in all the case studies, innovations did lead to more environmentally sound production processes. This went hand in hand (i) with financial benefits for the company, even though they were not always quantifiable and (ii) with enhancing the environmental image of companies, even though it was not often used as a marketing tool. Finally, heightened environmental awareness by staff was one of the most often cited benefits in all the case studies.

Differences were apparent between the three countries. Although in Germany there was concern about environmental issues as well as financial ones, sustainable improvements did not play a dramatic role. This can be explained by the relatively high level of environmental awareness already reached before EMAS was implemented. Social issues at the beginning of implementation were not prominent in the German case studies but did become more important in some cases as EMAS was applied. Only two case studies (Hannover, Lincoln) took social issues into account in the early stages of EMAS.

All six UK case studies were to a greater or lesser extent involved in the development of sustainability policies. The people responsible for EMAS implementation in companies and local authorities found it challenging to understand exactly what sustainable development meant for the way everyday business was carried out. They were familiar with the Brundtland definition of the concept and were also aware that it included the three interlinking elements of the social, economic and the environmental. Nevertheless, most sustainability strategies focused on environmental improvements. The social and community activities that some companies had been undertaking for some time were generally not included under the umbrella of sustainability policies, while with respect to local authorities they were.

In Greece, it was clear that environmental management systems (EMAS or ISO 14001) were a crucial tool for creating and increasing awareness of the environmental impacts of site activities. Although environmental management systems had been used mainly as a modern management and marketing tool for increasing sales and enhancing the site's profile in the eyes of their customers (and in some circumstances of the local government), savings were also anticipated in energy, water and raw materials, resulting in financial savings. However, at a later stage, by taking these issues seriously, environmental issues were specifically addressed.

10.2 Development of EMAS at the European Level

When negotiating EMAS in the early 1990s, sustainability was one of several concepts that underpinned the discussions. The European Commission had taken up ideas developed in the world of business for internal environmental audits but had faced opposition to the idea of a compulsory environmental audit. The 1993 regulation allowed for voluntary participation by companies in the manufacturing sector. However, sites had to take a number of internal steps before being validated by an external verifier and officially registered. Member states were required to establish systems for accreditation and supervision of environmental verifiers and for registering validated sites but this could be achieved with a high degree of discretion.[5]

In the Article 19 Committee and various networks associated with it, commission officials, national delegates and other relevant, non-government or semi-government players worked together in implementing the scheme. Stress should be laid on the open access relevant parties had to this decision- making arena. At the time this level of openness was unusual in DG XI and it served as an example for other directorates in DG XI. In working up the details, such as how to define the core terms of the scheme and how to operationalise them for practical application, the various participants were very active, thus allowing for a common European approach to the implementation of EMAS. At the same time, various elements of the system were evaluated, and in operationalising and evaluating the scheme a number of lessons were learnt which were to prove useful in the revision of the scheme. Many of these lessons influenced the drafting of the revised regulation (EMAS II), which was adopted early in 2001.

The revision of the EMAS directive began in 1997 through consultation with those involved in the Article 19 Committee. Further consultation with interested parties followed after a first informal proposal had been drafted in late 1997. The first official proposal was presented in autumn 1998. The core modifications proposed were a closer alignment of EMAS to the international standard for environmental management systems, ISO 14001, an extension of the scheme beyond the industrial sector, and improvement of stakeholder involvement and of the public profile of the scheme.

---

[5] It can be argued that EMAS explicitly addresses the "need for adaptability in EU environmental policy design and implementation" (Glachant 2001). On the one hand, the directive prescribes strict procedures for its application but, on the other hand, it offers member states enough discretion to achieve intended policy effects by recognising the divergent national conditions which are important for making the scheme work.

Following ratification of the Amsterdam Treaty, the European Parliament scrutinised the proposals under the co-decision procedure. In its first reading (April 1999), the EP proposed almost 60 modifications. For instance, it pushed for stronger involvement by interested parties (particularly employees) and for strengthening the credibility of the scheme through stronger demands for legal compliance, the use of BAT and stricter training for verifiers. Furthermore it demanded that the environmental statement should be presented in a printed version and that minimum requirements for its content be adopted by the Commission.

In its revised proposal the Commission accepted only a few of the Parliament's modifications. The Council, in working towards its common position, focused on the definition of the core terms of the scheme, which in implementing EMAS I had proved to be poorly defined. The same was true for the verification process. The regulation of many details was delegated to comitology, partly as new competencies and partly as the codification of activities which had already been fulfilled informally.

In its second reading in July 2000, the Parliament proposed 27 modifications to the common position. The main points again focused on the credibility of the scheme (strict legal compliance and tougher accreditation conditions), the involvement of interested parties and the publication of the environmental statement in printed form.

Since many of the modifications proposed were acceptable neither to the Commission nor to the Council, EMAS II had to go into conciliation procedure. After a period of tri-partite negotiation under pre-conciliation, a compromise was reached in November 2000. The Parliament was quite successful in pressing its demands.

Parallel to the negotiation of EMAS II which took much longer than initially planned, the Commission started tabling several guidance documents in the Article 19 Committee, most of them explicitly mentioned in the new regulation, so that the regulation would be operational from the moment it was adopted. Guidance was drafted on the entity to be registered, the content of the environmental statement, the use of the EMAS logo, employee participation and the identification of environmental impacts and assessment of their significance.

Some preliminary conclusions can be drawn on the interrelationship between participation and sustainability in the context of decision-making on EMAS at the European level.

Although the main conceptual origins of EMAS are older than the sustainability debate, EMAS I was considered to be an innovative policy tool at the time of its adoption because it reflected a number of ideas from that debate. Thus, EMAS can be seen as a tool for achieving (or at least working towards) sustainability at the company level.

In the course of implementing EMAS, shortcomings were revealed in the scheme in general as well as in terms of participation and sustainability. Participation by "holders" (interested parties) was given little emphasis and many of the environmental dimensions of sustainability (such as products, indirect environmental effects and supply chains) were not explicitly or adequately addressed.

The implementation of EMAS I in comitology allowed the majority of relevant "holder" groups to participate. Major players (member state governments, environmentalists, trade unions and industry) were given the opportunity to participate in the implementation, evaluation and reformulation of the scheme in the

Article 19 Committee. A wider network of informal structures, consultations, workshops and seminars offered access to other relevant players. By establishing this open, informal consultation process, the Commission offered opportunities for participation and influence and obtained a wide range of ideas from specific "holder" groups. This had a major impact on the implementation and revision of EMAS.

During the revision of EMAS, greater involvement in the scheme by "holders" (interested parties) was already included in the first Commission proposal. The European Parliament gave participation a high priority – be it by employees (and their representatives), "special holder" groups or by the general public. Great emphasis was put on defining more clearly who the relevant "holders" were and how they could be integrated into the system. In the end, the Parliament, due to its co-decision power, was relatively successful in achieving its aims. The Council of Ministers, on the other hand, by making definitions and annexes more precise and by delegating more detailed definition work to comitology, significantly improved expectations in relation to sustainability in the new regulation, for instance, by inserting a detailed list of indirect environmental effects (including biodiversity). The Parliament also gave sustainability a higher profile by pushing for changes relating to the credibility of the scheme.

In summary, EMAS developed at the EU level through a governance arrangement which can be characterised as reasonably participatory in terms of the involvement of and horizontal interaction with a broad spectrum of non-governmental actors. This can be linked to a relatively high degree of self-reflexivity (and learning) as well as a strengthening of the legal basis for the involvement of "interested parties" in implementing and applying EMAS at site level. Nevertheless, it was the European Parliament strengthened by the Amsterdam Treaty rather than the Commission or the Council, which was the crucial promoter of new participatory elements in EMAS II.

10.3 Implementing EMAS at National Level

One reason for choosing Britain, Greece and Germany for analysing the implementation of EMAS at the national and site levels was that these three countries exhibit very varying traditions and patterns of environmental regulation. Whereas in Britain the dominant approach has always placed emphasis on self-regulation and pragmatic negotiation between the regulators and their "audience", in Germany environmental policy has been characterised by a rigid regulatory command and control approach, combined with a federal structure of responsibilities. In Greece, there is a normative, legalistic regulatory policy combined with a more pragmatic approach in the administrative practices of a strongly centralised state. On environmental management, Britain was the first country to develop a national standard for environmental management systems (April 1992), whilst neither Germany nor Greece had experienced any national schemes for environmental management at the time when EMAS was adopted at the European level. It is argued that these differences, together with other issues, such as administrative structure, environmental awareness and economic situation, were influential on the way in which EMAS (as an innovative procedural tool) was accepted by these member states and transposed into national organisational structures, which were in turn accepted by the business community.

The first element in the implementation of EMAS was the establishment of systems for the accreditation and supervision of verifiers (by April 1995) and the registration of sites (by July 1994).

In Britain where EMAS was not seen as anything out of ordinary, no major problems occurred. The accreditation of verifiers was integrated into the existing National Accreditation Council for Certification Bodies (NACCB) later renamed the UK Accreditation Service (UKAS). Registration was located in central government at the Department of the Environment. A consultative body, with membership from industry, local government, the voluntary sector, the trade unions, the utilities, the accrediting body, an academic expert, a verifier, the National Rivers Authority and Her Majesty's Inspectorate of Pollution, was established to offer advice on EMAS to the Department of the Environment. In 1998 the role of the competent body was handed on to the Institute of Environmental Assessment (IEA).

In Germany, by way of contrast, EMAS was perceived as something new and unusual. A major dispute arose over the extent of government involvement in accreditation and registration with "environment" players (Ministry of the Environment, Environmental Agency, environmental groups and trade unions) favouring strong government involvement and "business" players (the Economics Ministry and business associations) opposing it. It was some time after the deadline in December 1995 that an agreement was reached so that the act could be adopted. Accreditation of verifiers was placed with a body supported by business (DAU) though guidelines for accreditation were developed by a committee (UGA) placed closely to the Ministry of the Environment. This committee comprised representatives of federal and Länder governments and of a range of interest groups, including business, trade unions, environmental groups and verifiers' professional associations.

In Greece, the EMAS machinery was not put in place until 1998, because of a lack of interest in the scheme and strong resistance by the administrative system to delegating responsibilities to non-governmental bodies. Accreditation is now carried out by the Greek Accreditation Council (ESYD), founded in 1994 and part of the Ministry for Development. Members of the Council are representatives from government ministries, scientific and social organisations as well as from industry. The competent body for the registration of sites in Greece was not set up until 1998. It comprises representatives from four different ministries.

In comparing the participation of "holders" in the accreditation and registration systems, we can identify a pragmatic approach in the UK, placing accreditation with a non-governmental accreditation body and creating an advisory body to the Department of the Environment which included most relevant "holder" groups. By way of contrast, in Germany, registration of sites was located close to business, whereas accreditation of verifiers was to be carried out by a non-governmental body under the close watch of government and given detailed, formal guidelines for their activities. Participation of relevant "holders" as well as of different territorial levels in this body (UGA) was formally guaranteed. In Greece, both systems were located very close to the state administration and little access was available to interested parties.

The comparison between the three countries on the means and intensity of participation in EMAS by companies revealed striking differences.

The low number of participating companies in Greece can be explained by three factors. First, the system for registration did not get established until 1998. Companies had been faced with uncertainty for a long period. Second, business, local authorities and social groups were not very interested in the scheme. Third, of those companies interested in environmental management systems, most preferred ISO 14001, which was up and running before EMAS and did not include a requirement to publish internal data, a requirement which many Greek companies did not see as providing competitive advantage.

The high number of EMAS sites in Germany came as a major surprise for many experts, because business associations had been strongly opposed to the scheme during the period of negotiation. There are a number of reasons why so many companies did go for EMAS (see Malek et al. 2000). The high regulatory burden in Germany made EMAS look like an instrument that could systematically address statutory obligations. At the same time, a trade off looked possible, particularly for larger sites in sectors such as the chemical industry, since some branches of government were offering the chance of regulatory relief (in relation to inspections and reporting requirements), if a large number of companies went for EMAS. Additionally, a high level of public environmental awareness made companies think that implementing EMAS would improve their image. Finally, it was the location of accreditation and registration bodies relatively "close to business" that made German firms trust the scheme. However, since 1998 there has been a decrease in newly registered sites in Germany, and about 10 per cent of sites have not re-registered. The absence of general popularity for the scheme and the low level of regulatory relief are likely explanations for this reduction in numbers.

Given the fact that EMAS was based on BS 7750, more enthusiasm could have been expected on the part of British companies. Even though growth of EMAS registered sites has been steady, the total number is modest. There are many companies that have developed environmental management systems, but most of them have preferred ISO 14001, mainly because they operate on the international market. The participation by local authorities in EMAS is in part due to support from central government but also by the backing given to it by the Local Government Association, the umbrella organisation of local authorities in England.

It can therefore be concluded that issues such as the level of environmental awareness of the public – of consumers and customers – as well as the marketing strategies of companies and the national traditions of environment policy, matter in trying to understood the national application of EMAS. In particular, the method of implementing EMAS in the member states is important. Not only is the timetable for making EMAS work through a new organisational setting of great relevance, but even more important, the culture of this organisational setting, i.e. its participatory features, seems to be important for the acceptance and credibility of EMAS to companies and local authorities as well as to other "interested parties" such as environmental NGOs, the general public and consumers.

10.4 General Conclusions and Policy Implications

All in all it seems that in competing with the international standard ISO 14001, EMAS does face difficulties, because it has to prove its added value. For some companies this added value is the enhanced involvement of employees and the requirement to communicate with an interested public, but this is unlikely to be the case for the vast majority of companies interested in environmental management systems. EMAS has been shown to be an instrument for a small elite, but not for large number of companies. Consequently the total effect on improving environmental performance is modest.

One of the objectives of the research team was to identify ways of improving the implementation of EMAS, and particularly to foster the participatory elements of the instrument. Three levels need to be distinguished: the European, the national and the organisation/site level. For these three levels we can summarise some policy implications for different kinds of actors: governmental organisations, non-governmental organisations (NGOs), trade unions and the business sector.

Table 10.1 Policy Implications for Key Actors at Different Territorial Levels

	Governmental Organisations	Non-Governmental Organisations	Trade Unions	Business
EU level	1	4	5	6
National level	2			
Organisation /site level	3			7

The numbers in the boxes refer to the following commentary.

1   In the *EU decision-making process* it can be demonstrated in the case of EMAS that comitology can lead to reasoned debate provided that such committees are opened up to a broad spectrum of actors. This involvement can improve the regulation in detail and this is important for its acceptability. However, the preferences of directly affected interests, such as companies, still have to be taken into account.

2   The same is true for decision-making at the *national governmental level*: To implement EMAS effectively, it is a wise move to include those actors who are important for its acceptance. Such inclusion can help to develop and improve new forms of *government-market* or *market-civil society governance*. In addition, the adoption of EMAS needs to be more positively encouraged and the benefits of the scheme more widely communicated.

3   *Local authorities and local NGOs* should use EMAS as an instrument for interacting with firms. This provides an opportunity to communicate with enterprises about their local environmental impacts. Furthermore, local

authorities can, like private enterprise, use EMAS as an instrument of environmental risk management as well as for innovations in products, production processes and organisational structure.

4 *International and national (as well as local) NGOs* may be concerned that environmental management systems (EMSs) can be used as an instrument of substitution with self-regulation replacing control of enterprises by governmental agencies. However, as such EMSs implemented anyhow, NGOs should promote EMAS instead of ISO 14001. It is a more stringent and publicly controlled scheme. It also leads to more detailed information about the specific application of the scheme at site level through the obligation to publish an environmental statement. This in turn provides opportunities to scrutinise and react to the environmental performance of the participating enterprises.

5 *Trade unions* at all levels should use EMAS as an instrument by which employees and their representatives can have more influence on innovation processes at site level. The influence brought about through EMAS can be used to improve working conditions in general and in particular to contribute to job enrichment through greater autonomy and higher qualifications.

6 *Umbrella organisations of private business as well as chambers of commerce at the international and national level* should recognise that transparency is of "benefit" in their lobbying activity and where they are directly involved in the implementation of EMAS. Any reduction of control of enterprises by governmental agencies (substitution) can only be expected if the scheme and its application at site level secures a reputation for being reliable and trustworthy. This kind of legitimacy has to be linked to proper rules and procedures. The same applies to the acceptance of the scheme in the market by consumers and customers. The success of the scheme as a marketing instrument depends crucially on its reputation that environmental management and auditing are taken seriously by the participating organisations.

7 Enterprises should recognise that EMAS can be used as an instrument for innovation in the broader context of organisational change. This is demonstrated by the fact that in most of the case studies (and in all cases which can be highlighted as "good practice"), environmental and quality management go hand in hand. Environmental improvements and risk management should provide the starting point but the objectives should go beyond them. For instance, financial savings are likely to occur only in the longer term. The crucial point is that EMAS can be used as an instrument for knowledge management and as a vehicle for the development of new kinds of corporate procedures. In this respect the participation of employees leading to continuous learning and improvement becomes a focal point – and should be taken seriously by all organisations interested in EMAS.

References

Alexandropoulos, S. (1990) 'Trends of corporatist representation and the Greek experience', *Parliamentary Review*, 4, 64–79.

Alexopoulou, S./Nastouli, C. (1998) 'The pilot application of EMAS in small and medium-sized dairy companies', paper presented at the Conference on Environmental Management and Ecological Control, Athens: Technical Chamber of Greece, 4 May.

Andrew, J. (2001) 'Britain joins the eco-tax warriors', *The Independent*, 4 April.

ASU/UNI (1997) 'Öko-Audit in der mittelständigen Praxis – Evaluierung und Ansätze für eine Effizienzsteigerung von Umweltmanagementsystemen in der Praxis', *UNI/ASU-Umweltmanagementbefragung 1997*, Bonn.

Audit Commission (1997) 'It's a Small World: Local Government's Role as a Steward of the Environment', London: Audit Commission.

Audit Commission (1999) 'It's a Small World: A Review of Progress in Environmental Stewardship', London: Audit Commission.

Ball, A./Owen, D./Gray, R. (2000) 'External transparency or internal capture? The role of third-party statements in adding value to corporate environmental reports', *Business Strategy and the Environment*, 9, 1–23.

Benchmark Environmental Consulting and the European Environment Bureau (1995) ISO 14001: 'An Uncommon Perspective: Five Public Policy Questions for Proponents of the ISO 14001 Series', EEB, Brussels.

Beuermann C. (1998) 'Local Agenda 21 in Germany', in O'Riordan, T./Voisey, H. (eds.) Agenda 21. *The Transition to Sustainability*. The Politics of Agenda 21 in Europe, London: Earthscan, 250–262.

BMWi (Bundesministerium für Wirtschaft) (1994) 'Bericht der Unabhängigen Expertenkommission zur Vereinfachung und Beschleunigung von Planungs- und Genehmigungsverfahren: Investitionsförderung durch flexible Genehmigungsverfahren', Baden-Baden.

Böhm-Altmann, E. (1997) 'Umweltpakt Bayern: Miteinander die Umwelt schützen', in *Zeitschrift für Umweltrecht*, 4, 178–182.

Bohne, E./Wagner, H. (1999) 'Öko-Audit und Deregulierung in innerstaatlichen Recht auf Gesetzes- und Vollzugsebene nach der Verordnung (EWG) 1836/93', Forschungsbericht 29718086 im Auftrag des UBA, Speyer.

Bohnen, H. (1995) 'Das Umweltauditgesetz im Streit zwischen Bundesrat und Bundestag: Deregulierungsprobleme?', in *Betriebs-Berater*, 35, 1757–1759.

Bovince Ltd (2000) Bovince ... The Sustainable Journey, September.

Braun, B./Geibel, J./Glasze, G. (2001) 'Umweltkommunikation im Öko-Audit-System – von der Umwelterklärung zum Umweltforum', in *Zeitschrift für Umweltrecht und – politik*, 2, 299–318.

Breuer, R. (1997) 'Zunehmende Vielgestaltigkeit der Instrumente im deutschen und europäischen Umweltrecht – Probleme der Stimmigkeit und des Zusammenwirkens', in *Neue Zeitschrift für Verwaltungsrecht*, 16, 833–845.

Brooking, A. (1999) *Corporate Memory: Strategies for Knowledge Management*, London: International Thomson Business Press.

Bültmann, A./Wätzold, F. (2000) 'The implementation of national and European environmental legislation in Germany: Three case studies', Leipzig: UFZ-Bericht No. 20.

Bundesverband der Deutschen Industrie (1998) 'Stellungnahme zu den Vorschlägen der EU-Kommission für die Novellierung der EG-Öko-Audit-Verordnung (EMAS)', Köln.

Bundesverband der Deutschen Industrie (1999) 'Pressemitteilung des BDI-Präsidenten Henkel: Öko-Audit braucht einen Schub nach vorne', Köln.

Carter, N. (1998) 'The Environment', in Lancaster, S. (ed.) *Developments in Politics 9*, Ormskirk: Causeway Press, 116–140.

Carter, N./Lowe, P. (1998) Britain: 'Coming to Terms with Sustainable Development?', in Hanf, K./Jansen, A.-I. (eds.) *Governance and Environment in Western Europe: Politics, Policy and Administration*, Harlow: Longman, 17–39.

Central and Local Government Environment Forum (1993) 'A Guide to the Eco-Management and Audit Scheme for UK Local Government', London: HMSO.

Christoforidis, A./Metaxas, M./Psimmenos, S./Zacharis, N. (1995) 'Results of the pilot application of EMAS in Greece: Priorities and perspectives', paper presented at the 2nd International Exhibition and Conference of Environmental Technology, Heleco '95 Proceedings, Volume II, Athens: Technical Chamber of Greece, 323–330.

Christoforidis, A./Zacharis, N. (1998) 'The application of EMAS in companies of the ready made clothing sector', paper presented at the Conference on Environmental Management and Ecological Control, Athens: Technical Chamber of Greece, 4 May.

Connelly, J./Smith, G. (1999) *Politics and the Environment: From Theory to Practice*, London: Routledge.

Dalton, A.J.P. (1998) *Safety, Health and Environmental Hazards at the Workplace*, London: Cassell.

Demmke, C. (1997) 'Implementation and Enforcement in the Member States: Internal Management of European Environmental Policy', in Demmke, C. (ed.) *Managing European Environmental Policy. The Role of the Member States in the Policy Process*, Maastricht: European Institute of Public Administration.

Department of the Environment (1990) 'This Common Inheritance: Britain's Environmental Strategy', Cm 1200, London: HMSO.

Department of the Environment (1992a) 'Explanatory Memorandum on European Community Legislation 5218/92', 11 May.

Department of the Environment (1992b) 'Development Plans and Regional Planning Guidance, Planning Policy Guidance Note 12', London: HMSO.

Department of the Environment (1994a) 'The Environmental Appraisal of Development Plans: A Good Practice Guide', London: HMSO.

Department of the Environment (1994b) 'Sustainable Development: The UK Strategy', Cm 2426', London: HMSO.

Department of the Environment (1994c) 'Climate Change: The UK Programme, Cm 2427', London: HMSO.

Department of the Environment (1994d) 'Biodiversity: The UK Action Programme, Cm 2428', London: HMSO.

Department of the Environment (1994e) 'Sustainable Forestry: The UK Programme, Cm 2429', London: HMSO.

Department of the Environment (1997) 'This Common Inheritance: UK Annual Report 1997, Cm 3556', London: The Stationery Office, February.

Department of the Environment, Transport and the Regions (1998a) 'Sustainable Local Communities for the 21st Century – Why and How to Prepare an Effective Local Agenda 21 Strategy', London, January.

Department of the Environment, Transport and the Regions (1998b) 'Opportunities for Change: Consultation Paper on a Revised UK Strategy for Sustainable Development', London: DETR, February.

Department of the Environment, Transport and the Regions (1998c) 'Full Government Response to the Second Annual Report of the UK Round Table on Sustainable Development', March 1997, London: DETR, February.

Department of the Environment, Transport and the Regions (1998d) 'Modern Local Government: In Touch with the People, Cm 4014', London: The Stationery Office, July.

Department of the Environment, Transport and the Regions (1998e) 'A New Deal for Transport: Better for Everyone, Cm 3950', London: The Stationery Office, July.

Department of the Environment, Transport and the Regions (1998f) 'Sustainability Counts: Headline Indicators', London: DETR.

Department of the Environment, Transport and the Regions (1998g) 'News Release: New EMAS Competent Body', Announced, 10 March.

Department of the Environment, Transport and the Regions (1999) 'A Better Quality of Life: A Strategy for Sustainable Development for the United Kingdom, Cm 4345', London: The Stationery Office, May.

Department of the Environment, Transport and the Regions (2000a) 'Explanatory Memorandum on European Community Legislation 9612/99', 21 January.

Department of the Environment, Transport and the Regions (2000b) 'Regional Planning, Planning Policy Guidance Note 11', London: The Stationery Office, October.

Department of the Environment, Transport and the Regions (2000c) 'Good Practice Guide on Sustainability Appraisal of Regional Planning Guidance', London: The Stationery Office, October.

Department of the Environment, Transport and the Regions (2000d) 'Local Quality of Life Counts', London, July.

Department of the Environment, Transport and the Regions (2000e) 'Transport 2010: The Ten Year Plan', London, July.

Department of the Environment, Transport and the Regions (2000f) 'Power to Promote or Improve Economic, Social or Environmental Well-being: Draft Guidance to Local Authorities', London, December.

Department of the Environment, Transport and the Regions (2000g) 'Preparing Community Strategies: Government Guidance to Local Authorities', London, December.

Department of Trade (1982) 'Standards, Quality and International Competitiveness, Cmnd 8621', London: HMSO, July.

Department of Trade and Industry, Department of the Environment (1993) 'Consultation Paper on the Implementation of the EC Eco-Management and Audit Regulation and Accreditation Arrangements for Certification to BS7750', London, July.

Department of Transport, Local Government and the Regions (2001) 'Strong Local Leadership – Quality Public Services, Cm 5327', London: The Stationery Office, December.

Deputy Prime Minister and Secretary of State for the Environment, Transport and the Regions (1998) 'The Government's Response to Environmental Audit Committee Report on the Greening Government Initiative, Cm 4108', London: The Stationery Office, November.

Deputy Prime Minister and Secretary of State for the Environment, Transport and the Regions (2000a) 'Government Response to the Royal Commission on Environmental Pollution's Twenty-First Report, Setting Environmental Standards, Cm 4794', London: The Stationery Office, July.

Deputy Prime Minister and Secretary of State for the Environment, Transport and the Regions (2000b) 'The Government's Response to the Environment, Transport and Regional Affairs Committee's Report: Audit Commission, Cm 4931', London: The Stationery Office, November.

Deputy Prime Minister and Secretary of State for the Environment, Transport and the Regions (2001) 'The Government's Response to Environmental Audit Commission (sic) Report – Environmental Audit: The First Parliament, Cm 5098', London: The Stationery Office, March.

Deutscher Naturschutzring (1999) 'Stellungnahme zu EMAS II', Bonn.

Dimadama, Z. (1998) *Environmental management and audit: A case study*, MA Dissertation, Department of Urban and Regional Development, Panteion University, February.

Doherty, B. (1999) 'Paving the way: the rise of direct action against road-building and the changing character of British environmentalism', *Political Studies*, 47 (2), June, 275–291.

Dyllick, T. (1995) 'Die EU-Verordnung zum Umweltmanagement und zur Umweltbetriebsprüfung (EMAS-Verordnung) im Vergleich mit der geplanten ISO-Norm 14001. Eine Beurteilung aus Sicht der Managementlehre', in *Zeitschrift für Umweltpolitik und Umweltrecht*, 18, 299–339.

Edvinsson, L./Malone M. S. (1997) *Intellectual Capital: Realising your Company's True Value by Finding its Hidden Brainpower*, New York: Harper Collins.

ENDS Daily (2000a) 'British PVC firms agree new green standards', 18 April.

ENDS Daily (2000b) 'EMAS certification draws in UK telecoms giant', 18 January.

European Commission (1992a) 'Proposal for a Council Regulation Allowing Voluntary Participation by Companies in the Industrial Sector in a Community Eco-Audit Scheme', COM (91) 459 final, 6 March.

European Commission (1992b) 'Towards Sustainability: A European Community Programme of Policy and Action in relation to the Environment and Sustainable Development', COM (92) 23 final, Volume II, 27 March.

European Commission (1996a) 'General Guidance Document for Accredited Verifiers under Regulation 1836/93 and on the Verification and Validation Approach', Brussels.

European Commission (1996b) 'Progress Report from the Commission on the Implementation of the European Community Programme of Policy and Action in relation to the Environment and Sustainable Development', COM (95) 624 final, 10 January.

European Commission (1997a) 'Informal Draft on the Revision of Regulation No 1836/93 (EMAS)', Brussels, 3 December.

European Commission (1997b) 'Report on Ad-Hoc EMAS Review Working Group Meeting', Amsterdam, Brussels, 2–3 June.

European Commission (1998) 'Proposal for a Council Regulation (EC) allowing Voluntary Participation by Organisations in a Community Eco-Management and Audit Scheme', COM (1998) 622 final 3 March.

European Commission (1999) 'Amended Proposal for a European Parliament and Council Regulation Allowing Voluntary Participation by Organisations in a Community Eco-Management and Audit Scheme (EMAS)', COM (1999) 313 final, 13 June.

European Commission (2000a) 'Opinion of the Commission pursuant to Article 251(2)(c) of the EC Treaty, on the European Parliament's amendments to the Council's common position regarding the proposal for a Regulation of the European Parliament and of the Council allowing voluntary participation by organisations in a Community eco-management and audit-scheme (EMAS)', COM (2000) 512 final, Brussels.

European Commission (2000b) 'Guidance on Organisational Structures Suitable for Registration to EMAS (Draft)', Brussels.

European Commission (2000c) 'Guidance on the Identification of Environmental Effects and Assessment of their Significance (Draft)', Brussels.

European Commission (2000d) 'EMAS Guidance on the Environmental Statement (Draft)', Brussels.

European Commission (2000e) 'Guidelines of the EU Commission on Employee Participation within the Framework of EMAS II (Draft)', Brussels.

European Communities (1993) 'Resolution of the Council and the Representatives of the Governments of the Member States, meeting within the Council of 1 February 1993 on a Community Programme of Policy and Action in relation to the Environment and Sustainable Development', *Official Journal of the European Communities*, No. C 138/1, Brussels.

European Environmental Bureau (2000) 'EMAS – EEB Suggestions for the Second Reading in the Parliament', Brussels, February.

European Parliament (1992) 'Opinion of the Committee on Economic and Monetary Affairs and Industrial Policy for the Committee on the Environment, Public Health and Consumer Protection on voluntary participation by companies in the industrial sector in a Community eco-audit scheme', PE 201.794/final, Brussels, 30 September.

European Parliament (2000a) 'Recommendation for second Reading on the Council common position for adopting a European Parliament and Council regulation allowing voluntary participation by organisations in Community Eco-Management and Audit Scheme (EMAS)', A5–0165/2000 final, 20 June.

European Parliament (2000b) 'Results of Conciliation Procedure on EMAS Regulation (98/0303/COD)' reached on 22 November (draft), Brussels.

Everett, T. (1998) Minutes of Evidence: 'The Greening Government Initiative, Environmental Audit Committee', House of Commons Paper 517–vi (1997–98), London: The Stationery Office, 28 April.

Falk, H./Wilkinson, D. (1996) 'The EU's Eco-Audit Regulation: The Role and Accreditation of Environmental Verifiers in Six Member States', London: Institute for European Environmental Policy.

Franke, J.F./Wätzold, F. (1996) 'Voluntary Initiatives and Public Intervention: The Regulation of Eco-auditing', in Lévêque, F. (ed.) *Environmental Policy in Europe. Industry, Competition and the Policy Process*, Cheltenham: Edward Elgar, 175–199.

Garner, R. (1996) *Environmental Politics*, London: Prentice Hall/Harvester Wheatsheaf.

German Statistical Office (Statistisches Bundesamt) (1997) *Reihe 4*, Wiesbaden: Metzler-Poeschel.

Getimis, P. (1993) 'Social Policy and Local State', in Getimis, P./Gravaris, S. (eds.) *Welfare State and Policies*, Athens: Themelion, 91–121.

Getimis, P. (1998) 'Urban Politics and Social Movements in Cities', in Oikonomou, D./Petrakos G. (eds.): *Urban Politics and Social Movements in Cities*, Volos: Thessaly University Editions, 359–370.

Getimis P./Giannakourou G./Dimadama Z. (2001) 'The EMAS (and ISO 14001) Case Studies in Greece', in Heinelt, H./Malek, T./Smith, R./Töller, A.E. (eds.) *European Union Environment Policy and New Forms of Governance*, Aldershot: Ashgate, 341–343.

Giannakis, C. (1998) 'The EMAS and small and medium-sized enterprises', paper presented at the Conference on Environmental Management and Ecological Control, Athens: Technical Chamber of Greece, 4 May.

Giannakourou, G. (1992) 'Private interests' legitimisation methods in the Greek urban administration', in Topos. *Review of Urban and Regional Studies*, 4, 23–40.

Giannakourou, G. (1994) 'Spatial planning and administrative judge: from the control of law towards the formulation of spatial planning policy?', in *Perivallon kai Dikaio*. Semestrial Review of Territorial Sciences, 1, 23–40.

Giannakourou, G. (1996) 'The implementation of European environmental policy in Greece: Trends towards the Europeanisation of national policy making', paper presented at the workshop "The European Policy Process" organised for the Human Capital and Mobility Programme Network at the European University Institute, Florence, 20–22 March.

Glachant, M. (2001) 'The need for adaptability in EU environmental policy design and implementation', *European Environment*, 11, 239–249.

Glasbrenner, G./Riglar, N. (1997) 'The Eco-Management and Audit Scheme (LA-EMAS): costs, problems and benefits', *EG: Local Environment News*, 3 (8), September, 11–12.

Glatzner, L. (1998) 'Die Stellung des Umweltgutachters im Licht der bisherigen Erfahrungen', *Ökologisches Wirtschaften* No. 3–4, 21–22.

Gouldson, A./Murphy, J. (1998) *Regulatory Realities: The Implementation and Impact of Industrial Environmental Regulation*, London: Earthscan.

Grafe-Buckens, A./Beloe, S. (1998) 'Auditing and communicating business sustainability', in *Eco-Management and Auditing*, 5, 101–111.

Gray, T.S. (1995) (ed.) *UK Environmental Policy in the 1990s*, Basingstoke: Macmillan, 1–10.

Grote, J. R./Gbikpi, B. (eds.) (2002) *Participatory Governance: Political and Societal Implications*, Opladen: Leske & Budrich.

Hall, D. (2000) 'Industrial harm to an ecological dream', *Corporate Environmental Strategy*, 7, 53–61.

Hamer, L. (2001) 'Comment...', in *EG Magazine*, 7 (5), May, 1–2.

Harrison, M. (2000) 'Brown's green tax measures attacked by BMW', in *The Independent*, 16 November.

Hartkopf, G./Bohne, E. (1983) Grundlagen, Analysen, Perspektiven, *Umweltpolitik* Vol. 1. Opladen: Westdeutscher Verlag.

Heinelt, H. (2002a) 'Preface', in Grote, J. R./Gbikpi, B. (eds.) *Participatory Governance: Political and Societal Implications*, Opladen: Leske & Budrich, 13–16.

Heinelt, H. (2002b) 'Civic Perspectives on a Democratic Transformation of the EU', in Grote, J. R./Gbikpi, B. (eds.) *Participatory Governance: Political and Societal Implications*, Opladen: Leske & Budrich, 97–120.

Heinelt, H./Malek, T. (1999) 'Öko-Audits in deutschen Betrieben. Zum Ausmaß und zu den Hintergründen einer Erfolgsstory – auf der Basis einer schriftlichen Befragung', *Zeitschrift für Umweltpolitik und Umweltrecht*, 4, 541–566.

Heinelt, H./Malek, T./Staeck, N./Töller, A.E. (2001) 'Environmental Policy: The European Union and a Paradigm Shift', in Heinelt, H./Malek, T./Smith, R./ Töller, A.E. (eds.) *European Union Environment Policy and New Forms of Governance: A Study of the Implementation of the Environmental Impact Assessment Directive and the Eco-Management and Audit Regulation in Three Member States*, Aldershot: Ashgate, 1–32.

Heinelt, H./Getimis, P./Kafkalas, G./Smith, R./Swyngedouw, E. (eds.) (2003) *Participatory Governance in Multilevel Context: Concepts and Experience*, Opladen: Leske & Budrich.

Helm, D. (1998) 'The assessment: environmental policy – objectives, instruments and institutions', *Oxford Review of Economic Policy*, 14 (4), 1–19.

Héritier, A./Mingers, S./Knill, C./Becka, M. (1994) *Die Veränderung von Staatlichkeit in Europa. Ein regulativer Wettbewerb: Deutschland, Großbritannien und Frankreich in der Europäischen Union*, Opladen: Leske & Budrich.

Héritier, A./Knill, C./ Mingers, S. (1996) *Ringing the Changes in Europe: Regulatory Competition and the Transformation of the State*, Berlin: de Gruyter.

Herremans, I./Welsh, C./Bott, R. (1999) 'How an environmental report can help a company 'learn' about its own environmental performance', in *Eco-Management and Auditing*, 6, 158–169.

Hey, C./Brendle, U. (1994) *Umweltverbände und EG. Strategien, Politische Kulturen und Organisationsformen*, Opladen: Westdeutscher Verlag.

Hildebrandt, E./Schmidt, E. (1994) *Umweltschutz und Arbeitsbeziehungen in Europa*, Berlin: edition sigma.

Hillary, R. (1995) 'Environmental reporting requirements under the EU: Eco-Management and Audit Scheme (EMAS)', *The Environmentalist*, 15, 293–299.

Hillary, R. (1997) 'EU environmental policy, voluntary mechanisms and the Eco-Management and Audit Scheme', in Hillary, R. (ed.) *Environmental Management Systems and Cleaner Production*, Chichester: Wiley, 129–142.

Hillary, R. (1998a) 'Pan-European Union assessment of EMAS implementation' *European Environment*, 8 (6), November-December, 184–192.

Hillary, R. (1998b) 'An Assessment of the Implementation Status of Council Regulation (No 1836/93) Eco-management and Audit Scheme in the Member States (AIMS-EMAS)', Final Report, London, June.

Hillary , R. (1998c) 'Environmental auditing: concepts, methods and developments', in *International Journal of Auditing*, 2, 71–85.

Holdgate, M. (1994) 'Energy Policy and the Environmental Challenges and Options at the Turn of the Century', in Helm, D. (ed.) *British Energy Policy in the 1990s: The Transition to the Competitive Market*, Oxford: Oxera Press, 84–91.

Honkasalo, A. (1999) 'Environmental management systems at the national level', in *Eco-Managing and Auditing*, 6, 170–173.

Hood, C. (1991) 'A public management for all seasons?', *Public Administration*, 69 (1), 3–19.

House of Commons Environment, Transport and Regional Affairs Committee (2000) 'The Audit Commission, tenth report, House of Commons Paper 174–I', London: The Stationery Office, June.

House of Commons Environmental Audit Committee (1998a) 'The Greening Government Initiative, minutes of evidence, House of Commons Paper 517–vi', London: The Stationery Office, April.

House of Commons Environmental Audit Committee (1998b) 'The Greening Government Initiative, second report, House of Commons Paper 517–I', London: The Stationery Office, June.

House of Commons Environmental Audit Committee (1999a) 'The Greening Government Initiative 1999, sixth report, House of Commons Paper 426–I', London: The Stationery Office, June.

House of Commons Environmental Audit Committee (1999b) 'Energy Efficiency, seventh report, House of Commons Paper 159–I', London: The Stationery Office, July.

House of Commons Environmental Audit Committee (2000a) 'The Greening Government Initiative: First Annual Report from the Greening Ministers Committee, fifth report, House of Commons Paper 341', London: The Stationery Office, March.

House of Commons Environmental Audit Committee (2000b) 'Energy Efficiency: Government Response and Follow-Up, House of Commons Paper 571–i', London: The Stationery Office, June.

House of Commons Environmental Audit Committee (2000c) 'Budget 2000 and the Environment, sixth report, House of Commons Paper 404', London: The Stationery Office, July.

House of Commons Environmental Audit Committee (2000d) 'Environmental Audit: The First Parliament, first report, House of Commons Paper 67–I', London: The Stationery Office, December.

House of Commons Environmental Audit Committee (2001) 'Government Response to the Second Report from the Committee in the Last Parliament Session 2000–01, on the Pre-Budget Report 2000: Fuelling the Debate, first special report, House of Commons Paper 216', London: The Stationery Office.

House of Commons Trade and Industry Committee (1998) 'Energy Policy, fifth report, House of Commons Paper 471–I', London: The Stationery Office, June.

House of Lords (1993) 'Industry and the Environment, eighteenth Report of the Select Committee on the European Communities (1992–93) (HL Paper 73)', London: HMSO, 30 March.

Hucke, J. (1998) 'Kommunale Umweltpolitik', in Wollmann, H./Roth, R. (eds.) *Kommunalpolitik. Politisches Handeln in den Gemeinden*, Bonn: Bundeszentrale für politische Bildung, 645–661.

Hughes, P. (1998) 'Minutes of Evidence: The Greening Government Initiative, Environmental Audit Committee, House of Commons Paper 517–vi (1997–98)', London: The Stationery Office, 28 April.

International Chamber of Commerce (ICC) (1989) 'Umweltschutz-Audits, Publication No. 435', Paris: ICC.

Institute of Environmental Management (1995) 'EMAS, BS7750 and ISO14001. How do they all relate?', *Institute of Environmental Management Journal*, 3 (3), December, 6–15.

Institute of Environmental Management (1998) 'Survey 1998) ISO14001 and EMAS', *Institute of Environmental Management Journal*, 5 (4), October, 2–24.

Ioannidou , E. (1999) 'The Eco-management and audit scheme, EMAS', in EMSNET conference: The Environmental Management Systems in Greece, Thessaloniki: Mimeo.

Jacobs, M./Levett, R. for the Department of the Environment (1993) 'A Guide to the Eco-Management and Audit Scheme for UK Local Government', London: HMSO.

Jänicke, M./Weidner, H. (1997) 'Germany', in Jänicke, M./Weidner, H. (eds.) *National Environmental Policies: A Comparative Study of Capacity Building*, Berlin: Springer, 133–155.

Jordan, A. (2000) 'Environmental policy', in Dunleavy, P./Gamble, A./Holliday, I./ Peele. G. (eds.) *Developments in British Politics 6*, Basingstoke: Macmillan, 257–275.

Jordan, G./Schubert, K. (1992) 'A preliminary ordering of policy network labels', *European Journal of Political Research*, 21, 7–27.

Kähler, M./Rotheroe, N. (1999) 'Comparison of the British and German approach towards the European Eco-Management and Audit Scheme (EMAS)', *Eco-Management and Auditing*, 6, 115–127.

Kara, A. (1999) 'It's a small world – but how should we measure it?', *EG: Local Environment News*, 5 (10), November/December, 2–4.

Karaiskou, E. (1998) 'The pilot application of EMAS in the municipalities of middle-sized cities: the case of Volos, Larissa and Patras (LIFE96 ENV/GR/

000559)', paper presented at the Conference on Environmental Management and Ecological Control, Athens: Technical Chamber of Greece, 4 May.

Katsakiori, M. (1994) *Non-Governmental Oriented Organisations in Greece, Europe and the Mediterranean Basin*, Thessaloniki: The Goulandris Natural History Museum – Greek Biotope/Wetland Centre.

Keller, B. (1997) *Einführung in die Arbeitspolitik*, fifth edition, München: Oldenburg.

Kioukias, D. (1993) 'Forms of people's organisation in the Greek political system', *Parliamentary Review*, 9–10, 80–101.

Kiser, L./Ostrom, E. (1983) 'The Three Worlds of Action: A Metatheoretical Synthesis of Institutional Approaches', in Ostrom, E (ed.) *Strategies of Political Inquiry*, London: Sage, 179–222.

Knill, C. (1998) 'European policies: the impact of national administrative traditions', in *Journal of European Public Policy*, 18 (1), January-April, 1–28.

Knill, C./Lenschow, A. (1998a) Coping with Europe: the impact of British and German administrations on the implementation of EU environmental policy, *Journal of European Public Policy*, 5 (4), December, 595–614.

Knill, C./Lenschow, A. (1998b) 'Change as appropriate adaptation: administrative adjustment to European environmental policy in Britain and Germany', paper presented at the Workshop: Europeanization and Domestic Political Change, European University Institute, 19–20 June.

Koch, A. (1993) 'Umweltmanagementsysteme – der Airbag für den Umweltschutz', in *UmweltWirtschaftsForum*, 1 (3), 28–30.

Kooiman, J. (2002) 'Governance: A Social-Political Perspective', in Grote, J. R./ Gbikpi, B. (eds.) *Participatory Governance: Political and Societal Implications*, Opladen: Leske & Budrich, 71–96.

Kottmann, H. (1998) 'Das Verhältnis zwischen EMAS und ISO 14001', in Umweltbundesamt (ed.) Umweltmanagement in der Praxis. Teilergebnisse eines Forschungsvorhabens (Teile V und VI) zur Vorbereitung der 1998 vorgesehen Überprüfung des gemeinschaftlichen Öko-Audit-Systems, Berlin, 169–183.

Krisor, K. (1998) 'Aus den Erfahrungen lernen. Die Position der EU-Kommission zur Revision der EMAS-Verordnung', *Ökologisches Wirtschaften*, No. 3–4, 26–27.

Lahmeyer ERM International (1998) Fachwissenschaftliche Bewertung des EMAS-Systems (Öko-Audit) in Hessen. Endbericht zum Forschungsvorhaben, Hessisches Ministerium für Umwelt, Energie, Jugend, Familie und Gesundheit (ed.), Neu-Isenburg.

Laskowski, S. (1998) 'Die funktionelle Privatisierung staatlicher Überwachungsaufgaben – Überwachungsmodell zwischen unternehmerischer Eigenverantwortung und staatlicher Gewährleistungsverantwortung', in Gusy, C. (ed.) *Privatisierung von Staatsaufgaben: Kriterien – Grenzen – Folgen*, Baden-Baden: Nomos, 312–327.

Lean, G. (2001) 'Country air is more polluted than in cities', in *The Independent*, 4 February.

Lindblom, C. (1965) *The Intelligence of Democracy. Decision Making through Mutual Adjustment*, New York: The Free Press.

Lipietz, A. (1991) 'Die Beziehung zwischen Kapital und Arbeit am Vorabend des 21. Jahrhunderts', *Leviathan*, 1, 78–101.

Local Government Association 2001) *Delivering Wellbeing: A Handbook for Sustainable Decision Making*, London: Local Government Association.

Lowe, P./Ward, S. (1998) *British Environmental Policy and Europe*, London: Routledge.

Lübbe-Wolff, G. (1994) 'Die EG-Verordnung zum Umwelt-Audit', *Deutsches Verwaltungsblatt*, 7/1994, 361–374.

Lübbe-Wolff, G. (1995) 'Beschleunigung von Genehmigungsverfahren auf Kosten des Umweltschutzes. Anmerkungen zum Bericht der Schlichter-Kommission', *Zeitschrift für Umweltrecht*, 6, 57–73.

Lübbe-Wolff, G. (1996a) 'Das Umweltauditgesetz', in *Natur und Recht*, 5, 217–227.

Lübbe-Wolff, G. (1996b) 'Öko-Audit und Deregulierung', *Zeitschrift für Umweltrecht* 4, 173–180.

Lusser, H. (1996) 'Green shoots', *Municipal Journal*, 15–21 March, 29.

Macrory, R. (1999) 'The Environment and Constitutional Change', in Hazell, R. (ed.) *Constitutional Futures*, Oxford: Oxford University Press, 178–195.

Majone, G./Wildavsky, A. (1979) 'Implementation as Evolution', in Pressman, J.L./Wildavsky, A. (eds.) *Implementation: How Great Expectations in Washington are Dashed in Oakland*, Berkeley: University of California Press, 177–195.

Malek, T./Töller, A. E. (2001) 'The Eco-Management and Audit Scheme (EMAS) Regulation', in Heinelt, H./Malek, T./Smith, R./Töller, A.E. (eds.) *European Union Environment Policy and New Forms of Governance: A Study of the Implementation of the Environmental Impact Assessment Directive and the Eco-Management and Audit Regulation in Three Member States*, Aldershot: Ashgate, 43–55.

Malek, T./Töller, A.E./Heinelt, H. (2000) 'Zwischen Bremsen und Sprinten: Deutschland und die EU-Umweltpolitik – am Beispiel der Entwicklung der Öko-Audit-Verordnung', in Knodt, M./Kohler-Koch, B. (eds.) *Deutschland zwischen Europäisierung und Selbstbehauptung*, Frankfurt: Campus, 381–411.

Malunat, B. M. (1994) 'Die Umweltpolitik der Bundesrepublik Deutschland', *Aus Politik und Zeitgeschichte*, 29, 15–28.

Matthews, P. 2001) 'A changing climate for local government', *EG Magazine*, 7 (3), March, 9–10.

Mayntz, R. (ed.) (1980) *Implementation politischer Programme: Empirische Forschungsberichte*, Königstein/Taunus: Anton Hain.

Meacher, M. (1998) 'EC Eco-Management and Audit Scheme', House of Commons Hansard, Written Answers, 10 March, column 100.

Mertins, K. (eds.) (2001) *Knowledge Management: Best Practices in Europe*, Berlin: Springer.

Metaxas, M./Zacharis, N. (1995) 'Combined application of environmental management systems in the industry', paper presented at the 2nd International Exhibition and Conference of Environmental Technology, Heleco '95 *Proceedings*, Volume II, Athens: Technical Chamber of Greece, 287–294.

Morphet, J. (1991) 'Environmental auditing – an emerging role for green guardians', *Town and Country Planning*, 60 (6), June.

Müller-Brandeck-Bocquet, G. (1995) *Die institutionelle Dimension der Umweltpolitik. Eine vergleichende Untersuchung zu Frankreich, Deutschland und der Europäischen Union*, Baden-Baden: Nomos.

Myers, P. S. (ed.) (1996) *Knowledge Management and Organizational Design*, Boston: Butterworth Heinemann.

National Accreditation Council for Certification Bodies (1994) 'Draft Criteria: Accreditation of Environmental Management Certification Bodies and Environmental Verifiers-BS7750 and the EMAS Regulation', April.

National Accreditation Council for Certification Bodies (1995a) 'Environmental Accreditation Criteria', January.

National Accreditation Council for Certification Bodies (1995b) 'The Accreditation of Environmental Verifiers for EMAS', July.

Nelson, G. (1996) 'Trading standards and the EMAS experience', in *Trading Standards Review*, November.

NIFES (1999) SCEEMAS, Altrincham: Nifes.

Nisse, J./Jury, L. (2000) 'Greenpeace gets in bed with its foes', *The Independent on Sunday*, 15 October.

Nock, M. (2001) 'Local strategic partnership – can they *really* be local?', *EG Magazine*, 7 (3), March, 3–5.

O'Riordan, T./Voisey, H. (1998) 'The Political Economy of the Sustainability Transition', in O'Riordan, T./Voisey, H. (eds.) Agenda 21. *The Transition to Sustainability – The Politics of Agenda 21 in Europe*, London: Earthscan, 3–30.

OECD (2000) *Environmental Performance Reviews – Greece*, Paris: Organisation for Economic Co-operation and Development.

Piore, M. J./Sabel, C. J. (1984) *The Second Industrial Divide. Possibilities for Property*, London: Basic Books.

Porritt, J./Levett, R. (1999) 'A better quality of strategy? Assessing the White Paper on sustainable development', in *EG: Local Environment News*, 5 (6), June, 4–7.

Probst, G. (eds.) (1999) *Wissen managen: wie Unternehmen ihre wertvollste Ressource optimal nutzen*, 3rd edition, Frankfurt/M: Frankfurter Allgemeine Zeitung.

Riglar, N. (1996) 'Environmental management systems: getting started', *EG: Local Environmental News*, 2 (5), May, 2–4.

Riglar, N. (1997) 'Another year, another EMAS survey', *EG: Local Environmental News*, 3 (9), October, 7–9.

Rose, C. (1990) *The Dirty Man of Europe: The Great British Pollution Scandal*, London: Simon and Schuster.

Royal Commission on Environmental Pollution (1998) 'Setting Environmental Standards, 21st report, Cm 4053', London: The Stationery Office, October.

Sachverständigenrat für Umweltfragen (SRU) (1994) *Umweltgutachten 1994. Erreichtes sichern – neue Wege gehen*, Stuttgart: Metzel-Poeschel.

Sachverständigenrat für Umweltfragen (SRU) (2000) *Umweltgutachten 2000. Schritte ins nächste Jahrtausend*, Stuttgart: Metzel-Poeschel.

Sanchez, R./Heene, A. (1997) *Strategic Learning and Knowledge Management*, Chichester: Wiley.

Scharpf, F. W. (1988) 'Verhandlungssysteme, Verteilungskonflikte und Pathologien der politischen Steuerung', in Schmidt, M. G. (ed.) *Staatstätigkeit. International und historisch vergleichende Analysen (PVS Sonderheft 19)*, Opladen: Westdeutscher Verlag, 61–87.

Schink, A. (1996) 'Die Entwicklung des Umweltrechts im Jahr 1995 – Erster Teil', *Zeitschrift für Umweltrecht*, 9, 357–378.

Schlichter, O. (1995) 'Investitionsförderung durch flexible Genehmigungsverfahren. Eine Darstellung der Untersuchungsergebnisse der Unabhängigen Expertenkommission zur Vereinfachung und Beschleunigung von Planungs- und Genehmigungsverfahren', *Deutsches Verwaltungsblatt*, 110, 173–179.

Schmalz-Bruns, R. (2002) 'The Normative Desirability of Participatory Governance', in Heinelt, H./Getimis, P./Kafkalas, G./Smith, R./Swyngedouw, E. (eds.) *Participatory Governance in Multilevel Context: Theoretical Debate and the Empirical Arena*, Opladen: Leske & Budrich.

Schmitter P. C. (2000) *'Participatory governance' in the context of 'Achieving sustainable and innovative policies in a multi-level context'*, European University Institute, October, revised version.

Schmitter, P. C. (2002) 'Participation in Governance Arrangements: Is there any Reason to Expect it will Achieve "Sustainable and Innovative Policies in a Multilevel Context"?', in Grote, J. R./Gbikpi, B. (eds.) *Participatory Governance. Political and Societal Implications*, Opladen: Leske & Budrich, 51–69.

Schnutenhaus, J. (1995) 'Die Umsetzung der Öko-Audit-Verordnung in Deutschland', *Zeitschrift für Umweltrecht ZUR* No. 1, 9–13.

Schröter, E./Wollmann, H. (2001) 'New Public Management', in Blanke, B. et al. (eds.) *Handbuch zur Verwaltungsreform*, second edition, Opladen: Leske & Budrich, 71–82.

Scott, W. R. (1994) 'Institutions and Organisations: Towards a Theoretical Synthesis', in Scott, W. R./Meyer, J. W. (eds.) *Institutional Environment and Organisations: Structural Complexity and Individualism*, Thousand Oaks: Sage, 55–80.

Secretary of State for the Environment et al. (1995) 'This Common Inheritance: UK Annual Report 1995', Cm 2822, HMSO: London.

Secretary of State for the Environment, Transport and the Regions et al. (2000) 'Climate Change: The UK Programme, Cm 4913', London: The Stationery Office, November.

Seel, B./Paterson, M./Doherty, B. (eds.) (2000) *Direct Action in British Environmentalism*, London: Routledge.

Sellner, D./Schnutenhaus, J. (1993) 'Umweltmanagement und Umweltbetriebsprüfung (Umwelt-Audit) – ein wirksames, nicht ordnungsrechtliches System des betrieblichen Umweltschutzes?', *Neuere Zeitschrift für Verwaltungsrecht*, 12, 928–934.

Sheldon, C. (1997) 'BS7750 and Certification: The UK Experience', in Hillary, R. (ed.) *Environmental Management Systems and Cleaner Production*, Chichester: Wiley, 165–172.

Siouti, G. (1994) 'Basic principles for planning control in the case law of the State Council', in *Perivallon kai Dikaio. Semestrial Review of Territorial Sciences*, 1, 11–22.

Skordilis, A./Filippeou, M. (1995) 'Eco-audit as a tool for the advancement of small and medium-sized enterprises', paper presented at the 2nd International Exhibition and Conference of Environmental Technology, Heleco '95 *Proceedings*, Volume II, Athens: Technical Chamber of Greece, 302–307.

Skourtos, M. (1995) 'Economic Instruments and Environmental Protection in Greece', in Skourtos, M./Sofoulis, K. (eds.) *Environmental Policy in Greece*, Athens: Typothito, 221–256.

Souflis, I. (1998) 'EMAS/ BS 7750/ ISO 14000: A comparative evaluation, paper presented at the Conference on Environmental Management and Ecological Control', Athens: Technical Chamber of Greece, 4 May.

Spanou, C. (1995) 'Public Administration and the Environment: the Greek experience', in Skourtos, M./Sofoulis, K. (eds.) *Environmental Protection in Greece*, Athens: Typothito, 115–175.

Spanou, C. (1996a) 'On the regulatory capacity of the Hellenic state: a tentative approach based on a case study', *International Review of Administrative Sciences*, 62, 219–237.

Spanou, C. (1996b) 'Administrative responses to Greek EU membership: impulses and inhibitions', paper presented at the ECPR Joint Sessions of Workshops, Oslo, 29 March – 3 April.

Steger, U. (1998) 'Executive Summary Final Report "Empirically Based Evaluation and Policy Recommendation for the Revision of EMAS"', edited second draft, Hohenheim.

Strachan, P. (1999) 'Is the Eco-Management and Audit Scheme (EMAS) regulation an effective strategic marketing tool for implementing industrial organisations?', *Eco-Management and Auditing*, 6, 42–51.

Streeck, W. (1984) 'Co-determination: the fourth decade', *International Yearbook of Organizational Democracy 2*, 391–422

Sustainable Development Commission 2001) 'End of Term Review: Headlining Sustainable Development', London: November.

Taeger, J. (ed.) (1999) *Edition Umweltrecht*, Hannover: Schlütersche Verlagsanstalt

Taschner, K. (1998) 'Environmental Management Systems: the European Regulation', in Golub, J. (ed.) *New Instruments for Environmental Policy in the EU*, London: Routledge.

Tindale, S./Hewett, C. (1998) 'New Environmental Policy Instruments in the UK', in Golub, J. (ed.) *New Instruments for Environmental Policy in the EU*, London: Routledge.

Töller, A.E. (1998) 'The Article 19 Committee: The Regulation of the Environmental Management and Audit Scheme', in van Schendelen, M.P.C.M. (ed.) *EU Committees as Influential Policymakers*, Aldershot: Ashgate, 179–206.

Töller, A.E. (2002) *Komitologie. Theoretische Bedeutung und praktische Arbeitsweise von Durchführungsausschüssen in der Europäischen Union – am Beispiel der Umweltpolitik*, Dissertation, Opladen: Leske &Budrich.

Töller, A.E./Hofmann, H. C. H. (2000) 'Democracy and the Reform of Comitology', in Andenas, M./Türk, A. (eds.) *Delegated Legislation and the Role of Commitees in the EC*, The Hague: Kluwer Law, 25–50.

Töller, A.E./Taeger, J. (2000) 'Zu den Gründen der relativ hohen Beteiligung deutscher Betriebe an EMAS', in Heinelt, H. (ed.) *Prozedurale Umweltpolitik der EU*, Opladen: Leske & Budrich, 174–182.

Tope, G. (1998) 'Minutes of Evidence: The Greening Government Initiative', Environmental Audit Committee, House of Commons Paper 517–vi (1997–98), London: The Stationery Office, 28 April.

Transport and General Workers Union (1999) *Workplace Pollution Reduction*, London: T&GWU.

Tuxworth, B. (2000) 'Introduction to special issue', *EG: Local Environment News*, 6 (4), April, 2–3.

UBA (1998) 'Umweltmanagement in der Praxis – Teilergebnisse eines Forschungsvorhabens zur Vorbereitung der 1998 vorgesehenen Überprüfung des gemeinschaftlichen Öko-Audit-Systems', Berlin.

UGA (1997) 'Unternehmensbefragung des UGA', Bonn.

Van Schendelen, M.P.C.M. (1998) 'Prolegomena', in Van Schendelen, M.P.C.M. (ed.) *EU Committees as Influential Policymakers*, Aldershot: Ashgate, 3–22.

Van Waarden, F. (1992) 'Dimensions and types of policy networks', *European Journal of Political Research*, 21, 29–52.

Voelzkow, H. (1996) *Private Regierungen in der Techniksteuerung. Eine sozialwissenschaftliche Analyse der technischen Normung*, Frankfurt/New York: Campus.

Vogel, D. (1986) *National Styles of Regulation: Environmental Policing in Great Britain and the United States*, Ithaca, NY: Cornell University Press.

Voisey, H. (1998) 'Local Agenda 21 in the UK', in O'Riordan, T./Voisey, H. (eds.) *Agenda 21. The Transition to Sustainability – The Politics of Agenda 21 in Europe*, London: Earthscan, 235–249.

Wätzold, F./Bültmann, A. (2000) 'The implementation of EMAS in Europe: A case of competition between standards for environmental management systems', *UFZ-Bericht* No. 16, Leipzig.

Waskow, S. (1994) *Betriebliches Umweltmanagement. Anforderungen nach der Audit-Verordnung der EG*, Heidelberg: CF Müller Juristischer Verlag.

Waskow, S. (1997) 'Normung und Umweltaudit', in Rengeling, H.-W. (ed.) *Umweltnormung. Deutsche, europäische und internationale Rechtsfragen*, fünf Osnabrücker Gespräche zum deutschen und europäischen Umweltrecht am 12/13 Juni, Köln, 55–72.

Weale, A. (1996) 'Environmental Regulation and Administrative Reform in Britain', in Majone, G. (ed.) *Regulating Europe*, London: Routledge, 106–130.

Weale, A. (1997) 'Great Britain', in Jänicke, M./Weidner, H. (eds.) *National Environmental Policies: A Comparative Study of Capacity-Building (in collaboration with H. Jörgens)*, New York: Springer, 89–108.

Weidner, H. (1991) 'Umweltpolitik auf altem Weg zu einer neuen Spitzenstellung', in Süß. W. (ed.): *Die Bundesrepublik in den Achtziger Jahren: Innenpolitik, politische Kultur, Außenpolitik*, Opladen: Leske & Budrich, 137–152.

Weidner, H. (1997) 'Performance and characteristics of German environmental policy. Overview and expert commentaries from 14 countries', WZB-Paper, Berlin.

Weizsäcker , E. U. von/Lovins, A. B./ Lovins, L. H. (1995) *Faktor vier: doppelter Wohlstand, halbierter Naturverbrauch. Der neue Bericht an den Club of Rome,* München: Droemer Knaur.

Welford, R./Gouldson, A. (1993) *Environmental Management and Business Strategy,* London: Pitman.

Wild, A./Marshall, R. (1999) 'Participatory practice in the context of Local Agenda 21: a case study evaluation of experience in three English local authorities', *Sustainable Development,* 7, 151–162.

World Commission on Environment and Development (1987) *Our Common Future,* Oxford: Oxford University Press.

Young, S. (2000) 'Participation Strategies and Environmental Politics: Local Agenda 21', in Stoker, G. (ed.) *The New Politics of British Local Governance,* Basingstoke: Macmillan, 181–197.

Index

For Product Safety Concerns and Information please contact our EU
representative GPSR@taylorandfrancis.com Taylor & Francis Verlag GmbH,
Kaufingerstraße 24, 80331 München, Germany

Printed and bound by CPI Group (UK) Ltd, Croydon, CR0 4YY

01/05/2025

01858474-0001